Microprocessors & Microcontrollers

About the Author

Dr. Santanu Chattopadhyay, is a Professor at the Department of Electronics and Electrical Communication Engineering, Indian Institute of Technology (IIT) Kharagpur. Currently, he is working as the Director, Indian Institute of Information Technology, Kalyani. He has been the Associate Dean of the Faculty of Engineering and Architecture, IIT Kharagpur. He was also the Chairman of the Kalpana Chawla Space Technology Cell, IIT Kharagpur. He has been a Senior Member of the IEEE Computer Society.

He has authored/co-authored several text and reference books in the domains of Cellular Automata, Compilers, Embedded Systems, VLSI Testing, Network-on-Chip etc. Apart from the reputed national publishers, his books have been published by the international publishers like IEEE Computer Society Press and CRC Press. The book on Compiler Design has been translated in Chinese also. He has more than 200 publications/presentations to his credit in reputed refereed international journals (IEEE, ACM, Science Direct and Springer etc.) and conferences.

Professor Chattopadhyay has successfully handled several research projects sponsored by both public and private agencies. He has been the principal investigator in many of the projects sponsored by the Department of Science & Technology, Ministry of Information Technology, Synopsys Inc. USA etc. His research interests include Microcomputer Architectures, Embedded Systems, Low-Power VLSI Design and Test, Hardware Security.

Microprocessors & Microcontrollers

Santanu Chattopadhyay
Professor
Department of Electronics and
Electrical Communication Engineering
IIT Kharagpur

McGraw Hill Education (India) Private Limited

Published by McGraw Hill Education (India) Private Limited
Registered Office: Anjana Complex No: 5/90 A, Butt Road, St. Thomas Mount, Chennai – 600016

Microprocessors & Microcontrollers

Print Edition:
ISBN (13): 978-93-5532-651-5
ISBN (10): 93-5532-651-3

E-Book Edition:
ISBN (13): 978-93-5532-863-2
ISBN (10): 93-5532-863-X

1 2 3 4 5 6 7 8 9 D103074 28 27 26 25 24

Printed and bound in India.

Typeset at PHi Business Solutions Limited, # 98-99, Sub-City Centre, Sector 34-A, Chandigarh
and Printed at Rajkamal Electric Press, Kundli, Haryana - 131 028.

Cover Designer: Neeraj Dayal
Cover Image Source: Shutterstock

Visit us at: www.mheducation.co.in
Write to us at: support.india@mheducation.com
CIN: U80302TN2010PTC111532
Toll Free Number: 1800 103 5875

To

SANTANA, MY WIFE

My inspiration

and

SAYANTAN, OUR SON

Our hope

Preface

Processor is an essential part of any computing system. With advancements in electronic design automation and VLSI technology, processors and computing systems have been implemented as monolithic integrated circuit chips. A single-chip processor realization is called a *microprocessor*. Realization of a complete computing system in a chip results in a *microcontroller*. The developments in this domain started way back in the 1980s. However, the field has grown exponentially with rapid developments in allied areas like computer architecture, operating systems, and improved optimization techniques available with modern compilers. Performance demand from the user community poses newer challenges to processor designers. Numerous vendors have come up with a multitude of microprocessors and microcontrollers. Over the years, the complexity of such processors has increased manifold. An understanding of their features is essential to select the proper alternative for an application. Processor architectures exhibit increasing order of complexity. For a system design, a number of peripheral devices are also to be interfaced with the processor to interact with the environment. Processor complexity has increased in two directions – computational power and communication abilities. This book is an effort to introduce the readers to the important processor architectures. For microprocessors, the book starts with one of the simplest processors, 8085, followed by gradual advancements – 8086, 8088, Intel 286, 386, 486, and Pentium. For microcontrollers, the discussion starts with 8051 and is followed by the advancements, including PIC, AVR, TI, and ARM microcontrollers. Apart from the traditional I/O interfaces (keyboard, display, ADC, DAC, stepper motor, etc.), many microcontrollers support advanced communication primitives (SPI, I2C, IEEE-488, Current Loop, Bluetooth, Zigbee, Wi-Fi, etc.). Interfacing circuitry using these primitives has been discussed. Advanced processor features like CISC, RISC, Cache, and Virtual Memory, along with an overview of Internet of Things (IoT), have been included. Tremendous effort has been made to keep the volume of the book within a reasonable size that can be covered in a single semester.

Organization of the Book

This book is suitable for students pursuing undergraduate courses in engineering in colleges and universities across India. It consists of twelve chapters.

Chapter 1 introduces the basics of microcomputer system architecture. Block diagram level enumeration of the architecture of a typical CPU has been presented. Execution of machine instructions by the processor has been explained. Typical structure of a microcontroller along with its difference from a microprocessor has been detailed.

Chapter 2 is on 8085 microprocessor architecture. It starts by enumerating the important features of 8085. The pin layout of the chip has been presented along with a grouping of pins based on their functionality. The internal operation of the 8085 CPU has been explained with the help of a block diagram detailing the functional blocks present in it. The CPU blocks are responsible for executing the machine instructions. The process has been elaborated with the help of timing diagrams corresponding to memory read and write operations. Interfacing of memory and I/O chips with 8085 has been discussed. Bulk data transfer using *Direct Memory Access* (*DMA*) has been presented.

Chapter 3 enumerates the machine instruction set of 8085 and the associated assembly language programming. Individual instructions work on operands that may be of different types. The addressing modes of the processor has been listed. This has been followed by discussion on the machine instructions, grouped by their functionality. Fundamentals of assembly language programming has been illustrated through several programming examples.

Chapter 4 discusses the interrupt facilities available with 8085. The interrupts have been classified into different categories and their triggering modes enumerated. Usage of software instructions and service routines for processing of interrupts have been detailed. Processor responds to concurrently occurring interrupts based on their priority. The basic priority policy of 8085 has been detailed. The dedicated programmable interrupt controller chip 8259 has been introduced, which can be utilized in system design to handle more complex interrupting mechanisms.

Chapter 5 illustrates the hardware and software requirements to interface basic input-output devices with 8085. A generic port structure in an 8085-based system has been illustrated. Interfacing of simple I/O devices, such as switches, LEDs, keyboard and display units, LCD, etc., has been included. The powerful programmable peripheral interface chip 8255 structure and programming have been enumerated. The 8255 chip has been used to connect to Analog-to-Digital and Digital-to-Analog converters. Many applications require facilities for precision timer operations. For this, the usage of chip 8254 has been explained. To interface devices supporting serial data transfer, 8085 provides two dedicated pins, Serial Input Data (SID) and Serial Output Data (SOD) lines, along with software instructions RIM and SIM. The usage of the serial data transfer policy has been enumerated.

Chapter 6 introduces the readers to the next advanced microprocessor – 8086. The architecture of the 8086 CPU has been explained. The pin layout of the chip has been presented. The organization of physical memory for 8086 has been illustrated. Timing diagrams corresponding to the two modes of operation (*minimum* and *maximum*) of 8086 have been explained. The internal structure of the 8086 CPU has been enumerated with the help of a block diagram. Different addressing modes and instructions available with 8086 have been listed. The interrupt structure of 8086 has been illustrated.

Chapter 7 deals with the assembly language programming techniques for 8086 and the strategies for I/O interfacing. Interfacing of basic input-output devices like keyboard, seven-segment displays, ADCs and DACs have been included. Further, a stepper motor interface has been taken up for illustration of I/O operations.

Chapter 8 introduces one of the most widely used microcontrollers – 8051. The chapter illustrates the pin layout of the chip and its internal organization at a block diagram level. The distribution of memory regions into various functional zones has been illustrated. The structure of built-in I/O ports has been enumerated. To accommodate programs with code or data sizes more than the available internal memory, 8051 can be interfaced with external ROM and RAM chips. The connection procedure for external memory interfacing has been explained.

Chapter 9 illustrates the instruction set and assembly language programming for 8051. Various addressing modes have been enumerated, along with illustrations of groups of instructions. Several example programs have been written in the assembly language of 8051. To facilitate the programmers, compilers are available to target C language programs to 8051. The registers and other CPU resources can be directly accessed in such programs, giving rise to *embedded C*. Applications have been developed using the timers, counters, interrupts, and serial communication facilities of 8051.

Chapter 10 discusses the interfacing of basic I/O devices with 8051. In the process, hardware and software modules have been reported for connecting LEDs, keyboard, seven-segment display, LCD, ADC and DAC. The connection of 8255 chip with 8051 has been presented as well. Using the ports, a stepper motor has been interfaced with 8051.

Chapter 11 covers the communication techniques available in contemporary microprocessors and microcontrollers. In the process, it acquaints the reader with communication standards like RS-232, SPI, I2C, IEEE-488 (GPIB), Current Loop, Zigbee, and Wi-Fi. For each of them, an example microprocessor/ microcontroller supporting the protocol has been discussed. Many interfacing chips are also available through which the traditional processors can be made to provide such communication protocols. Such chips have been enumerated for the understanding of the reader.

Chapter 12 enumerates several advanced processors that are being used extensively in electronic systems. These can be grouped to belong to the CISC or RISC architecture family. Generic CISC and RISC features have been explained. Memory hierarchy in a typical computing system has been illustrated. The operation of Cache and virtual memory have been explained. Advanced microprocessors like 80286, 386, 486, and Pentium have been detailed. Advanced microcontrollers, such as PIC, AVR, MSP, and ARM, have been introduced. The integration of these processors in IoT has been taken up to gain a good understanding of the domain.

Instructor Resources

The text is supported by online resources for instructors, which include PowerPoint presentations and solutions to end-of-chapter questions. The detailed instructions for accessing the online resources are printed on the cover.

Acknowledgements

I must acknowledge the contribution of my teachers, who taught me subjects like Digital Electronics, Computer Architecture, Microprocessors, Operating Systems, Compilers, VLSI Design, and so forth. The clear discussions in those classes helped me to consolidate my knowledge in those domains and develop an eagerness to explore contemporary processor architectures. The findings have been compiled into this book. I feel deeply blessed to have such a nice group of teachers. I am also thankful to all the students I have taught this subject to. They all helped me to identify the gaps and mistakes in the manuscript.

My source of inspiration for writing this book is my wife Santana, whose relentless wish and pressure forced me to bring the book in its present shape. Over this long period, she has sacrificed a lot in the family front to allow me to have time to continue writing, taking all other responsibilities onto herself. I am also thankful to my son Sayantan for his encouragements in writing this book.

I acknowledge the authors of the books and research papers that have served as reference material for writing this book. Thanks are also due to the publishers, McGraw Hill, and the editorial and production teams for providing me with the necessary support to see my thoughts in the form of a printed book.

If the book can be fruitfully utilized by the students, for whom it has been written, I shall consider my efforts to be successful. All constructive suggestions for improving the content will be welcomed.

Santanu Chattopadhyay

PUBLISHER'S NOTE

McGraw Hill Education (India) invites suggestions and comments from you, all of which can be sent to support.india@mheducation.com (kindly mention the title and author name in the subject line). Piracy-related issues may also be reported.

Programme and Course Outcomes

Outcome based curriculum and assessment enable the evaluators to determine whether the students have achieved the outlined standard, specific and measurable outcomes. *Programme Outcomes* (*POs*) are statements that describe what students are expected to know and be able to do upon graduating from the program. As per the *National Board of Accreditation* (*NBA*), the following are the 12 POs for an Engineering degree graduate. These relate to the skills, knowledge, analytical ability, attitude and behavior that students acquire through the program.

PO1 – Engineering knowledge: Apply the knowledge of mathematics, science and engineering fundamentals and engineering specialization to solve the complex engineering problems.

PO2 – Problem analysis: Identify, formulate, review research literature, and analyze complex engineering problems reaching substantiated conclusions using first principles of mathematics, natural sciences, and engineering sciences.

PO3 – Design/development of solutions: Design solutions for complex engineering problems and design system components or processes that meet the specified needs with appropriate consideration for the public health and safety, and the cultural, societal, and environmental considerations.

PO4 – Conduct investigations of complex problems: Use research-based knowledge and research methods including design of experiments, analysis and interpretation of data, and synthesis of the information to provide valid conclusions.

PO5 – Modern tool usage: Create, select, and apply appropriate techniques, resources, and modern engineering and IT tools including prediction and modeling to complex engineering activities with an understanding of the limitations.

PO6 – The engineer and society: Apply reasoning informed by the contextual knowledge to assess societal, health, safety, legal and cultural issues and the consequent responsibilities relevant to the professional engineering practice.

PO7 – Environment and sustainability: Understand the impact of the professional engineering solutions in societal and environmental contexts, and demonstrate the knowledge of, and need for sustainable development.

PO8 – Ethics: Apply ethical principles and commit to professional ethics and responsibilities and norms of the engineering practice.

PO9 – Individual and team work: Function effectively as an individual and as a member or leader in diverse teams and individual, and as a member or leader in diverse teams, and in multidisciplinary settings.

PO10 – Communication: Communicate effectively on complex engineering activities with the engineering community and with society at large, such as, being able to comprehend and write effective reports and design documentation, make effective presentations, and give and receive clear instructions.

PO11 – Project management and finance: Demonstrate knowledge and understanding of the engineering and management principles and apply these to one's own work, as a member and leader in a team, to manage projects and in multidisciplinary environments.

PO12 – Life-long learning: Recognize the need for, and have the preparation and ability to engage in independent and lifelong learning in the broadest context of technological change.

The course corresponding to the present book "*Microprocessors and Microcontrollers*" is a part of the graduate program in *Electronics Engineering*. The following are the *Course Outcomes* (*COs*) for the subject. By the end of the course the students will be able to:

CO1 – Understand and compare fundamentals of microprocessors and microcontrollers.

CO2 – Illustrate the architecture of 8085 microprocessor.

CO3 – Enumerate the instructions of 8085 and use them in programming.

CO4 – Utilize the interrupts to notify external events to the 8085.

CO5 – Interface input-output devices with the 8085.

CO6 – Illustrate the architecture of 8086 microprocessor.

CO7 – Enumerate the instructions of 8086 and their use in programs.

CO8 – Illustrate the architecture of 8051 microcontroller.

CO9 – Enumerate the instructions of 8051 and use in programs.

CO10 – Interface input-output devices with 8051.

CO11 – Enumerate the communication primitives available with contemporary microprocessors and microcontrollers.

CO12 – Illustrate features of advanced microprocessors, microcontrollers and Internet of Things.

Mapping of Course Outcomes with Programme Outcomes to be done according to the matrix given below:

Course Outcomes	Expected Mapping with Programme Outcomes (1-Weak Correlation; 2-Medium Correlation; 3-Strong Correlation)											
	PO1	PO2	PO3	PO4	PO5	PO6	PO7	PO8	PO9	PO10	PO11	PO12
CO1	3	2	1	1	1	1	1	1	2	2	3	3
CO2	3	2	1	1	1	1	1	1	2	2	1	3
CO3	3	3	2	2	1	1	1	1	2	2	1	3
CO4	3	3	2	3	2	1	1	2	2	2	2	3
CO5	3	3	3	3	1	1	1	2	2	2	2	3
CO6	3	2	1	1	1	1	1	1	2	2	1	3
CO7	3	3	2	2	1	1	1	1	2	2	1	3
CO8	3	2	1	1	1	1	1	1	2	2	1	3
CO9	3	2	3	2	2	1	1	1	2	2	1	3
CO10	3	3	3	3	2	1	1	2	2	2	2	3
CO11	3	2	2	2	3	2	1	1	2	2	3	3
CO12	3	2	2	2	3	2	1	2	2	2	3	3

Brief Contents

Contents

1

Microcomputer Fundamentals

LEARNING OUTCOMES

After reading this chapter, students will be able to:

LO 1 List the components in a computer system

LO 2 Draw block diagram explaining basic computer organization

LO 3 State the structure and functioning of a CPU

LO 4 Compare between microprocessors and microcontrollers

LO 5 Distinguish between high- and low-level languages

LO 6 Enumerate the role of tools present in an Integrated Development Environment

1.1 Introduction

LO 1, 2

A computer consists of a computing element at its heart that can perform a set of arithmetic and logic operations. The computing unit is typically called the *Central Processing Unit* (*CPU*), or simply the *processor*. The processor needs to be instructed about the sequence of operations that it has to perform, may be repetitively. The set of instructions that the processor is asked to carry out for accomplishing some task is called a *program*. Further, the operations are carried out on some *data* values. The data may be embedded into the program itself or provided separately via some *input device*. The program (and data as well) is kept in a storage element, called *memory*. The memory may be an integrated store for both the program and the data or may have separate storages – *program memory* and *data memory*. The processor, via execution of program instructions, may compute some value which may be stored back in the memory or output to an *output device*. The input and the output devices are jointly known as *peripheral devices*. Figure 1.1 shows a block-level interconnection between the components of a computer. Such computer, also

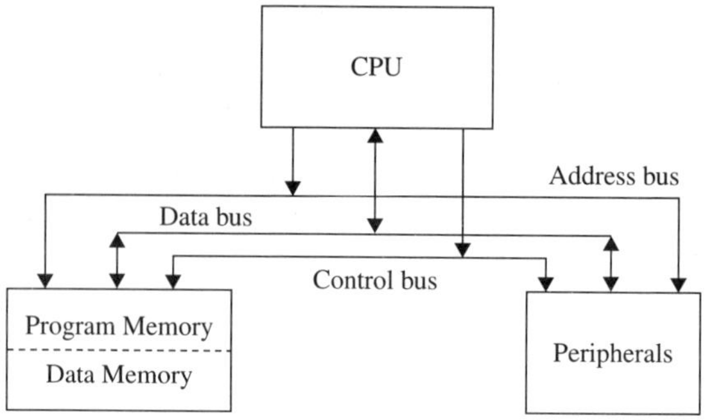

FIGURE 1.1 A typical computer system

called *stored-program computer*, has a program stored in memory. If the program memory and the data memory are put together, it is called *von Neumann* architecture. On the other hand, if the program and data memories are separate, it is known as *Harvard* architecture.

1.2 Buses in a Computer System

> A *bus* is set of parallel conducting lines carrying electrical signals. Typical buses are 8-bit to 256-bit wide. Much faster than serial bit-by-bit transfer.

The CPU needs to communicate with the memory and the peripheral devices to get the assigned computational task carried out satisfactorily. The communication takes place via buses. A *bus* is a set of parallel signal lines, often realized by drawing parallel wires onto the PCB (Printed Circuit Board) or by a strip of wires connecting the chips. A bus may be of various widths, such as 8-bit, 16-bit or more. The parallel wires aid in transferring parallel data of size equal to its width in one go. The bus may be unidirectional or bidirectional in nature. A unidirectional bus transfers data from, say, device *A* to device *B* (where *A* and *B* are the two devices connected by the bus) only. On the other hand, a bidirectional bus will allow data transfer from *A* to *B* as well as *B* to *A*. A computer system has three major buses – *address bus, data bus* and *control bus*. Out of these, in general, address bus and control bus are unidirectional, while the data bus is bidirectional.

1.2.1 Address Bus

> The *address bus* is output from CPU, input to memory. It carries the address of the memory location CPU accesses.

This is a unidirectional bus running from the CPU to the memory and peripherals. A memory chip consists of a number of locations, called *words*. A word may be of size equal to 1, 2 or more number of bytes. Each word location is identified by an *address*. A typical memory chip contains number of locations which is a power of 2 (for example, 1 kilobyte [2^{10} bytes], 2 kilobyte [2^{11} bytes] etc.). That is, the total number of addresses in a memory chip is a power of 2. The *address bus* is used by the CPU to provide the address of the location it wants to access. If the CPU address bus has n bits in it (that is, of width n bit), it can be used to access 2^n number of memory locations. However, it is not mandatory that the size of address bus be same as the number of address lines for the memory chip. For a CPU with n-bit address bus, if it connects to a memory chip with m address lines ($m < n$), the m least significant bits of the address bus are connected to the address lines of the memory chip. For example, if the CPU has 16-bit address bus and it is interfaced with a memory chip of size 4 kilobyte, the least significant 12 bits of the address bus are connected to the memory chip (since 4 kilo = 2^{12}). The remaining 4 bits of the address bus remain unutilized.

A k-to-2^k decoder has k inputs, 2^k outputs. Based on k-bit input value, only one output is 1, rest 0 (or vice versa).

It may be noted that if it is required to have the maximum amount of memory connected to the CPU using memory chips of lesser capacity, a number of memory chips need to be interfaced. If the address bus width is n and the memory chips have m address lines, $2^n/2^m = 2^{n-m}$ number of memory chips are needed to be connected to have the full address space covered. Individual memory chips possess a *chip select* line. For a memory chip to be enabled for access, its chip select line must be activated. The availability of this chip select line facilitates interfacing a number of memory chips to the CPU. For this, the higher order (n–m) bits are connected to a *decoder*. The decoder output lines are connected to the chip select lines of individual memory chips. The lower order m bits of the address bus are connected to all the memory chips. For an n-bit address generated by the CPU, one of the memory chips gets selected depending upon the higher order (n–m) bits. From this chip, depending upon the lower order m bits, one location gets selected. The situation has been explained in Example 1.1.

EXAMPLE 1.1

Consider a CPU with 16-bit address bus. With 16-bit, it can access 2^{16} = 65,536 different locations. Now, if the individual memory chips are of size 16 kB, four such memory chips are needed to be connected. Each memory chip has 14 address lines. Assuming that the 16-bit CPU address lines are identified as $A_{15}A_{14}A_{13}\ldots A_1A_0$, the least significant 14 bits, A_{13} through A_0, are to be connected to all the four memory chips. The higher order two bits, A_{15} and A_{14}, are connected to a 2-to-4 decoder. The four decoder outputs are connected to the chip select lines of four memory chips. The connection pattern has been shown in Figure 1.2. For addresses in the range 0000H-3FFFH, memory chip 1 gets selected. Similarly for address range 4000H-7FFFH, 8000H-BFFFH and C000H-FFFFH, memory chips 2, 3 and 4 are accessed, respectively.

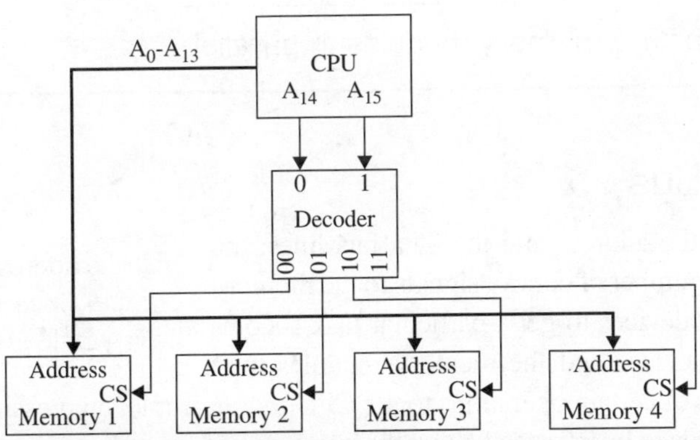

FIGURE 1.2 Memory interfacing with CPU

It may be noted that instead of connecting memory chips at each of the decoder outputs, some peripheral device can also be connected (that has some type of chip select facility). This will of course affect the memory capacity of the computer system.

1.2.2 Data Bus

This is a bidirectional bus that is used to transport data between CPU and memory or I/O devices. If the CPU data bus is of width *n* bits, *n* bits of data can be transferred at a time. The individual memory chips may have *m* bit data lines. Typically, *n* and *m* are equal; however, if $m = n/2$, two memory chips can be connected in parallel to provide *n*-bit data to the CPU. Processing power of the CPU is often guided by the size of its data bus. If the data bus size is *n*-bits, the CPU works with *n* bit data. All operations, such as addition and subtraction, work on *n* bit data. Parallel connection of memory chips has been illustrated in Example 1.2. If two such chips are connected in parallel, the address bus lines connect to both of them and one chip select output from the decoder selects both the memory modules simultaneously.

EXAMPLE 1.2

Figure 1.3 shows a memory interface in which the CPU has 8-bit data bus, whereas a memory chip possesses 4-bit data lines. Individual locations of the memory chip are 4-bit wide. Thus, to match with the 8-bit data bus size, two such memory chips are connected in parallel. A particular address generated by the CPU selects two locations – one in each memory module. These two 4-bit nibbles are joined together to give the 1-byte word to the CPU.

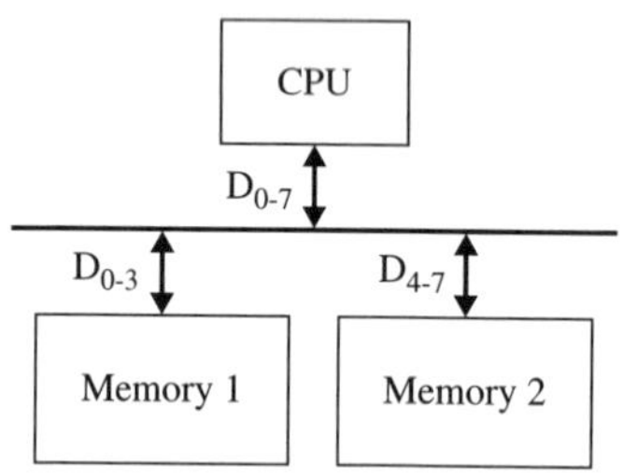

FIGURE 1.3 Connecting memory modules in parallel

1.2.3 Control Bus

Apart from connecting the address and the data bus lines, the CPU needs to provide a number of control signals to the memory chips and the peripheral devices to get a particular task accomplished. Similarly, the devices and the memory modules may need to inform the status of certain operations to the CPU. For example, on pressing a key in the keyboard by the user, the CPU needs to be informed about the event. These control and status bit lines together form

> Unlike address and data buses, individual lines of a control bus may carry data in different directions.

the *control bus*. Since the majority of the signals are from the CPU to the memory or peripheral devices, the control bus is often shown as unidirectional. The number of lines in the control bus is dependent on the chips used to realize CPU, memory modules and I/O devices. However, two very common control signals used in such interfaces are *Read* and *Write*. For most of the processors, these lines are active low and are represented as $\overline{RD}$ and $\overline{WR}$, respectively.

In general, there are two types of memory chips used in a computer system – *Read Only Memory* (*ROM*) and *Random Access Memory* (*RAM*). Though there are different variants of ROM, it is used to hold information that is not going to change during the operation of the system and also does not get erased, even if the system is turned off. Data into a ROM is written with the help of special *programming* devices. On the other hand, the RAM content is volatile, can be written and read during the system operation and gets erased if the system is turned off. In a computer system, ROM is used to hold semi-permanent data, such as operating system code and other program code/data that need not be changed frequently. For the temporary program code and data, RAM chips are used. Thus, in system operation, the ROM content is only read, while the RAM content may be read and written as well. As a result, the $\overline{RD}$ control of CPU is connected to both ROM and RAM chips, while the $\overline{WR}$ control is connected to the RAM chips only. The situation has been shown in Figure 1.4. Example 1.3 discusses about a memory interfacing.

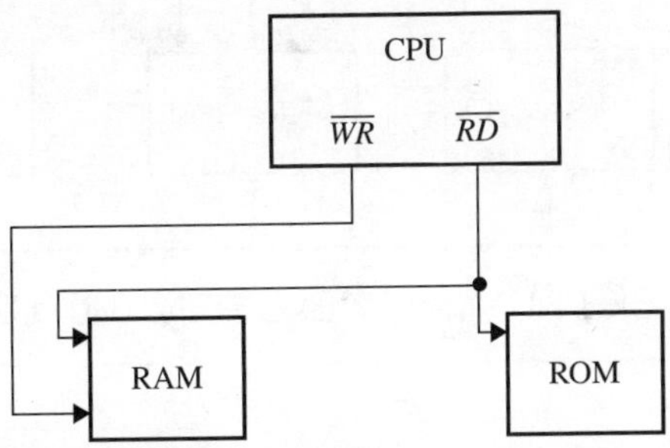

FIGURE 1.4 Read, write signals to memory

EXAMPLE 1.3

A typical computer system with the details of its memory interfacing has been shown in Figure 1.5. The CPU has 16-bit address bus and 8-bit data bus. Thus, it can connect to a 64 kB of memory space. In the figure, that entire memory space has been attached to the CPU. Out of 64 kB, 16 kB is the ROM space and 48 kB RAM space. Individual RAM and ROM chips are of capacity 16k×4-bits each. Since the CPU data bus size is 8 bit, two memory chips are to be activated in parallel to provide 8-bit data. Thus, a total of 2 ROM chips and 6 RAM chips are needed to provide the 64 kB memory with the desired ROM-RAM distribution. The CPU address lines A_0 to A_{13} have been connected to the address lines of all the memory chips. A 2-to-4 decoder has been used with inputs coming from the CPU address bits A_{14} and A_{15}. Individual decoder outputs have been used to select two parallel memory modules, at a time. Out of two modules selected, one memory chip provides data for bits D_4-D_7 of the CPU data bus, while the other module gives the D_3-D_0 bits.

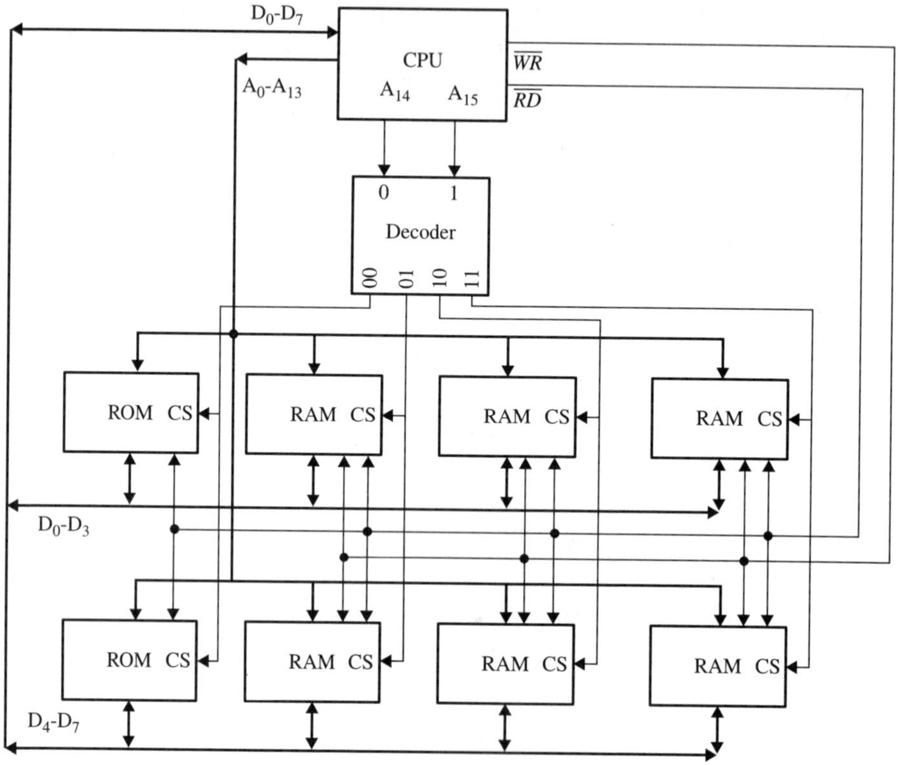

FIGURE 1.5 Interfacing 16kB ROM and 48kB RAM with CPU

1.3 Inside the CPU

LO 3

Figure 1.6 shows the block diagram of a typical CPU. Internally, the CPU consists of a set of *registers*, an *Arithmetic Logic Unit* (*ALU*) and a *controller* to coordinate the overall operation. The components inside a typical CPU have been detailed in the following.

1. **Registers:** These are temporary storages for data values or memory addresses. The data may be used in the computation process, while the addresses are used as pointers to memory locations that may contain the data for operation. Each register can be of width 8 bits, 16 bits, 32 bits etc. As the registers are located inside the CPU, access to data stored there is much faster than from the memory chips. However, their number is limited as a large number of registers will imply bigger chip size and higher power consumption for the CPU. There are two types of registers present in any CPU. The first group is known as the *General Function Registers* (*GFRs*), while the second one is *Special Function Registers* (*SFRs*).

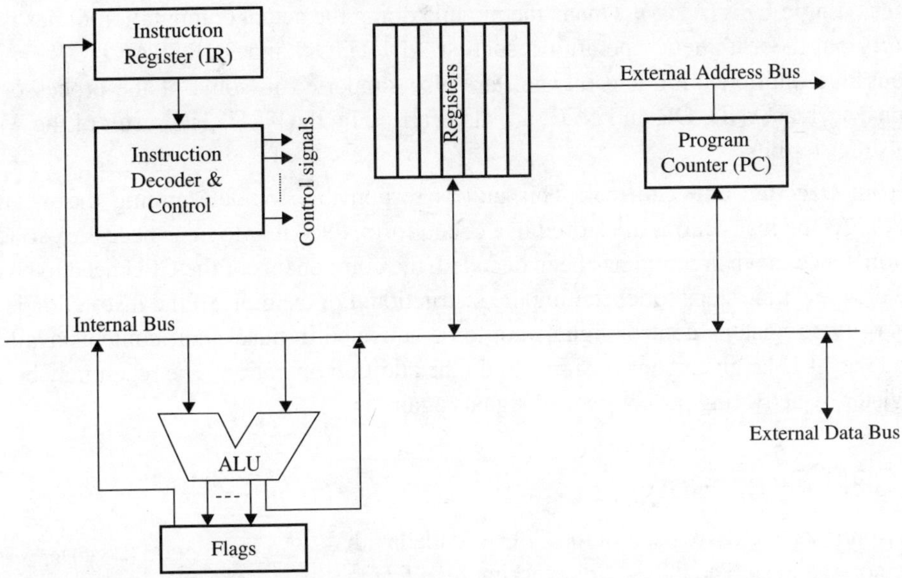

FIGURE 1.6 Structure of a typical CPU

a. **General Function Registers (GFRs):** These registers are available to the users to store temporary data of their programs. The registers can participate in arithmetic and logic operations carried out by the CPU. The storage locations marked as "Registers" in Figure 1.6 are the GFRs.

b. **Special Function Registers (SFRs):** Apart from the GFRs, there are some more registers with dedicated functionalities. The following are some of the commonly available SFRs in the processors.

   i. **Program Counter (PC):** This register contains the address of the memory location having the next instruction to be executed by the processor. When the processor is reset, the PC is loaded with a fixed value set by the designer of the processor. The PC content is put onto the address bus of the CPU to select the memory location containing the instruction. After fetching an instruction from this memory to CPU, the PC content is incremented by the instruction size, to point to the next instruction. In some of the processors, the PC is also named *Instruction Pointer* (*IP*).

   ii. **Instruction Register (IR):** This register holds one complete instruction after it has been fetched from memory to the CPU. Since the instruction is stored in a coded form, it needs to be decoded before execution. The IR feeds the decoder with the instruction.

   iii. **Flags:** This register holds the status information corresponding to the last operation carried out by the ALU. Though there are some variations in the information stored in this register across the processors, some of the important bit-level information are named ZERO, CARRY etc. The ZERO bit is set to 1 if the result of the last ALU operation was zero. Similarly, the CARRY bit is set to 1 if the ALU operation produced a result beyond the range of numbers that could be stored in the registers. The flags help in implementing the branching operation as well.

2. **Arithmetic Logic Unit (ALU):** This is the module doing the actual computation. It has the capability to carry out the arithmetic operations, such as ADDITION and SUBTRACTION. The complex operations like multiplication and division are also supported in some of the processors. Logical operations, such as AND, OR and NOT, are also carried by the ALU. The result of the ALU affects the *Flags* register bits.
3. **Instruction Decoder and Control:** This unit is responsible for determining the meaning of an instruction. As the instructions are stored in a coded form, decoding logic is necessary to identify the instruction. Once the instruction has been decoded, the components of the CPU need to be controlled over the next few time steps for executing the instruction. For example, if the instruction is to add the contents of two registers, control signals are to be activated to make their contents available to the ALU and the ALU be given control signal to do the addition operation. The result may be stored in a third register by activating proper control signals again.

1.3.1 CPU Operation

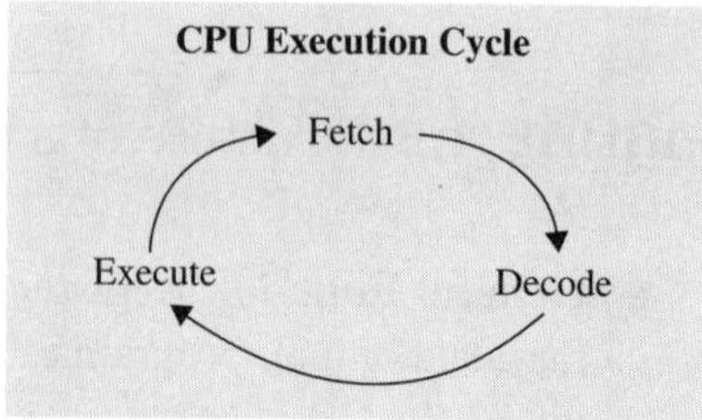

Upon power-on/reset, the PC register of the CPU is loaded with a preset memory address. The CPU starts working by *fetching* the first instruction from this memory address to the IR. The instruction is then *decoded* with the help of the instruction decoder. Once decoded, the controller instructs the units within the CPU to *execute* the operation desired by the instruction. In the meantime, the PC is incremented by the size of the fetched instruction, so that it points to the next instruction in sequence. If the current instruction asks the program control to be transferred to a different address (a branch instruction), the PC is loaded with the targeted address. In either case, by the time the execution of the current instruction finishes, the PC has been loaded with the address of the next instruction to be executed. CPU goes to fetch the next instruction followed by its decoding and execution. Thus, the CPU continually performs the "*fetch → decode → execute*" cycles, till the processor executes a HALT instruction to stop all processing, or a reset signal is received to restart from the reset address. The operations have been summarized in Table 1.1.

TABLE 1.1 Instruction execution steps of CPU

Fetch	1. PC content (address of next instruction) put on the address bus. 2. $\overline{RD}$ signal activated. 3. Instruction register gets the next instruction from memory. 4. PC gets incremented pointing to the next instruction.
Decode	5. Decoder decodes the instruction and decides the sequence in which various control signals are to be activated to accomplish the instruction.
Execute	6. Operands for the instruction brought from memory, if required. 7. Operation carried out in ALU, flags updated. 8. Result stored back in register/memory.

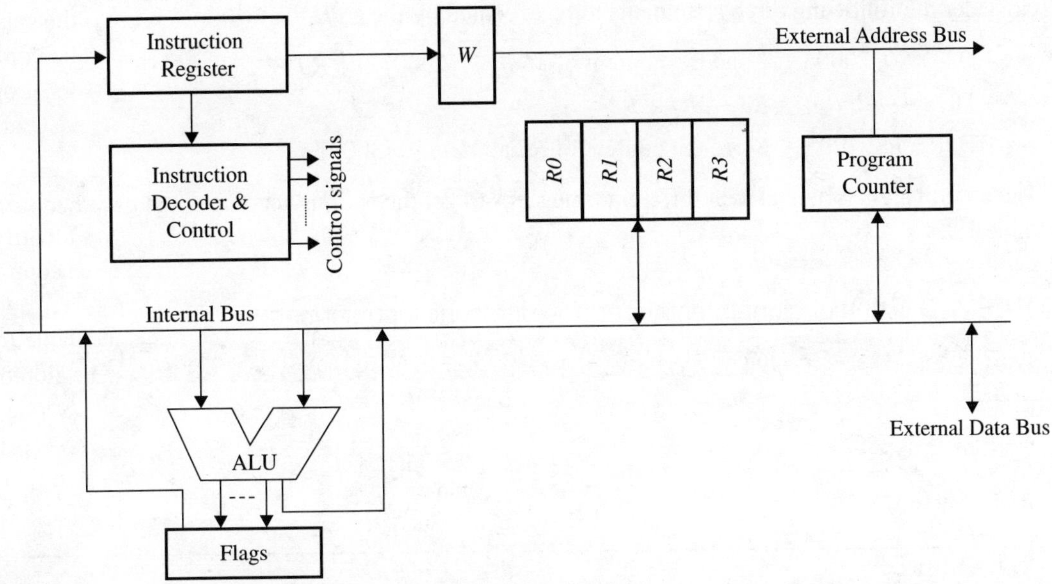

FIGURE 1.7 A hypothetical CPU

To illustrate the operation of the CPU further, consider a hypothetical CPU shown in Figure 1.7. It contains four general-purpose registers – *R0*, *R1*, *R2* and *R3*. There is a special register *W* that can hold an address. It is connected to the *IR* register, so that for instructions specifying an address in it (for example, "LOAD <*address*>"), the address part could be copied directly from the *IR* to *W* and made available in the address bus for memory access.

For the controller to control the operation of registers, ALU etc., each of the components need a few control signals. The control signals are generated by the *Instruction Decoder & Control* unit and applied to individual components. In general, each register possesses a load signal that loads the content available at its input to its storage, and an enable signal that allows its storage content to be put onto its output. The ALU needs control signals to indicate the intended operation. Table 1.2 lists the control signals for this CPU.

TABLE 1.2 Control signals for the CPU

Component	Control Signals
Instruction Register	*Load_IR*, *Enable_IR*
W Register	*Load_W*, *Enable_W*
R0, R1, R2, R3 Registers	*Load_R0*, *Load_R1*, *Load_R2*, *Enable_R0*, *Enable_R1*, *Enable_R2*, *Load_R3*, *Enable_R3*
ALU	*Add*, *Sub*, *And*, *Or*, *Not*
PC Register	*Load_PC*, *Enable_PC*, *Increment_PC*

Consider the following three statements to be executed by the CPU.

- LOAD *R0*, 2000; Load the content of *Memory*[2000] into *R0*
- ADD *R1*, *R2*, *R3*; $R3 \leftarrow R1 + R2$
- STORE *R3*, 3000; Store the content of *R3* at *Memory*[3000]

The control signals activated at different time steps to get the actions accomplished have been shown in Table 1.3.

TABLE 1.3 Control signals activated for different instructions

Instruction	Time Step	Control Signals
LOAD *R0*, 2000	t_1	*Enable_PC, RD*
	t_2	*Load_IR*
	t_3	*Decode, Increment_PC*
	t_4	*Load_W, Enable_IR*
	t_5	*Enable_W, RD, Load_R0*
ADD R1, R2, R3	t_1	*Enable_PC, RD*
	t_2	*Load_IR*
	t_3	*Decode, Increment_PC*
	t_4	*Enable_R1, Enable_R2, Add_ALU, Load_R3*
STORE *R3*, 3000	t_1	*Enable_PC, RD*
	t_2	*Load_IR*
	t_3	*Decode, Increment_PC*
	t_4	*Load_W, Enable_IR*
	t_5	*Enable_W, WR, Enable_R3*

In Table 1.3, it may be noted that the steps t_1 and t_2 correspond to the fetch stage, t_3 stands for the decoding stage and the rest of the steps are for executing the instruction.

1.4 Microprocessor and Microcomputer

LO 4

The processor architecture, discussed in Section 1.3, when put into a single chip, is called a *microprocessor*. In a single chip, it includes the ALU, the registers and the control unit along with all internal buses. It may be noted that the components inside the CPU could be realized by using logic gates and flip-flops as stand-alone components with connections between them. This strategy for processor design was followed till late 1960s. However, due the presence of discrete components and off-chip connections, the devices used to be quite big and too slow to use in applications. In the early 1970s, with the phenomenal change in the semiconductor technology and the chip manufacturing process, entire CPU could be housed into a single chip, leading to the introduction of microprocessors. The size of a microprocessor

> There is nothing called a miniprocessor.

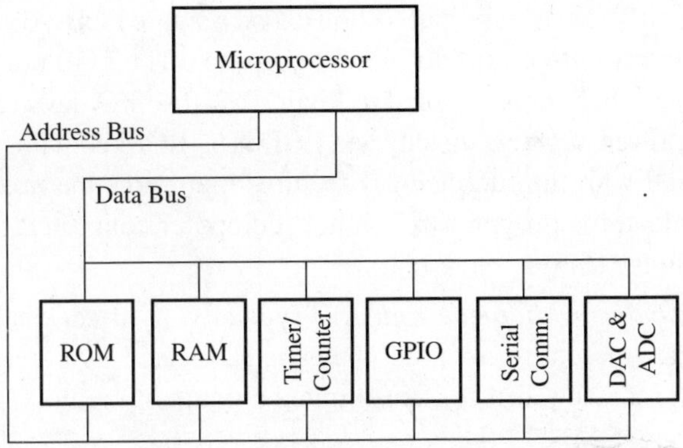

FIGURE 1.8 Generic microcomputer system

became several thousand times smaller than the discrete component-based designs. The speed improved by at least several hundred times. The dramatic improvement in speed came from the fact that in the semiconductor technology, on-chip communication is much faster (10–100 times) than the off-chip communication. In 1971, the first microprocessor, Intel's 4004, was introduced. It was a 4-bit processor (4-bit internal data bus) with 10-bit address bus and 4-bit data bus. It could operate at a frequency of 108 kHz. Intel 4004 was quickly succeeded by 8008 (8-bit data bus), 8080 and 8085. The microprocessor 8085 was introduced in the year 1976. It has 8-bit data bus and 16-bit address bus and can operate at a clock frequency of 3 MHz.

A computing system built around a microprocessor is commonly known as a *microcomputer*. With the advancement in the microprocessor technology, such systems have found applications in almost any domain of scientific and business computations. Apart from the microprocessor, a microcomputer system contains memory (both ROM and RAM), timers, digital input-output lines for parallel communication (GPIO – General Purpose Input Output), serial communication interface, analog-to-digital and digital-to-analog interfaces to interact with the analog environment and so on. The timers can also be used as external event counters. The components of a generic microcomputer system along with its address and data buses have been shown in Figure 1.8.

A general classification of microcomputer systems into different categories is as follows.

1. *Personal Computer (PC)/Desktop*: These are mostly single-user systems that can perform a wide range of computations. The applications include office automation, finance, internet browsing and limited amount of scientific computations. The systems generally contain a large number of application packages.
2. *Workstation*: These are high-performance desktop systems dedicated to some applications. These systems can perform certain engineering/scientific computations efficiently. Examples of such applications include Computer-Aided Design (CAD), Computer-Aided Manufacturing (CAM), and Computer-Aided Engineering (CAE). The microprocessors used in such systems are high-end ones with large primary and secondary storage, high-resolution screen and specialized I/O devices (for example, plotter and digitizer).

3. *Single-Board Computers*: These are computers housed on a single PCB. Also known as trainer boards, such computers are primarily used for educational purposes. The PCB contains the microprocessor, limited RAM, ROM and other chips as noted in Figure 1.8. The I/O devices are very simple, such as matrix keyboard, LED, seven-segment display and LCD. The ROM contains a monitor program which is very simple and small with the rudimentary features that allow the user to enter the program in machine code or download the program from a host computer connected to it, execute the program and, if necessary, debug the same.
4. *Single-Chip Microcomputers* (*Microcontrollers*): Typically used in small applications involving a controlling task. Here, the entire microcomputer is fabricated as a single chip. The chip, apart from containing a full processor, also contains limited amount of memory, timers, digital and analog input-output facilities.

1.5 Microcontroller

LO 4

The typical microcontroller structure has been shown in Figure 1.9. These are single-chip microcomputers containing processor, memory and other peripherals, clock generation circuitry, timers, data converters and so on. However, to keep the area under control and the power consumption within limit, the processor is generally much simpler than the advanced microprocessors, ROM and RAM spaces are limited. Thus, only small programs can be held there. However, almost all microcontrollers support extended memory with externally connected ROM and RAM chips. Microcontrollers are typically used

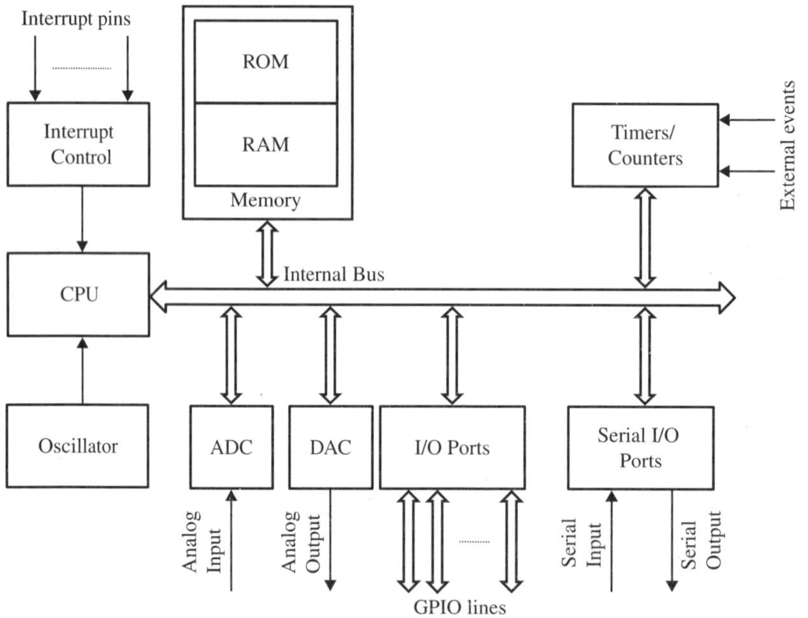

FIGURE 1.9 Block diagram of a typical microcontroller

in dedicated applications, known as *Embedded Systems*, such as mobile phone, digital camera, security alarm and plant control.

The 8-bit microcontroller, 8051, introduced by Intel in the year 1981, has 128-byte RAM, 4kB ROM, two timers, one serial port and four 8-bit parallel ports. Several variants of the basic 8051 are also available, namely 8031 and 8052. Several other microcontrollers also got popularity, such as Zylog's Z8, PIC, AVR and ARM to name a few.

1.6 Selecting Microprocessor/Microcontroller for Application

With so many different microprocessors and microcontrollers being available from various vendors, it is extremely important to choose the most suitable one for an application. An electronic design may have different types of constraints applied on it – some of which are scientific and some are financial. Scientific constraints include issues, such as area, performance and power, whereas the financial constraints may be on the designated price range of the product, time-to-market etc. Each processor has its own architecture, unique instruction set and in-built peripherals. Thus, choosing the proper device for an application becomes a challenge. The following are the major concerns in selecting a candidate processor.

1. Whether the highest available speed of the device is sufficient for the application or not. The processors vary widely on the range of frequency at which they can operate. Thus, choosing the correct processor is guided by the speed requirement of the application.
2. The size of the chip matters, as it decides the PCB size for the system. For example, the chip may have a 40-pin dual-inline packaging (DIP) and 40-pin quad-flat packaging (QFP). The packaging also determines the assembling and prototyping of the product.
3. Power consumption is also an important issue, as it determines the battery life for battery-operated devices. With increasing power consumption, heat generation also increases which may necessitate cooling arrangements. Now, depending on the design of the final product, the cooling facility may or may not be feasible.
4. For microcontrollers, the amount of on-chip RAM/ROM space available should ideally be able to hold the entire program for the application. Otherwise, external memory is needed to be interfaced, subject to the availability of such facility with the microcontroller. On-chip memory access may be 10–100 times faster than the off-chip external memory access.
5. Availability of other on-chip resources, such as number of I/O pins, number of timers and their sizes, availability of ADCs and DACs along with the number of channels and resolution, may play crucial roles in deciding the processor. These must cater to the requirements of the application. Most of the microcontrollers possess some such resources within them, whereas, for microprocessors, availability of such companion chips may be an issue.
6. Cost of a single processor chip often becomes a deciding factor in the overall cost of the system.
7. Ease of upgradation to higher performance and/or lower power consumption variants may be a factor guiding processor selection.

8. Availability of good development platform is a very essential issue. Each processor has its own instruction set. Mastering all of them is very difficult, if not impossible, for any system designer. To solve this problem, the vendors of these chips normally come up with a set of software to help in programming the microcontroller. It is often the case that the development platform allows the program to be written in some high-level language, such as C. Processor-specific compiler and assembler translate the program to machine language. User-friendliness of such development platform often biases the selection of microcontroller by the designer.
9. Availability of processor chip and its companions in the market is another guiding factor. Many of the processors are now available as *Intellectual Property Core* (*IP Core*) to be integrated in single-chip *System-on-Chip* (*SoC*) designs. With the advancement in VLSI design, this domain has progressed a lot.

1.7 Low and High Level Languages LO 5

Any CPU understands the instructions coded in binary form only. For a program to be executed by the processor, it must be available in this format. A program in binary form that is understandable by the processor is called a *machine language program* for that processor. Though easily understood by the processor, it is too cumbersome for any user to write a program in machine language directly. At the same time, the user may wish to work close to the actual CPU hardware (in terms of CPU registers, ALU operations etc.) to have a much optimized version of his/her program. To enable it, processor designers have created an intermediary language for the valid instruction set of a processor. This is called the *assembly language* of the processor. For every instruction supported by the processor, the designer has a fixed, binary-coded instruction in the machine language. For example, the assembly language statement “ADD B” of 8085 has a fixed machine language code 87H. A user may write the program in assembly language in a text file. It is converted to a machine-level binary file by a tool (often provided by the processor designer), called *assembler*. The task of an assembler has been shown in Figure 1.10. Programs written in assembly language can refer to the CPU registers, I/O ports, memory locations and other system resources directly. Thus, the size of the program and the time needed for its execution can be determined by the programmer by consulting the user manual of the processor.

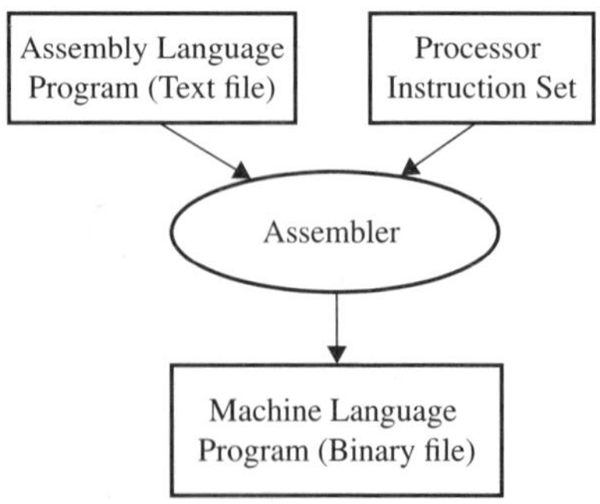

FIGURE 1.10 Assembly language vs. machine language

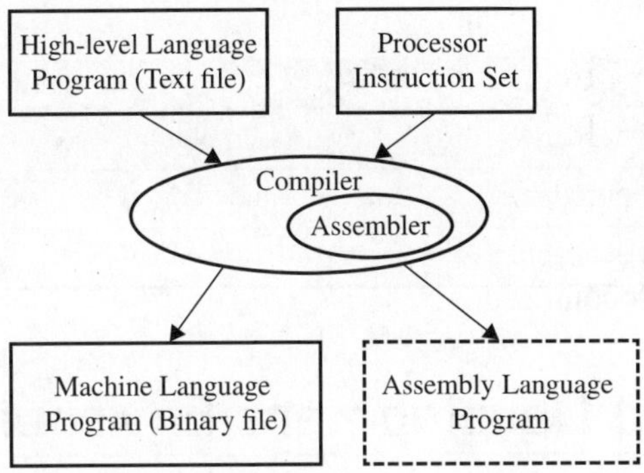

FIGURE 1.11 High-level language compilation

Machine language is the lowest level at which a program may be written for a processor. Assembly language is the next higher level providing facility to have a program in human-readable form. For a program of medium-to-high complexity, even assembly language programming is quite difficult. This is particularly so because of the absence of programming constructs like *loop*, *if-then-else*, and *switch-case*. It is cumbersome to express the program logic elegantly in assembly language. In particular, readability of the program code and understanding by a third person becomes an issue. Also, the maintainability, debugging and upgradation of assembly language program are difficult. Even with a moderate degree of documentation of program code, it is very difficult for a third person to understand an assembly language program. In contrast, programs written in languages like C/C++ may utilize all the programming facilities. Such languages are known as *high level languages*. A program written in a high-level language is converted to machine level by means of tools known as *compilers*. Compilers often generate code in some intermediary language, which is then converted to the assembly language of the targeted processor. An assembler of the target processor converts the assembly language program to the final machine code, understood by the processor. For programs written in high-level languages, updating and migrating to other processors become very easy. The compiler may be asked to generate the assembly language program also, as a by-product. This has been shown in Figure 1.11. It should be understood that due to the automated translation of programs written in high-level languages, the machine code generated by a compiler may not be highly optimized, compared to the assembly language program written by an experienced system programmer. A summary of feature comparison between low-level and high-level languages has been shown in Table 1.4.

TABLE 1.4 Feature comparison between low-level, assembly-level and high-level languages

Feature	Low-Level Language	Assembly Language	High-Level Language
Programmer friendly	No	Moderate	Yes
Machine friendly	Yes	Yes	No
Debugging	Difficult	Fair	Easy
Maintenance	Difficult	Fair	Easy

(Continued)

TABLE 1.4 *Continued*

Feature	Low-Level Language	Assembly Language	High-Level Language
Compiler/Interpreter	Not needed	Assembler	Compiler
Porting across platforms	Not portable	Can be ported	Can be ported
Library functions	Not available	Not available	Available

1.8 Integrated Development Environment (LO 6)

Developing a complete application, starting with the program written in some high-level language and targeting a microprocessor/microcontroller, requires a set of tools (such as compilers) specific to that processor. To make such system design task easy, the vendors often come up with an integrated environment that aids in the entire development cycle. This complete tool is known as an *Integrated Development Environment* (*IDE*) that contains all the tools necessary for system design. An IDE typically contains a code editor, a compiler and a debugger. A *Graphical User Interface* (*GUI*) is also provided. It may be noted that IDEs are also available in other domains. For example, *Microsoft Visual Studio* provides an environment for developing computer programs, supporting various programming languages.

The tools commonly available in an IDE targeted to microprocessor/microcontroller systems are as follows.

1. *Editor*: This is the first tool in the development chain. The tool helps in writing program code in some high-level language. The code is often termed *source code.*
2. *Compiler*: The compiler is used to translate the source code into machine code, also called *object code.* The IDE may be running on a PC with Pentium processor, while the compiler in the IDE is producing code for some other processor (say 8085), the *target processor*. Such a compiler running on a system and producing code for some other target processor is called a *Cross Compiler*.
3. *Assembler*: System designer may prefer to develop the application using the assembly language of the processor. As noted earlier, an assembler translates the assembly language program to machine language. The assembler is also included as a step in the compilation process.
4. *Linker*: This tool combines a number of object code files into a single executable. This provides an avenue for a team of coders to design parts of the program and then combine into one module later. Additionally, an IDE may contain predesigned libraries to be included in the executable. Figure 1.12 shows the details.
5. *Debugger*: While a compiler or assembler can detect the grammatical errors (often called syntax errors), they cannot detect logical errors in the program. For this, a debugger tool is necessary. A debugging software running inside the IDE can be used to diagnose logical errors in the code. There are two approaches followed by a debugger – *simulation* and *emulation*.
   a. *Simulation*: A simulator for a processor models its behaviour in software. However, it should be understood that a simulator cannot match the speed of a real processor. Simulating the external events (such as inputs from sensors arriving at desired time) is also very difficult. Thus, a simulator may not be able to mimic the behaviour of the overall system in the true sense.

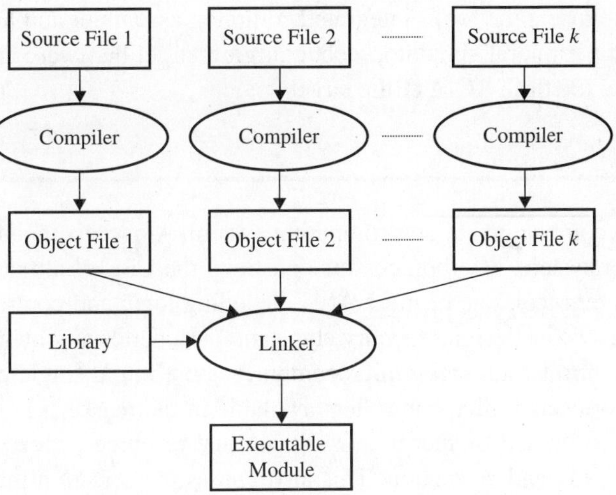

FIGURE 1.12 Linking multiple source and library modules

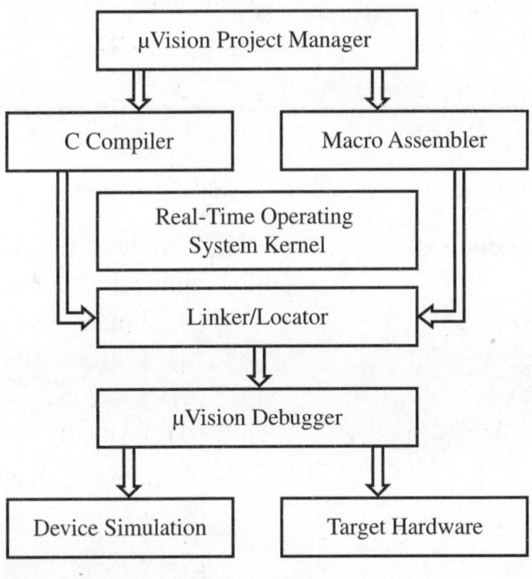

FIGURE 1.13 Components in Keil µVision IDE

b. *Emulation*: This is a piece of hardware that behaves exactly as the microprocessor/microcontroller chip. The functionality of the processor is emulated in real time. An emulator stores the program in RAM which behaves as the program memory. The RAM code can be modified quickly to aid in the debugging process. Once the designer is satisfied with the performance of the system in the emulator, the program may be downloaded into ROM/EPROM of the final board/chip.

Some of the well-known IDEs include *MPLAB* for PIC microcontrollers, *Arduino* for Arduino microcontrollers and *Keil* for range of microcontrollers like 8051, ARM7 and C16x/ST10. Figure 1.13 shows the structure and essential components of *Keil µVision* IDE. It has a full-featured source code editor, device database for configuring the tools, integrated compiler, assembler and linker, source- and assembly-level debugger with peripheral simulator, debugging on target hardware and flash programming to download final code into the flash ROM of the target device.

SUMMARY

Processors are the main components of any computing system. A microcomputer is a single-chip implementation of a processor/system. Microprocessors constitute the CPU of a microcomputer. A microprocessor contains a set of registers, one or more ALU, decoding logic and controller for operation. Apart from that, to complete a system design, memory chips and peripherals are interfaced with the processor. To qualify for a microcontroller, an entire microcomputer is put into a single chip. With several microprocessors, their accessories and microcontrollers available in the market, choice of the one to be used for an application is often guided by the processor's internal resources, speed, cost etc. Availability of good IDE aids in debugging and correction. This also guides the selection process. To understand the architecture and operation of different processors, in the next chapter, we shall look into one of the most popular microprocessors from Intel, the 8085. This is perhaps the most studied microprocessor used by the academic community to initiate understanding the operation of all other processors, reported so far.

EXERCISES

Multiple Choice Questions

1. The processor execution sequence is
   (a) Fetch-Decode-Execute
   (b) Decode-Fetch-Execute
   (c) Execute-Fetch-Decode
   (d) Execute-Decode-Fetch
2. The third phase of instruction execution is
   (a) Fetch
   (b) Decode
   (c) Execute
   (d) None of the above
3. The component which is definitely not part of any microprocessor is
   (a) ALU
   (b) Register
   (c) Bus
   (d) None of the above
4. Which of the following is NOT a criterion in choosing a processor?
   (a) Number of pins
   (b) On-chip debugging
   (c) Easy availability
   (d) None of the above
5. Component that may not be part of microcontrollers is
   (a) Registers
   (b) Timers
   (c) DAC, ADC
   (d) None of the above
6. A low-level language is friendly to
   (a) Programmer
   (b) Processor
   (c) Both Programmer and Processor
   (d) None of the above

7. Porting of programs across microcontrollers will be easy if the program is written in
   (a) Machine language
   (b) Assembly language
   (c) High-level language
   (d) None of the above
8. Simulator and emulator are
   (a) Both software
   (b) Software and hardware respectively
   (c) Hardware and Software respectively
   (d) Both hardware
9. The first tool in IDE tool flow is
   (a) Editor (b) Compiler
   (c) Assembler (d) Debugger
10. Executable code is produced by
   (a) Editor (b) Compiler
   (c) Linker (d) Debugger
11. A stored-program computer must have
   (a) Memory
   (b) Processor
   (c) Both Memory and Processor
   (d) None of the above
12. Size of a word can be
   (a) 1 byte
   (b) 2 bytes
   (c) 4 bytes
   (d) All of the above
13. In a memory system, address decoder outputs are connected to the memory signal line
   (a) Chip Select
   (b) RD
   (c) WR
   (d) None of the above
14. If the processor word size is 32 bits and data bus of individual memory chips are of 8 bits, the number of memory chips to be activated in parallel is
   (a) 4
   (b) 2
   (c) 8
   (d) 3
15. Which combination of RD and WR signals is NOT valid?
   (a) Only RD active
   (b) Only WR active
   (c) Both RD and WR active
   (d) None are active
16. The other name of Program Counter is
   (a) Instruction Pointer
   (b) Memory Counter
   (c) Instruction Counter
   (d) None of the above
17. GPIO stands for
   (a) General Purpose Input Output
   (b) General Port for Input Output
   (c) Global Port for Input Output
   (d) None of the above
18. High-resolution screen is often used in
   (a) Desktops
   (b) Workstations
   (c) Embedded computers
   (d) None of the above
19. QFP stands for
   (a) Quadrature Frequency Port
   (b) Quad-Flat Packaging
   (c) Quadrature Flat Packaging
   (d) Quad Frequency Port
20. Which memory access is faster?
   (a) On-chip
   (b) Off-chip
   (c) Both are same speed
   (d) Depends on size
21. An assembler generates machine language program from
   (a) High-level language
   (b) Multiple languages
   (c) Assembly language
   (d) Another machine language
22. The files input to a linker are
   (a) Source files
   (b) Object files
   (c) Both source and object files
   (d) None of the above

23. The output of linker is an
    (a) Object file
    (b) Source file
    (c) Executable file
    (d) None of the above
24. A debugger works on
    (a) Source file
    (b) Object file
    (c) Executable file
    (d) All of the above
25. The library modules are attached by
    (a) Compiler
    (b) Assembler
    (c) Linker
    (d) All of the above

Answers to Multiple Choice Questions

1. a	2. c	3. d	4. d	5. d	6. b	7. c	8. b	9. a	10. c
11. c	12. d	13. a	14. a	15. c	16. a	17. a	18. b	19. b	20. a
21. c	22. b	23. c	24. c	25. c					

Long Answer Type Questions

1. List down the CPU signal lines to be connected to memory for interfacing.
2. How does a decoder help in memory addressing?
3. When does the Program Counter get incremented during program execution?
4. A hypothetical CPU has 18-bit address bus and 8-bit data bus. Draw a diagram depicting the interfacing of 4kB RAM and 4kB ROM with it using memory chips of size 2k × 4-bits each. Show all control line connections as well.
5. With respect to hypothetical CPU shown in Figure 1.7, list down the control signals to be activated at different time steps to execute the instruction "SUB R1, R2, R3" that computes R3 ← R1 – R2.
6. Explain how the availability of large number of GFRs may help in program execution.
7. How does the instruction decoder aid in program execution by the processor?
8. Study about the typical flag bits in the FLAGS register of a processor.
9. How does a microcomputer system differ from a general computer system?
10. Identify the typical components of a microcontroller.
11. How does a microcontroller differ from a microprocessor?
12. Compare the features of low- and high-level languages.
13. Identify the components of an IDE.
14. Study the Keil μVision IDE for the microcontroller 8051 and list its features.
15. How does simulation differ from emulation?

2

The 8085 Microprocessor Architecture

LEARNING OUTCOMES

After reading this chapter, students will be able to:

LO 1 State the important features of 8085

LO 2 Draw pin diagram of 8085

LO 3 Group pins of 8085 as per their functionalities

LO 4 Draw and explain the functional block diagram of 8085

LO 5 Draw and illustrate the machine cycles corresponding to 8085 operations

LO 6 Interface memory and I/O devices with the 8085

LO 7 State the procedure to interface DMA chip with the 8085

2.1 Introduction

LO 1

> An 8-bit processor performs operations primarily on 8-bit data. However, there may be some operations on lesser or higher sized operands, as well.

The *Intel 8085* (also pronounced as *eighty-eighty-five*) is one of the very popular microprocessors. It is an 8-bit microprocessor, introduced in March 1976. It is fully compatible (*software-binary compatible*) with its predecessor 8080. This means that the machine language programs of 8080 will run without any modification or reassembly, by the 8085 as well. Compared to 8080, there are only two minor changes – interrupt services enhanced (5 interrupt lines in 8085, compared to only 1 in 8080) and serial input/output feature has been introduced. Also, the 8085 requires only a +5V power supply compared to +5V, –5V and +12V required for 8080. The 8085 comes in a 40-pin DIP package with 16-bit address and 8-bit data lines. The lower order address bus and the data bus share the pins to multiplex their operation. Some important features of the 8085 are as follows.

1. The 8085 is an 8-bit processor that can access, process and provide 8-bit data simultaneously.
2. The clock has a 50% duty cycle.
3. The 8085 can operate at 3-MHz clock frequency; however, the variant 8085A-2 can operate upto a maximum frequency of 5 MHz.
4. Provides facility to connect upto 256 input-output ports.
5. A total of 74 machine instructions are supported.
6. Five different addressing modes are supported.
7. The arithmetic operations supported are addition, subtraction, increment and decrement. The logic operations include AND, OR, XOR, NOT and bit-shifting.
8. Enough status information is made available, so that external devices can identify the machine cycle the 8085 is in at any point of time.
9. Though only five interrupt pins are available, it is possible to increase the number of interrupt sources significantly.
10. The 8085 has the ability to share the system bus with the *Direct Memory Access* (*DMA*) controller. This helps in transferring huge chunk of data from the input-output device to the memory or vice versa at high speed bypassing the CPU.

2.2 The 8085 Chip

LO 2, 3

Figure 2.1 shows the pin diagram of the 8085 chip. As noted earlier, the chip is a 40-pin DIP package. Figure 2.1(a) lists the pins in order of their position in the package, while Figure 2.1(b) presents the pins in terms of their functional grouping. In the following, the pins have been described in terms of their functions.

DIP stands for *Dual In-line Package*. The pins of the chip are arranged in two parallel rows along the periphery. Another organization QFP (Quad Flat Package) arranges pins in all the four sides of the chip.

V_{CC}, V_{SS} (Pins 40, 20): The pin V_{CC} is the positive power supply of +5V needed by the chip. V_{SS} is the pin for ground connection.

X_1, X_2, CLK(OUT) (Pins 1, 2, 37): These pins are related to the clock generation and distribution. A crystal of frequency 6 MHz is to be connected between the pins X_1 and X_2 to feed the internal clock generation circuitry. The clock circuit divides the crystal frequency by 2 to generate smooth clock signal with 50% duty cycle. The clock generated in this manner is used by all the sequential logic elements within the chip. This clock is further output at pin 37, called CLK(OUT) for interfacing any device that may need to synchronize its operation with the 8085 CPU.

$\overline{\text{RESET IN}}$, RESET OUT (Pins 36, 3): The pin $\overline{\text{RESET IN}}$ is an input pin to the 8085. If the signal to this pin goes low, the processor is reset. This causes the processor to resume its operation by fetching instruction from memory address zero (the *Program Counter* (*PC*) register in the processor gets reset

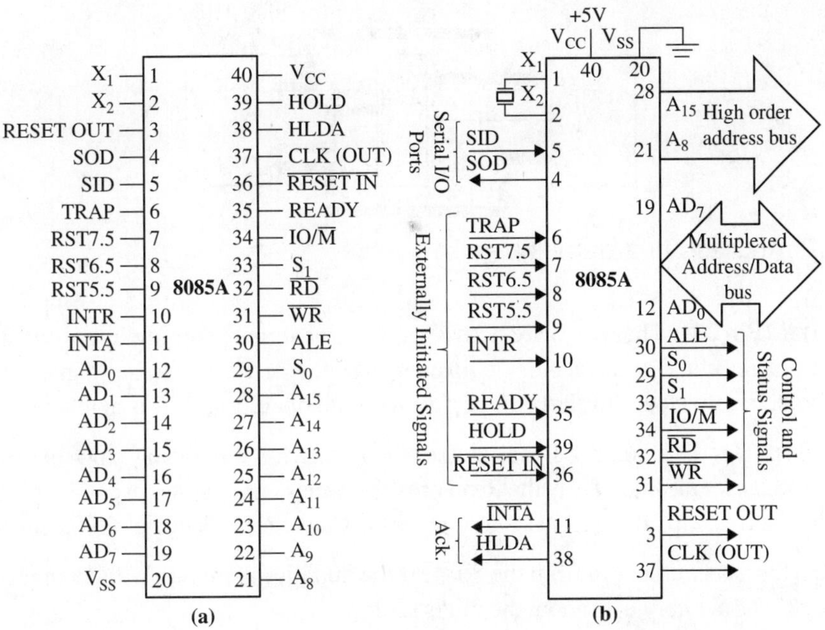

FIGURE 2.1 8085A processor **(a)** Pin diagram **(b)** Grouping of pins

to zero), the buses get tri-stated. Along with that, the output pin RESET OUT is made high. This can be used by the connected devices to reset their operation as well, when the processor is reset.

AD_0-AD_7 (Pins 12–19), A_8-A_{15} (Pins 21–28), ALE (Pin 30): These lines correspond to the address and data bus signals. The 8085 has 16-bit address bus A_{15}-A_0 and 8-bit data bus D_7-D_0. While there are explicit pins for the lines A_{15}-A_8, the lower order address bus (A_7-A_0) and the 8-bit data bus (D_7-D_0) are mapped onto same pins, 12–19. This is done to keep the number of pins required for the chip under control. For a memory access (both for instruction and data), the lines A_{15}-A_8 hold the higher order address bits continually for the entire duration of the memory access. However, the pins AD_7-AD_0 hold the lower order address bits in the early stages of memory access (first clock cycle only). Once the data is put into the data bus, these lines hold the data bits. It may be noted that the data is put onto the data bus by the 8085 for a *write* operation, while for a *read* operation, the data is put by the memory module. However, the duration for which the lower order address is available may not be sufficient for many of the memory chips. To solve this problem, the 8085 designers have provided an additional signal at pin 30, called ALE (*Address Latch Enable*), that is expected to be used by the system designers to latch the address by means of some external latch. This has been shown in Figure 2.2. The ALE signal is connected to the *enable* line (EN) of the 8-bit latch. When EN is activated, the values available at AD_7-AD_0 are stored in the 8-bit latch and are available at latch output. Now, even if the ALE gets deactivated, the latched output remain unaffected and are available to provide the lower order address inputs to the memory continually. It may be noted that the higher order address lines A_{15}-A_8 do not require any such latch as they are held steady for the entire access duration by the 8085 itself. The ALE signal is used to demultiplex the lower order address bus and the data bus lines by using an external latch.

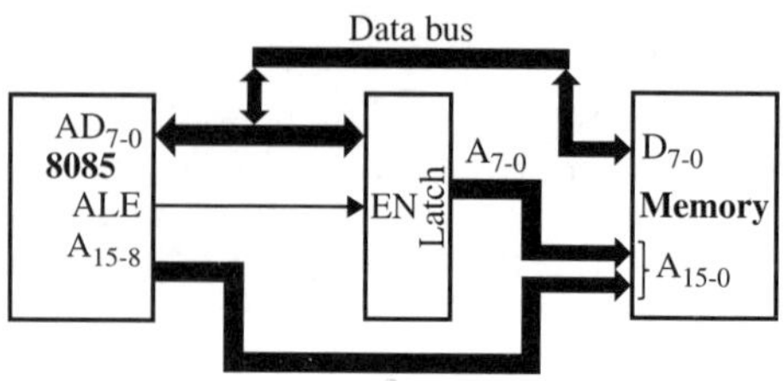

FIGURE 2.2 Address bus demultiplexing

***$\overline{RD}$* (Pin 32), *$\overline{WR}$* (Pin 31):** These two are active low control signals coming out of the 8085 chip. The signals are used to interface with memory chips and I/O devices. For reading content from memory/devices, $\overline{RD}$ signal is activated. Similarly, to perform a write operation, $\overline{WR}$ is actuated.

READY (Pin 35): This is a control signal to the 8085 used for interfacing slow peripheral modules. Activating this line causes the 8085 delay the Read or Write cycles until a slow-responding peripheral device is ready to send or accept data. If the signal is low, the 8085 waits for clock cycles till the line becomes high.

IO/$\overline{M}$(Pin 34): This is an output pin from the 8085. If the 8085 is interacting with an input-output device, the pin is set high. For memory operation, the pin is set low.

S_0 (Pin 29), S_1 (Pin 33): These two are status signals like IO/$\overline{M}$. The values of these lines combined with IO/$\overline{M}$ give information about the operation currently carried out by the 8085. The execution of an instruction consists of a set of *machine cycles*. One machine cycle may refer to a memory read, an instruction fetch, a memory write, an I/O read/write etc. Table 2.1 shows the values of the pins IO/$\overline{M}$, S_1 and S_0 for different machine cycles.

SOD, SID (Pins 4, 5): These two pins are for serial communication through the 8085 chip. The serial communication transmits/receives one bit of data at a time. Thus, a complete byte is transmitted over 8 clock cycles, compared to the single clock cycle parallel communication through the data bus lines.

TABLE 2.1 IO/$\overline{M}$, S_1, S_0 values for machine cycles

Machine Cycle	Status Signals			Control Signal
	IO/$\overline{M}$	S_1	S_0	
Fetch	0	1	1	$\overline{RD}$
Memory Read	0	1	0	$\overline{RD}$
Memory Write	0	0	1	$\overline{WR}$
I/O Read	1	1	0	$\overline{RD}$
I/O Write	1	0	1	$\overline{WR}$
InterruptAcknowledgement	1	1	1	$\overline{INTA}$
Halt	Z	0	0	$\overline{RD} = \overline{WR} = Z$
Hold	Z	X	X	$\overline{INTA} = 0$
Reset	Z	X	X	

The transmission takes place via the pin *Serial Output Data* (*SOD*) while the reception happens via the *Serial Input Data* (*SID*) pin.

HOLD, HLDA (Pins 39, 38): These two pins are useful in interfacing another bus master with the 8085. A typical such bus master may be a chip acting as *DMA* controller. The address-, data- and control-bus connections to the memory modules may be driven by the 8085 and the DMA controller. The DMA controller is responsible for transferring large chunk of data from peripheral devices to the memory, without the intervention of the 8085. For this, the DMA controller sends a request to the 8085 by activating the HOLD signal, requesting the control of buses. If the 8085 is ready to release the buses, it relinquishes the control of the buses and acknowledges via the HLDA line to the DMA controller.

TRAP, RST7.5, RST6.5, RST5.5., INTR, $\overline{\text{INTA}}$ (Pins 6–11): These pins correspond to *interrupt* operations for the 8085. In its normal course of operation, the 8085 executes the instructions one-by-one, by fetching them from the memory sequentially. However, it may be necessary to notify the occurrence of some external event to the processor. For example, a fire sensor interfaced with the 8085 may have to inform the occurrence of fire to the processor. Though the 8085 may read the sensor periodically, the occurrence of fire is a rare event. Thus, periodic reading leads to wastage of the processor time. Rather, a peripheral device may interrupt the normal processing sequence of the 8085 to request it to take up appropriate action to handle the situation. A real-life analogy may be as follows. Consider the situation in which a person is reading a book in a room. The doorbell rings and the person needs to go to open the door. The visitor arriving to the house and pressing the doorbell switch generates an interrupt for the reading process by the reader. Upon hearing the doorbell, the person will first complete the line/paragraph being read, a bookmark would be kept to remember the page/line. The person will then go to open the door and interact with the visitor. Once the interaction is over, the person will come back and start reading from the point at which it was left. Here, going to open the door is equivalent to jumping to the routine to service the interrupt, interaction part is equivalent to the actual interrupt service and restarting the reading process is like resuming the original program being executed after serving the interrupt. The 8085 has a total of five such interrupt pins – TRAP, RST7.5, RST6.5, RST5.5. and INTR. Out of these, the interrupt TRAP cannot be inhibited (that is, masked off) while the rest are maskable. User may mask off some of the interrupts, so that, the 8085 does not respond to the occurrence of those interrupts until unmasked. However, the non-maskable interrupt line cannot be ignored. On occurrence of TRAP, RST7.5, RST6.5 and RST5.5, the 8085 branches to fixed memory addresses. However, on occurrence of INTR, the 8085 activates the $\overline{\text{INTA}}$ (*interrupt acknowledge*) line for the interrupting source to provide the starting memory address of the service routine to be executed. The process will be discussed in detail in Chapter 4.

2.3 Functional Description of the 8085

LO 4

Figure 2.3 shows a functional block diagram of the 8085 CPU. The internal architecture consists of the following – a set of General Purpose Registers (GPRs), a set of Special Function Registers (SFRs), ALU, Interrupt Control, Serial Control, Timing Control, Instruction Decoder, Buffers for the Address and Data buses.

GPR: These registers are used to store the temporary data pertaining to the arithmetic and logic operations. The individual registers are 8-bit wide and are named as A (Accumulator), B, C, D, E, H and L.

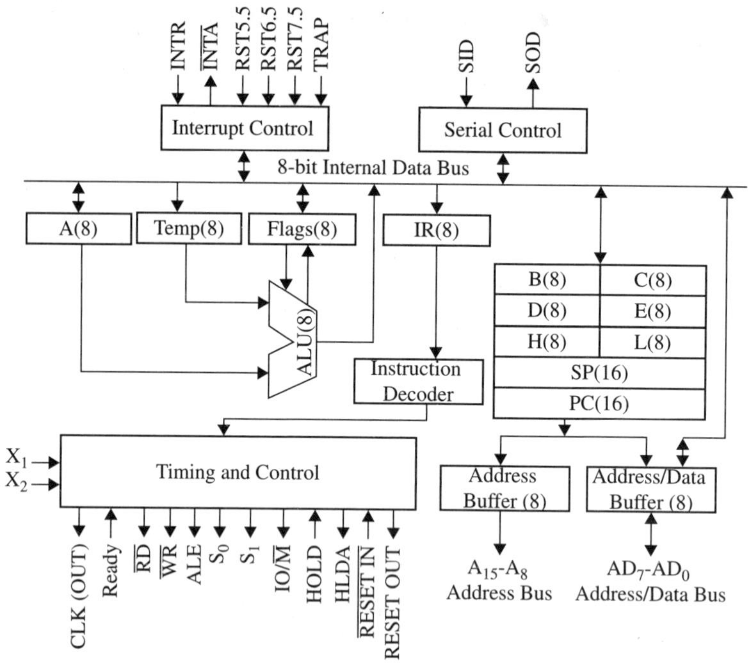

FIGURE 2.3 Internal architecture of the 8085A

Out of these, the A register is different in functionality in the sense that it acts as the first source operand and also the destination for all ALU operations. Though the individual B, C, D, E, H, L registers are 8-bit wide, a pair of them is often used as 16-bit operand in many instructions. The 8-bit registers B and C may be accessed as a 16-bit B-register. Similarly, the registers D and E, H and L form the 16-bit D- and H-register, respectively.

SFR: The 8085 contains a number of registers with some specific functionalities. As mentioned previously, the A register itself has got some special role in arithmetic and logic operations. Apart from that, the following registers also have some special functions.

- *Program Counter* (*PC*): This is a 16-bit register that contains the address of the next instruction to be executed by the 8085.
- *Stack Pointer* (*SP*): A program may contain a special subprogram within it, which is called from several places of the main program. On getting a subprogram call instruction, control branches to the subprogram. On finishing the called subprogram, control returns to the point in the main program from where it was called. Subprograms may have several levels of nesting – one subprogram may call further subprograms from its body. To implement subprogram call and return, a part of RAM space is defined as *stack*. Register values can be saved into the stack. In particular, when a subprogram is called, the address of the instruction, just after the call instruction (current PC content) is saved onto stack. While returning from the subprogram, the address is retrieved from the stack and copied into the PC, so that main program can resume from the position it branched to the subprogram. The stack grows in the upward direction and the SP register points

to the memory address till which the stack is full at present. For example, if the stack is implemented to grow from the RAM address 2000H, the SP is initialized to 2001H. If a subprogram call happens at this time, the current PC value is saved at memory address 2000H and 199FH. The new SP value becomes 199FH. After the subprogram has finished executing, contents of locations 2000H:199FH are copied to the PC, while the SP becomes 2001H.

- *Temporary Register* (*Temp*): This is an 8-bit register, not directly available to the user. The register holds the second operand for ALU operations.
- *Instruction Register* (*IR*): This is an 8-bit register that stores the instruction, as fetched from the memory. It feeds the Instruction Decoder that is used to decode the meaning of the operation code (opcode) part of the instruction.
- *Flags*: This is an 8-bit register that holds information about the outcome of the last ALU operation. The structure of this register has been shown in Figure 2.4. The meanings of the individual bits of the register are as follows.

D7	D6	D5	D4	D3	D2	D1	D0
S	Z		AC		P		CY

FIGURE 2.4 Flags register

  - *Z* (*Zero*): Set to 1 if the result of the last ALU operation is zero, reset otherwise.
  - *CY* (*Carry*): Set to 1 if the last ALU operation generated a carry bit at the most significant bit position, reset otherwise.
  - *S* (*Sign*): Set to 1 if the result of the last ALU operation is negative, reset otherwise.
  - *P* (*Parity*): Set to 1 if the 8-bit result of the last ALU operation has an even number of 1s in it, reset otherwise.
  - *AC* (*Auxiliary Carry*): This bit is set to 1 if in the last ALU operation a carry was generated from the bit-position 3 to the bit-position 4, reset otherwise. This flag bit is used in conjunction with *BCD* (*Binary Coded Decimal*) operations.

Arithmetic Logic Unit (ALU): This is the main component in the CPU that performs the computational tasks. In the 8085, the ALU is an 8-bit one, that is, the operands and result are 8-bit in size. Of the two inputs to the ALU, the first operand comes from the accumulator register, A. The second operand comes from the temporary register. The result is stored in the A register. The operations that could be carried out by the ALU are addition, subtraction, increment, decrement and logical operations – AND, OR, XOR, NOT, Compare and bit shifting of A – left shift, right shift, rotate left and rotate right. As noted previously, the outcome of the ALU operation also affects the Flags register.

Serial Control: This module controls the SID and SOD pins of the 8085. One of the bits of the A register is output to the SOD line. Similarly, the data available at SID pin can be transferred to a bit of A register for serial input under program control.

Interrupt Control: This module is responsible for the interfacing of interrupt lines. The controller can be instructed to enable or disable a selected set of interrupts. Among all the interrupts, the TRAP cannot be disabled, while the others could be. The handling of interrupts will be discussed later in Chapter 4.

Buffers: For the two external buses A_{15}-A_8 and AD_7-AD_0, two buffers are used to have steady values available at the buses while the 8085 performs other computations.

Timing and Control: This module can be said to be the heart of the 8085. It is responsible for generation of control signals over the clock cycles to get the intended task done by the processor. The control signals emanating from the module have been already discussed in the last section, and their activation will be enumerated as we discuss about different instructions and mechanisms throughout the book.

2.4 Instruction Cycle of 8085 LO 5

The 8085 follows the *Fetch-Decode-Execute* cycle for executing any instruction. This is called the *Instruction Cycle* of the processor. The whole instruction cycle consists of several clock cycles, also known as *T-states*. It may be noted that the duration of a T-state is equal to the period of the clock generated by the processor from the crystal connected to X_1 and X_2 pins of the 8085. A group of control signals are activated in a T-state. The timing and control unit activates control signals relevant for current T-state and corresponding to the current instruction being executed. Individual instructions may need different number of T-states depending upon the volume of work to be accomplished. A set of T-states pertaining to specific subtask forms a *machine cycle*. Thus, an instruction cycle consists of a set of machine cycles, whereas a machine cycle can further be divided into a set of clock cycles or T-states.

The following are the different types of machine cycles performed by the 8085.

- Opcode fetch
- Memory read
- Memory write
- I/O read
- I/O write
- Interrupt acknowledge

We shall detail the interrupt acknowledge machine cycle while discussing about the interrupts in Chapter 4. The rest of the machine cycle types are discussed next.

2.4.1 Opcode Fetch Machine Cycle

The opcode fetch cycle generally consists of four T-states. Some instructions take six T-states also. A fetch cycle with four T-states has been shown in Figure 2.5. The higher order address bus A_{15}-A_8 contains the higher order byte of the memory address (that is, higher order byte of PC) for three cycles T_1 through T_3. The lines AD_7-AD_0 hold the lower order byte of the address bus (that is, lower order byte of PC) in the first clock cycle, T_1. The ALE signal is activated in the T_1 state, so that the lower order address lines could be latched by the external circuitry. The lines $IO/\overline{M}$, S_1 and S_0 are set to 0, 1 and 1, respectively, to indicate that the 8085 is now in the opcode fetch machine cycle. The $\overline{RD}$ signal is activated across the states T_2 and T_3 and the opcode from the memory is expected to be available in the data bus lines AD_7-AD_0. At state T_4, the action taken by the processor is not documented by the designers of the 8085.

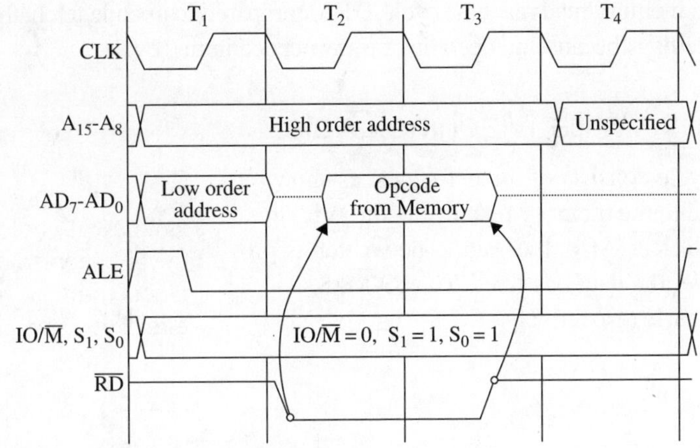

FIGURE 2.5 Opcode fetch machine cycle

However, it could be inferred that the 8085 uses this time to decode the opcode received to carry out the future operations for successful accomplishment of the task, intended by the instruction.

2.4.2 Memory Read Machine Cycle

The memory read cycle consists of three T-states, as shown in Figure 2.6. The operation is similar to the opcode fetch cycle, since both the fetch and the memory read operations read a byte from the memory address. However, unlike memory fetch, the memory address is not from the PC register. The address may come from some other registers in the register bank. The $IO/\overline{M}$, S_1 and S_0 lines are set to 0, 1 and 0, respectively, by the

> Memory read cycle is similar to opcode fetch, with the difference that opcode fetch cycle uses one more clock cycle to reserve time to decode the instruction opcode.

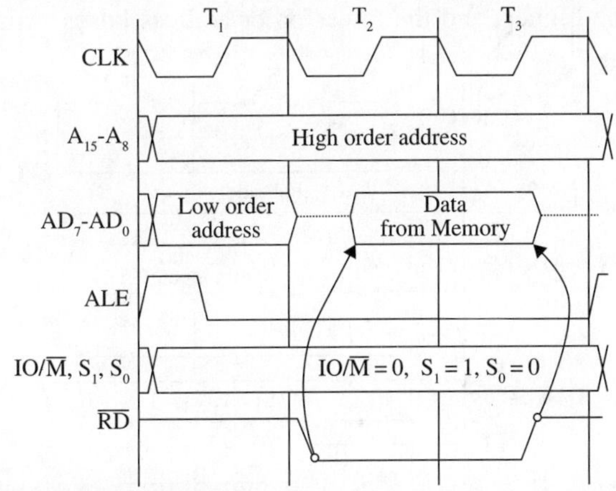

FIGURE 2.6 Memory read machine cycle

processor to indicate a memory read machine cycle. Also, compared to opcode fetch, this machine cycle has got one less T-state, perhaps because no decoding task is needed in this cycle.

2.4.3 Memory Write Machine Cycle

The memory write cycle consists of three T-states, as shown in Figure 2.7. Compared to the memory read cycle, the $\overline{WR}$ signal is activated, instead of $\overline{RD}$. Also, the data to be written is provided by the 8085 over the data bus lines. The IO/$\overline{M}$, S_1 and S_0 lines are set to 0, 0 and 1, respectively.

Unlike memory read cycle, in the write cycle, the content of AD_7-AD_0 never becomes undefined/tristated, since there is no reversal in the direction of the bus.

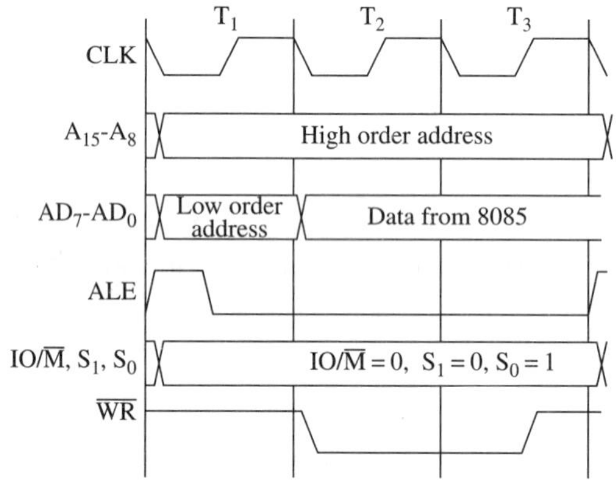

FIGURE 2.7 Memory write machine cycle

2.4.4 I/O Read and I/O Write Machine Cycles

The I/O read and I/O write machine cycles consist of three T-states each and have been shown in Figures 2.8 and 2.9, respectively. Since the I/O port addresses in the 8085 are only 8-bit wide, the same address is put onto both the higher- and the lower order address buses. Apart from that, the I/O read

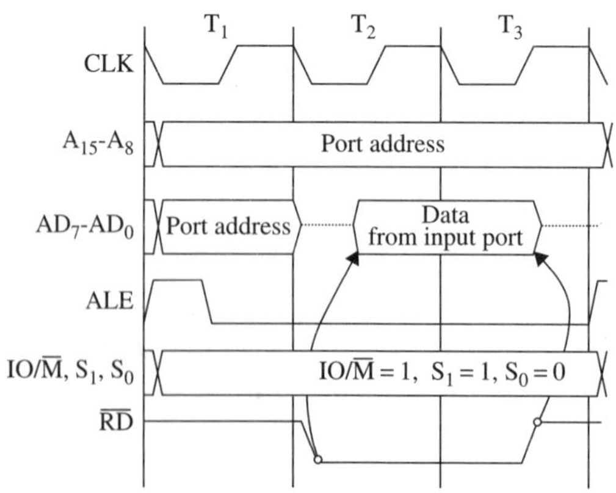

FIGURE 2.8 I/O Read machine cycle

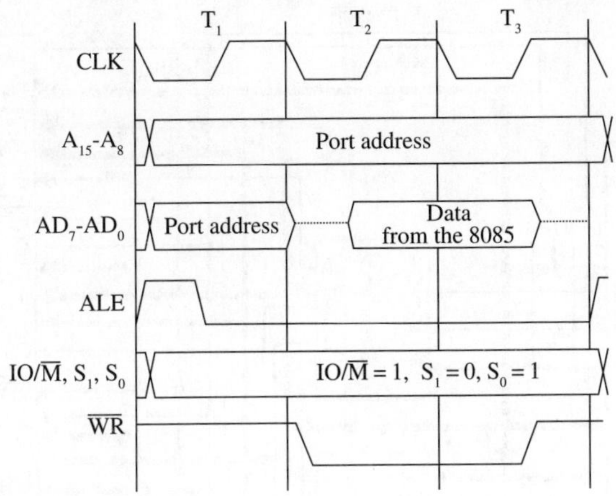

FIGURE 2.9 I/O write machine cycle

machine cycle is similar to the memory read machine cycle, whereas the I/O write machine cycle is similar to the memory write machine cycle with the $IO/\overline{M}$ bit being 1, indicating an I/O operation.

2.5 Memory and I/O Interfacing with 8085

LO 6

In Section 1.2, it has been shown how to interface memory chips with a CPU by connecting address, data and control lines through proper decoding logic. The same may be extended for the 8085 based system design as well. However, additionally, the 8085 provides some more signal lines, such as, ALE and $IO/\overline{M}$ that may be used by the system designers to interface the required number of memory chips and I/O devices. For this, it is assumed that the decoder chip used has an enable input (EN) which needs to be activated for the decoder to work. We shall discuss three different cases – interfacing only the memory modules, interfacing of I/O devices and finally a combined scheme of interfacing for systems containing both memory chips and I/O devices.

2.5.1 Interfacing Memory Modules

The 8085 has a 16-bit address bus. Thus, it can address upto 65536 (2^{16}) memory bytes. The total memory interfaced with the 8085, at a time, should be ≤65,536 bytes. Let, in a certain system configuration, the total memory to be interfaced with the 8085 be of size 2^m bytes, $m \leq 16$. If each memory chip is of size 2^p bytes ($p \leq m$), the number of such chips needed is 2^{m-p}. Since the individual memory chips are of size 2^p bytes, the p least significant address lines from the 8085 have to be connected to the p address lines of all memory chips. Furthermore, we should have a decoder with at least 2^{m-p} output lines, such that each output line can be used to select one memory module. This means that the decoder should have at least $(m-p)$ input lines. Depending upon the value of m, different cases may occur.

Case I, m=16. In this case, the entire memory space addressable by the 8085 has been utilized in the system. Figure 2.10 shows the case for memory chips with $p = 14$, that is, individual memory chips are

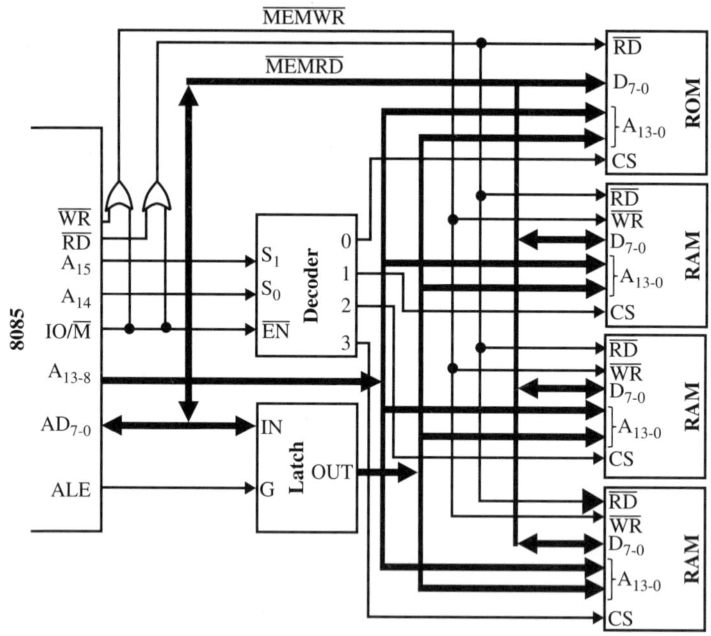

FIGURE 2.10 Interfacing 64 kB memory with 8085

of size 16 kB. Out of the total 64 kB, 16 kB is assumed to be ROM and the remaining 48 kB RAM. The ALE signal has been utilized to demultiplex the lower order address bus from the data bus. The IO/$\overline{\text{M}}$ line has been used to generate the enable signal for the address decoder and also to generate the memory read and write signals, in conjunction with the $\overline{\text{RD}}$ and the $\overline{\text{WR}}$ signals of the 8085.

Case II, m<16. In this case, the total memory interfaced is less than the full addressable capacity of 64 kB. Thus, all the 16 address lines of the 8085 may or may not be utilized for memory interfacing. Again, two situations can occur. In the first case, all 8085 address bits are utilized to either connect to the memory module or to the decoder input. The second case connects only the minimum required number of address lines to the decoder and p lines to the memory chips.

(i) If a decoder with $(16 - p)$ input lines is used, some of its outputs will remain unutilized, for the time being. In future, those outputs can be utilized to augment the memory capacity of the system. The situation has been shown in Figure 2.11 with $m = 14$ and $p = 13$. Here, the total memory interfaced is 16 kB with the individual chips being of capacity 8 kB. A 3-to-8 decoder has been used to generate the chip select signals for the memory chips. The first two outputs of the decoder have been connected to the memory chips and the remaining six outputs are available for future extension.

(ii) If $(m - p)$ address lines are used for decoding purpose, $(16 - m)$ most significant address lines from the 8085 remain unconnected to any memory chip or decoder. As a result, for an address generated by the 8085, the actual memory chip and location accessed becomes independent of those most significant $(16 - m)$ bits. For example, if $m = 14$, the 8085 address bits A_{15} and A_{14} remain unconnected. Memrory addresses generated by the 8085 have these two bits as don't cares, as a result,

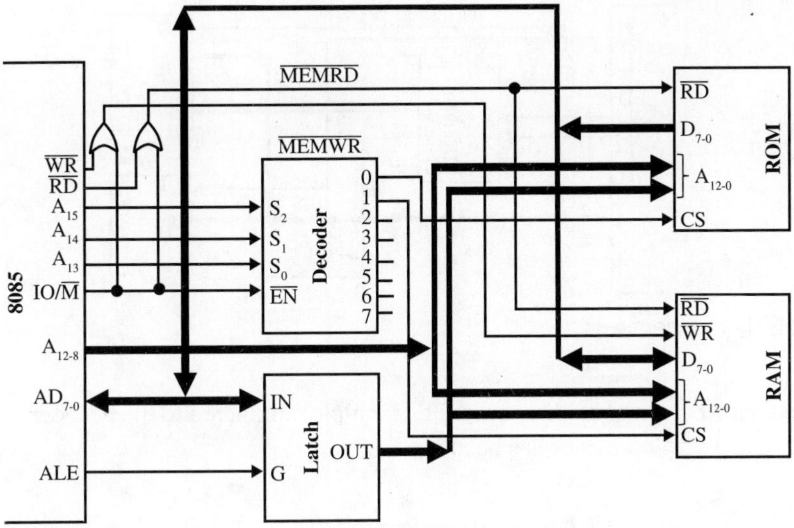

FIGURE 2.11 Interfacing 16 kB memory with 8 kB memory chips

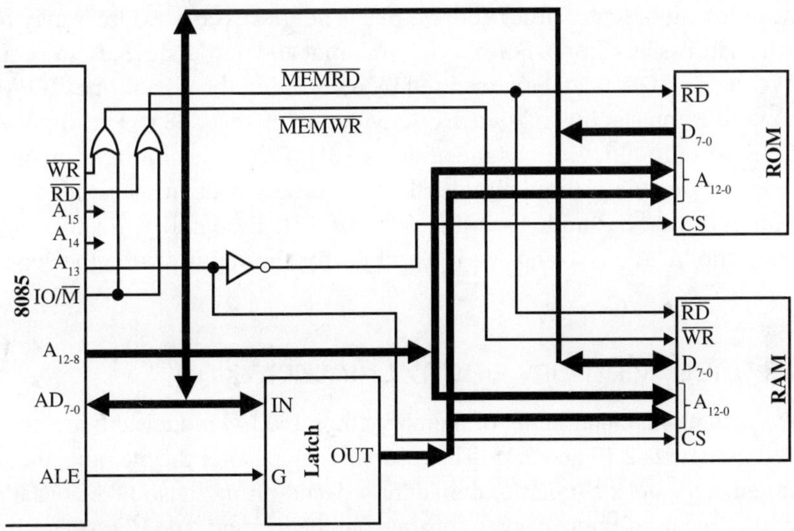

FIGURE 2.12 Memory folding

several memory addresses (differing in their don't care bits) get mapped onto some physical memory location. This phenomenon of multiple CPU-generated addresses being mapped onto the same physical memory location due to incomplete address decoding is called *memory folding*. For example, the addresses 0000H, 4000H, 8000H and C000H map to the same memory location – memory location 0 of the ROM. The addresses 2000H, 6000H, A000H and E000H map to the RAM address location 0. The last addressable location of the RAM has got multiple addresses – 3FFFH, 7FFFH, BFFFH and FFFFH. Folding restricts the extension of the memory capacity of the system. For example, unlike in Figure 2.11, for the memory system in Figure 2.12, no further memory chip can

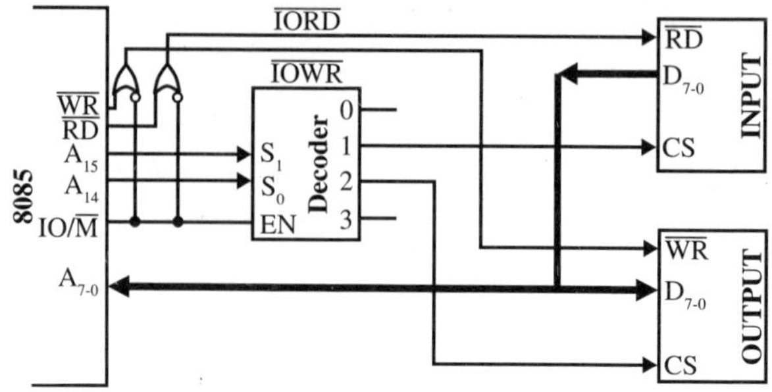

FIGURE 2.13 I/O interfacing

be connected. However, it saves hardware. For example, compared to the decoder in Figure 2.11, in Figure 2.12 only a single inverter has been used.

2.5.2 Interfacing I/O Modules

In 8085, for I/O access, the 8-bit address is available in both the higher order and the lower order address bus. Thus, demultiplexing of lower order address bus is not necessary. Address may be taken care of by the higher order address bus alone. For interfacing input and output devices, the control signals *IO read* ($\overline{IOR}$) and *IO write* ($\overline{IOW}$) can be generated by combining the signal lines IO/$\overline{M}$, $\overline{RD}$ and $\overline{WR}$. Figure 2.13 shows the interfacing of two I/O devices with the 8085. The input device has the address 40H while the output device has the address 80H. Principle similar to memory folding has been used to keep the decoding logic simple. Thus, the output device has addresses like 80H to BFH while the input device has addresses like 40H to 7FH. Essentially, for the output device, the requirement is that the bit A_{15} is 1 and A_{14} is 0, while for the input device the requirement is that A_{15} is 0 and A_{14} is 1.

2.5.3 Combined Memory and I/O Interfacing

Figure 2.14 shows a combined interfacing of memory chips and I/O modules. It is a combination of the schemes presented in Figures 2.12 and 2.13. The read/write signals for the memory and the I/O modules have been generated separately. Similarly, the address decoding part also takes place separately. The IO/$\overline{M}$ signal line has been used extensively to separate out the memory and I/O access parts. The address ranges for the memories and the I/O devices are same as those discussed in Sections 2.5.1 and 2.5.2, respectively. The memory and I/O module addressing patterns show folding in their ranges. The memory interface could use the scheme depicted in Figure 2.11 to avoid folding of memory space. Similarly, all the address lines A_8-A_{15} could be made to participate in the I/O address decoding to avoid the folding in the I/O address range.

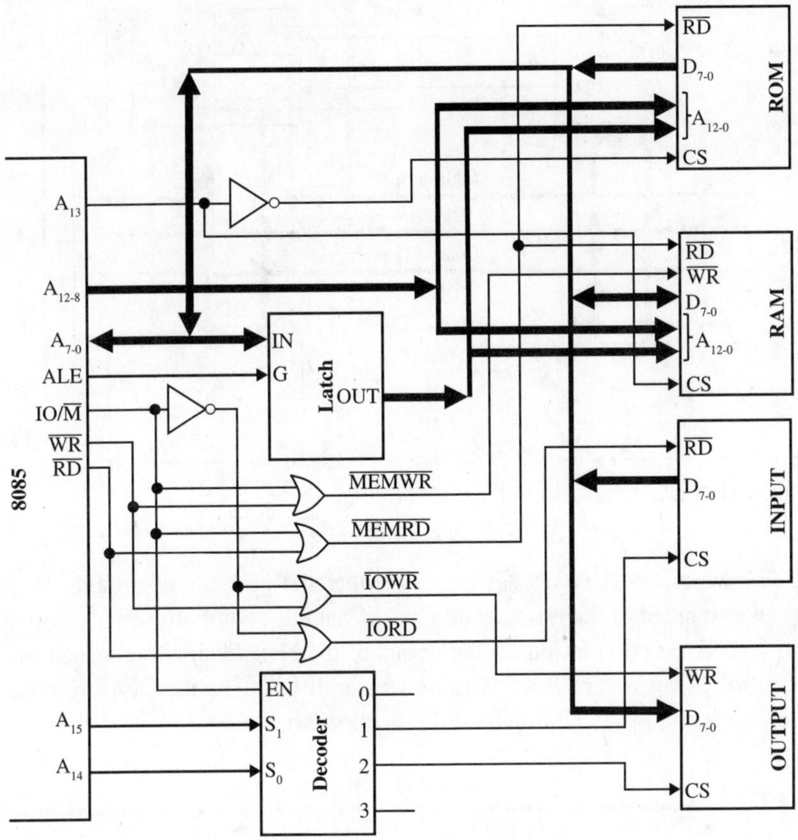

FIGURE 2.14 Combined memory and I/O interfacing

2.6 Direct Memory Access (DMA)

LO 7

The memory and I/O device interfacing techniques discussed in Section 2.5 can be utilized to transfer data from/to an I/O device to/from memory locations through the processor registers. Each byte transfer will necessitate one read operation, followed by a write operation. Two 8085 instructions are to be spent for each byte transfer. This makes the whole transfer process very slow. Moreover, in this transfer, the processor does not have any significant role to play, as no computation is involved in it.

A way out to save processor time and speed up the data transfer process is via *DMA*. In this, apart from the 8085, another bus controller chip (called DMA controller) is interfaced. Both the 8085 and the DMA controller have the capability to drive the address, data and control lines, as shown in Figure 2.15. Whenever an I/O device is ready for data transfer, it sends request to the DMA controller via the DRQ line. The DMA controller, in turn, sends a request to the 8085 by pulling the HOLD signal high. The 8085 relinquishes (turns to high impedance state) control of *Address*, *Data* and *Control* buses in the following machine cycle. Once the control has been relinquished, the 8085 pulls the acknowledgement

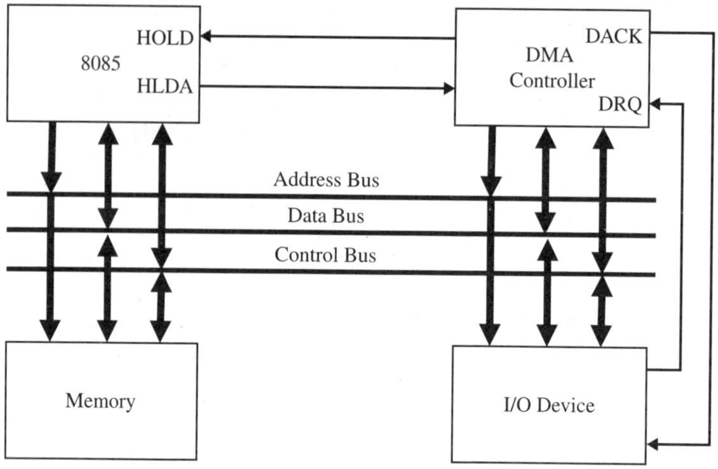

FIGURE 2.15 The 8085 and DMA controller

line HLDA high. Now the DMA controller acknowledges the peripheral through the DACK line. The DMA controller takes control of the buses and transfers data directly between memory and I/O devices, thus bypassing the 8085. At completion of data transfer, the HOLD signal is turned low, informing the 8085 to regain control of buses. The 8085 withdraws the HLDA. The detailed structure and interfacing with a DMA controller chip has been illustrated in Section 5.11.

SUMMARY

The 8085 is an 8-bit microprocessor, an improvement over its predecessor 8080. It is a 40-pin chip with 16-bit address and 8-bit data bus. The lower order address bus is multiplexed with the data bus. The 8085 uses the classical fetch-decode-execute instruction cycle. An instruction cycle consists of different types of machine cycles – opcode fetch, memory read/write, I/O read/write and interrupt acknowledge. Apart from parallel data transfer, 8085 also supports serial data transfer. The 8085 contains several general purpose and some special purpose registers in it. Memory interfacing with the 8085 may use either all or a subset of address lines, depending upon the total memory size. Incomplete decoding results in memory folding. Compared to data transfer through the CPU, DMA provides a much faster way of transfer between memory and peripheral. DMA bypasses the CPU altogether. This chapter has provided a detailed discussion on all these concepts. There are several interrupts supported by the 8085 which can be used to inform it about the occurrence of some external event. The interrupts have been discussed in detail in Chapter 4. The next chapter provides the details of the 8085 instruction set and its programming.

EXERCISES

Multiple Choice Questions

1. Number of interrupts in 8080 and 8085 are respectively
   (a) 1, 5 (b) 5, 1
   (c) 2, 5 (d) None of the above
2. Number of power supplies needed in 8080 is
   (a) 1 (b) 2
   (c) 3 (d) None of the above
3. On reset, program counter becomes
   (a) 0000H (b) FFFFH
   (c) 1000H (d) 1C00H
4. Duration of opcode fetch machine cycle can be
   (a) 3 (b) 4
   (c) 5 (d) 7
5. Destination in all arithmetic/logic operations in 8085 is
   (a) A (b) Immediate value
   (c) B (d) HL
6. Serial input of 8085 is
   (a) SID (b) SOD
   (c) SLD (d) None of the above
7. In 8085, compared to crystal frequency, processor frequency is
   (a) Same (b) Half
   (c) Double (d) None of the above
8. To get control of address and data buses from 8085, the line to be used is
   (a) HOLD (b) HLDA
   (c) RESET (d) None of the above
9. In 8085, the active state values of INTR and INTA signals are
   (a) Both high
   (b) INTR high, INTA low
   (c) INTR low, INTA high
   (d) Both low
10. Terminal Count register is of bit width
    (a) 8 (b) 14
    (c) 16 (d) None of the above
11. Power supply required for 8080 is/are
    (a) +5V (b) –5V
    (c) +12V (d) All of the above
12. Number of machine instructions in 8085 is
    (a) 56 (b) 74
    (c) 85 (d) None of the above
13. The duty cycle of the 8085 clock signal is
    (a) 20% (b) 30%
    (c) 40% (d) 50%
14. The Halt state of 8085 is identified by the values of S_1 and S_0
    (a) 0, 0 (b) 0, 1
    (c) 1, 0 (d) 1, 1
15. In 8085, the interrupt that cannot be inhibited is/are
    (a) TRAP (b) RST 7.5
    (c) RST 6.5 (d) All of the above
16. The number of 8-bit general-purpose registers in the 8085 is
    (a) 6 (b) 7
    (c) 8 (d) 9
17. The bit size of ALU in the 8085 is
    (a) 8 (b) 12
    (c) 4 (d) 16
18. If the last ALU operation in the 8085 has even number of 1s, the parity bit will be set to
    (a) 1 (b) 0
    (c) Not affected (d) Z
19. The serial data transmission happens from the 8085 register
    (a) H (b) D
    (c) A (d) PSW
20. To avoid memory folding
    (a) All address lines be used in decoding
    (b) Most significant address bit be used in decoding
    (c) Least significant address bit be used in decoding
    (d) None of the above
21. Memory folding provides
    (a) Enhanced memory space
    (b) Multiple address for one location

(c) Reduced Read/Write cycle time
(d) None of the above

22. DMA transfer happens
(a) Through the processor
(b) Bypassing the processor
(c) Interrupting the processor
(d) None of the above

23. In DMA transfer, data bus control is available via
(a) DMA controller
(b) CPU
(c) Both CPU and DMA controller
(d) None of the above

24. In DMA transfer, address bus control is available via
(a) Both CPU and DMA controller
(b) CPU
(c) DMA controller
(d) None of the above

25. In DMA transfer, control bus control is available via
(a) CPU
(b) DMA controller
(c) Both CPU and DMA controller
(d) None of the above

Answers to Multiple Choice Questions

1. a	2. c	3. a	4. b	5. a	6. a	7. b	8. a	9. b	10. c
11. d	12. b	13. d	14. a	15. a	16. b	17. a	18. a	19. c	20. a
21. b	22. b	23. a	24. c	25. b					

Long Answer Type Questions

1. How is the ALE signal used in demultiplexing of AD_7-AD_0 in the 8085?
2. Compare and contrast between the 8080 and 8085.
3. How do the status signals help in identifying the operation being carried out by the 8085?
4. Enumerate the interrupts present in the 8085.
5. Draw and explain the internal architecture of the 8085.
6. Show the structure of the PSW register in 8085.
7. Enumerate the conditions affecting the setting/resetting of bit patterns in PSW.
8. Enumerate the different types of machine cycles in the 8085.
9. Draw the timing diagrams of opcode fetch, memory read and memory write, I/O read and I/O write machine cycles of the 8085.
10. Explain the concept of memory folding.
11. In an address decoding strategy, the address bits A_{15}, A_{14} and A_{13} have not been utilized. Identify the folded address ranges for the interface.
12. Enumerate the advantages of DMA mode data transfer over other methods.
13. Explain the role of HOLD and HLDA signals of 8085 in DMA operation.
14. Enumerate the typical signals coming out of a DMA controller.
15. Draw an interfacing showing the connection between 8085 and a typical DMA controller.

3

The 8085 Instruction Set and Programming

LEARNING OUTCOMES

After reading this chapter, students will be able to:

LO 1 Enumerate the addressing modes of 8085

LO 2 List the data transfer instructions of 8085

LO 3 List the arithmetic instructions of 8085

LO 4 Enumerate the logical instructions of 8085

LO 5 Enumerate the branching instructions in 8085

LO 6 Explain the stack operations of 8085

LO 7 Draw timing diagrams of instruction execution in 8085

LO 8 Differentiate between memory-mapped I/O and I/O-mapped I/O operations

LO 9 Write assembly language programs for 8085 and evaluate their execution times

3.1 Introduction

The 8085 supports a number of machine instructions. Each instruction consists of an 8-bit field, called *mnemonic* (or *opcode*) and a few operands. The mnemonic part identifies the meaning of the instruction and is understandable by the instruction decoder within the 8085. The bit pattern corresponding to an instruction is fixed by the designers of 8085. For human understanding, each such mnemonic has been given a human readable name as well. A typical instruction may direct the 8085 to transfer the content of a register/memory location to some other register/memory location, perform some arithmetic/logic operation, to jump to a different memory address instead of executing the very next instruction pointed to by the Program Counter (PC) etc. Some example instructions are as follows:

> Processor understands bit-level coding of instructions. For human understanding, such instructions are written in textual format, giving rise to assembly language for the processor.

```
MOV A, B    ; A register gets content of B register
ADD B       ; A gets A + B
```

In 8085, the opcode part is 8-bit wide, which means that the decoder can differentiate between 256 different instructions, each instruction being encoded by an 8-bit pattern. However, the 8085 uses only 246 combinations that represent a total of 74 instructions. The operands may be of different types – an immediate value, a register identifier, a memory address etc. Depending upon the number of operands and their type, an instruction may be of size 1-byte, 2-bytes or 3-bytes. Thus, an 8085 instruction is as follows, in which only the *opcode* part is mandatory and the *operands* may be present depending upon the *opcode*.

< *opcode* > [*operand-1* [, *operand-2*]]

The ways in which the operands can be specified for a processor are known as its addressing modes. The 8085 instructions can be grouped into five classes, as noted in Table 3.1. The instructions will be discussed in detail after we look into the addressing modes in the next section.

TABLE 3.1 Classification of 8085 instructions

Type	Purpose	Examples

Data Transfer	Used to transfer data from source to destination. A register, memory location or an immediate value may act as source. Destination may be a register or a memory location.
	Restriction: Both source and destination cannot be memory locations simultaneously.
		`MOV B, A ; A register copied to register B`
		`MVI B, 40H ; B register gets 40H`
		`MOV A, M ; A gets Memory[address in HL]`
		`LXI H, 145AH ; H gets 14H, L gets 5AH`

Arithmetic Operations	Addition, subtraction, increment and decrement operations are supported.	`ADD C ; A gets A + C`
		`SUB B ; A gets A - B`
		`INR H ; H gets H + 1`

Logical Operations	Operations, such as, AND, OR, XOR, Rotate, Compare, Complement are supported.	`ANA C ; A gets A AND C`
		`CMP B ; Compare A with B, flags affected`
		`RLC ; Rotate accumulator left`

(Continued)

TABLE 3.1 *(Continued)*

Type	Purpose	Examples

Branching	Conditional and unconditional jump to an address, call subprogram, return from subprogram are supported	`JMP 2000H ; PC loaded with 2000H`
		`JC 3000H ; PC loaded with 3000H if CY = 1`
		`CALL 2000H ; Subprogram call`
		`RET ; Return from subroutine`

Machine Control	Some instructions affecting the operation of the CPU are supported.	`HLT ; Halt the CPU operation`
		`RST ; Software interrupt`
		`NOP ; Do nothing`

3.2 Addressing Modes

LO 1

The 8085 executes instructions to operate on some data and store the result at some destination. The ways in which the source and the destination operands could be specified in the 8085 are called its addressing modes. There are five different addressing modes supported by the 8085.

Addressing modes supported by the 8085 are fundamental in nature and are supported by almost all the later processors.

- Immediate addressing
- Direct addressing
- Register direct addressing
- Register indirect addressing
- Implicit addressing

In the following, each of the addressing modes has been discussed in detail.

1. **Immediate Addressing Mode.** In this case, the 8/16-bit data on which the 8085 operates is specified along with the instruction itself. An 8-bit data makes the instruction 2-bytes long while a 16-bit data makes the instruction 3-bytes long. The following are a few examples of the same.

```
MVI A, 76H    ; A gets the immediate value 76H
LXI B, 4575H  ; BC register pair gets 4575H (45H to B, 75H to C)
```

2. **Direct Addressing Mode.** The operand in this case is a memory location. The 16-bit address of the memory location is specified directly in the instruction. Some examples are as follows.

```
LDA 2000H     ; A gets Memory[2000H]
STA 3000H     ; Memory[3000H] gets A
```

3. **Register Direct Addressing Mode.** In this case also the operand is directly specified in the instruction; however, the operand is a register within the 8085 CPU. Some examples are as follows.

```
MOV B, C      ; B gets the content of C
ADD B         ; A gets A + B
```

4. **Register Indirect Addressing Mode.** In this case, the operand is a memory location. The address of the location is available in a register pair mentioned in the instruction. Some examples are as follows.

> Indirect addressing modes are generally used to implement pointers available in higher level languages.

```
LDAX B     ; A gets Memory[Content of BC-pair]
STAX D     ; Memory[Content of DE-pair] gets A
```

5. **Implicit Addressing Mode.** This mode is used by instructions in which the source and destination operands are fixed and need not be specified separately. Some such instructions are as follows.

```
CMA        ; Complement accumulator
STC        ; Set Carry flag
```

3.3 Instruction Set

As explained previously, instructions in the 8085 can be grouped into five different classes. In the following, instructions belonging to each class have been discussed in detail.

3.3.1 Data Transfer Instructions

LO 2

Instructions in this category are responsible for copying data from the source to the destination. The opcodes responsible of such data transfer are MOV, MVI, LXI, LDA, IN, OUT, LDAX, LHLD, STA, STAX, SHLD and XCHG. The instructions along with examples have been noted next.

1. **MOV.** This instruction copies the source operand to the destination operand. At least one of the two operands must be a register, while the other operand may also be a register or a memory location. For memory location, address is contained in the HL-pair. Content of the source operand is not affected. Some examples of the MOV instruction are as follows. It may be noted that the MOV instruction is useful for 8-bit data movement only. This is a 1-byte instruction.

> Memory-to-memory transfer not allowed.

```
MOV B, C     ; B gets content of C
MOV A, M     ; A gets content of Memory[(HL pair)]
MOV M, C     ; Memory[(HL pair)] gets content of C
```

2. **MVI.** The *Move Immediate* instruction is useful if the second operand happens to be an immediate value specified as part of the instruction itself. The first operand may be a register or a memory location (with address pointed to by the HL pair). This is a 2-byte instruction. Some examples are as follows.

```
MVI B, 56H   ; B gets 56H
MVI M, 45H   ; Memory[(HL pair)] gets 45H
```

3. **LXI.** This is the 16-bit variant of the MVI instruction and can be used to load the register pairs BC (named as B), DE (named as D), HL (named as H) or SP. This is a 3-byte instruction. Some examples are as follows.

```
LXI B, 2345H   ; BC-pair gets 2345H with B getting 23H and C 45H
LXI H, 1F2EH   ; HL-pair gets 1F2EH
```

4. **LDA.** This instruction loads the accumulator register A with the content of the memory location whose address is mentioned in the instruction. This is a 3-byte instruction. A typical example is as follows.

```
LDA 1020H      ; A register gets Memory[1020H]
```

5. **STA.** This is the opposite of LDA instruction. This 3-byte instruction is to store the A register content in some specified memory location. A typical example is as follows.

```
STA 1020H      ; Memory[1020H] gets content of A register
```

6. **IN.** The instruction reads the content of an input port into the accumulator. The 8-bit address of the input port is specified in the instruction. Thus, it is a 2-byte instruction. A typical instruction can be as follows.

```
IN 45H         ; A register gets content of the input port 45H
```

7. **OUT.** This is the reverse of IN instruction. The instruction outputs the content of A register to the specified output port. This is also a 2-byte instruction with the following as a typical example.

```
OUT 56H        ; A register content is output to port address 56H
```

8. **LDAX.** This instruction is an indirect way to load the A register from memory. The memory address may be specified by the B register (BC register pair) or D register (DE register pair). This is a 1-byte instruction. Typical examples are as follows.

```
LDAX D         ; A register gets Memory[(DE pair)]
LDAX B         ; A register gets Memory[(BC pair)]
```

It may be noted that the HL-pair is not allowed which may be realized by the "MOV A, M" instruction.

9. **STAX.** This is the counterpart of LDAX to save the A register content in memory. This is also a 1-byte instruction that uses 16-bit B and D registers as operand.

```
STAX D         ; Memory[(DE pair)] gets A register content
STAX B         ; Memory[(BC pair)] gets A register content
```

10. **LHLD.** This is a 3-byte instruction that contains a 16-bit address as part of the instruction. The HL register pair is loaded from the memory locations identified by the address mentioned in the instruction. The L register gets the content of memory location having the given address. Content of the next higher memory location is copied into the H register. The following is an example of the same.

```
LHLD 2040H     ; L register gets Memory[2040], H register gets
                 Memory[2041]
```

11. **SHLD.** This is the opposite of LHLD. The content of the L register is stored at the specified memory address while the H register is stored at the next higher address. For example,

```
SHLD 2040H     ; Memory[2040] gets L register content, Memory[2041]
                 gets H register
```

12. **XCHG.** This instruction is of single byte, containing only the mnemonic. The operands are implicit. Content of register pair HL gets exchanged with pair DE. The registers H and D get exchanged and also the registers L and E get exchanged. The format is as follows.

```
XCHG           ; HL and DE get exchanged
```

3.3.2 Arithmetic Instructions

LO 3

The 8085 supports a number of arithmetic instructions. The arithmetic operations supported are addition, subtraction, increment and decrement. The instructions are ADD, ADC, ADI, ACI, SUB, SBB, SUI, SBI, INR, INX, DCR, DCX, DAD and DAA. The instructions affect the status flags. In the following, the instructions have been discussed in detail.

Operations like multiplication and division not supported, to be implemented via software.

1. **ADD.** This instruction adds the A register with the specified operand. The operand may be a CPU register or a memory location. This is a 1-byte instruction. It affects the status flags. The following are a few examples.

```
ADD B       ; A gets A+B
ADD M       ; A gets A+Memory[(HL)]
```

2. **ADC.** The instruction stands for "add with carry". It adds the specified operand and the carry flag (CY) with the accumulator A. This is also a 1-byte instruction and affects all the flags. Some examples are as follows.

```
ADC B       ; A gets A+B+CY
ADC M       ; A gets A+Memory[(HL)]+CY
```

3. **ADI.** This is "add immediate" instruction. It is a 2-byte instruction. The 8-bit immediate operand specified in the instruction is added to the A register. All status flags are affected. A typical such instruction is as follows.

```
ADI 45H     ; A gets A+45H
```

4. **ACI.** This is the immediate version of the ADC instruction. It is a 2-byte instruction. The immediate operand is added to the accumulator along with the carry (CY) flag. A typical example is as follows. All flags are affected.

```
ACI 56H     ; A gets A+56H+CY
```

5. **DAD.** This is a 16-bit addition instruction that adds another 16-bit register content to the HL pair. It is a single-byte instruction. The operand may be B, D, H or SP. Only the CY

16-bit addition operation, normal ADD is over 8-bit data.

flag is set if the result is more than 16-bit wide. No other flags are affected. Some examples are as follows.

```
DAD B      ; HL gets (HL+BC)
DAD SP     ; HL gets (HL+SP)
DAD H      ; HL pair is doubled
```

It may be noted that the second example may be used to transfer the content of SP to HL by first clearing the content of HL pair and then executing the DAD instruction. The content available in the HL register may be used to save the content of current SP register at some place. The following instruction sequence carries out the transfer.

```
LXI H, 0   ; Clear the HL pair
DAD SP     ; Get SP content into HL
```

6. **SUB, SBB, SUI, SBI.** These instructions are for the subtraction operation of the 8085. The instructions SUB (subtract), SBB (subtract with borrow), SUI (subtract immediate) and SBI (subtract with borrow immediate) are similar to ADD, ADC, ADI and ACI instructions, respectively. The following are a few examples. All flags are affected.

```
SUB B      ; A gets A-B
SUB M      ; A gets A-Memory[(HL)]
SBB C      ; A gets A-C-CY
SUI 34H    ; A gets A-34H
SBI 45H    ; A gets A-45H-CY
```

7. **INR.** This is the increment instruction. It is a 1-byte instruction with a single operand. The operand may be a register or a memory location pointed to by the HL pair. All status flags are affected by the operation. Some examples are as follows.

> "INR B" is equivalent to
> ```
> MOV A, B
> ADI 1
> MOV B, A
> ```
> However, the content of A register is lost.

```
INR B      ; B gets B+1
INR M      ; Memory[(HL)] gets Memory[(HL)]+1
```

8. **DCR.** This is the decrement operation, opposite of INR. Rest of the features are same as INR. Some examples are as follows. All flags are affected.

```
DCR B      ; B gets B-1
DCR M      ; Memory[(HL)] gets Memory[(HL)]-1
```

9. **INX, DCX.** These are 16-bit variants of the INR and DCR instructions. The operand may be B, D, H and SP (Stack Pointer). However, a fundamental difference from the INR/DCR instructions is that INX/DCX instructions do not affect any of the status flags. Typical examples are as follows.

```
INX H      ; HL pair is incremented by 1
DCX D      ; DE pair is decremented by 1
INX SP     ; SP is incremented by 1
```

10. DAA. The *Decimal Adjust Accumulator* converts the accumulator content into decimal form, instead of binary. It is used in conjunction with other arithmetic operations that place the result in the accumulator in hexadecimal format. The DAA instruction operates on this result and converts the final result into decimal form. The decimal number is represented in BCD (Binary-Coded Decimal) format. In this format, each digit of the decimal number is converted to its 4-bit binary form. For example, the decimal number 38 is represented as 00111000. If 38 is added with 45 (01000101), the result available in the accumulator will be 01111101, that is 7DH. The DAA instruction working on this content will put 83 (10000011) into the accumulator which is the correct result of adding 38 and 45. The rules for the conversion are as follows.

- If the least significant hex digit in A is ≤ 9 and AC flag is 0, the digit is not modified.
- If the least significant hex digit in A is > 9 or if the AC flag is 1, 6 is added to the digit. In case of a carry from this operation, the most significant digit is also incremented. If this operation generates a carry, CY flag is set to 1.
- If the most significant hex digit in A ≤ 9 and CY flag is 0, the digit is not affected.
- If the most significant hex digit is > 9, or if CY is 1, 6 is added to the most significant hex digit of A and CY is set to 1.

3.3.3 Logical Instructions

LO 4

The 8085 supports a number of logic instructions. The instructions perform Boolean operations like AND, OR, XOR and NOT (complement). Apart from that, the accumulator content can be rotated left or right by a number of bits. The accumulator may also be compared with an operand and the result will affect the status flags. Program execution decisions may then be taken based upon the conditions of the status flags.

1. **ANA, ANI, ORA, ORI, XRA, XRI.** The instructions ANA and ANI are for bitwise and-operation of A register with the specified operand (a register or memory location for ANA and an immediate value for the ANI). Same is true for the instructions ORA, ORI (for or-operation) and XRA, XRI (for xor-operation). All the instructions affect the S, Z and P status flags, and the flag CY is reset while AC is set. Some examples of the instructions are as follows. The ANA, ORA and XRA are all 1-byte instructions while ANI, ORI and XRI are 2-byte instructions.

```
ANA B      ; A gets (A AND B)
ORI 45H    ; A gets (A OR 45H)
XRA M      ; A gets (A XOR Memory[(HL)])
```

2. **CMA.** This single-byte instruction complements the accumulator by inverting each of its bits. The instruction does not affect any of the status flags. The format of the instruction is as follows.

```
CMA        ; Complement accumulator bits
```

3. **CMP, CPI.** These instructions are to compare the accumulator register with the specified operand and affect the status flags accordingly. The CMP is a 1-byte instruction and the operand specified can

be a register or a memory location. The CPI is a 2-byte instruction with the 8-bit immediate operand specified as part of the instruction itself. Some examples are as follows.

```
CMP B      ; Compare A with B register
CMP M      ; Compare A with Memory[(HL)]
CPI 23H    ; Compare A with 23H
```

The status flags are affected as follows.

- If A < operand: Carry flag (CY) is set.
- If A = operand: Zero flag (Z) is set.
- If A > operand: Both Carry (CY) and Zero (Z) flags are reset.

4. **STC, CMC.** These are single-byte logical instructions affecting the carry (CY) flag only. The STC instruction sets the carry bit, while the CMC complements the carry. No other flag gets affected. It may be noted that there is no single instruction to clear the carry flag. The following is a sequence of instructions that can be used to clear the CY.

```
STC        ; Set Carry flag, CY
CMC        ; Complement CY
```

5. **RLC, RRC, RAL, RAR.** These are single-byte instructions affecting the content of the accumulator A and the carry flag CY. No other operand needs to be mentioned. The accumulator is rotated left (in RLC and RAL) or right (in RRC and RAR) by one bit. In RLC, each bit of A is shifted left by 1 position. The bit D_7 is placed in D_0 and also in the CY flag. In RAL, the bit D_7 reaches the CY, while the previous CY reaches D_0. The RRC and RAR are similar to RLC and RAL respectively, with D_0 content reaching CY, instead of D_7. Figure 3.1 shows the execution of the four instructions. Some examples are as follows:

```
RLC        ; Rotate accumulator left
RRC        ; Rotate accumulator right
RAL        ; Rotate accumulator left through carry
RAR        ; Rotate accumulator right through carry
```

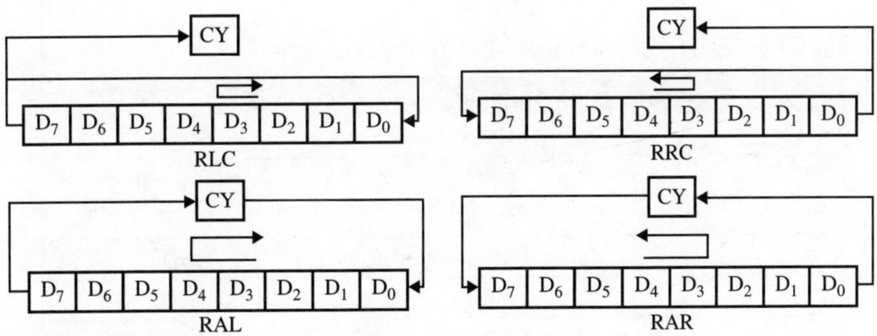

FIGURE 3.1 Rotate instructions in 8085

3.3.4 Branching Instructions

LO 5

In its normal course of action, the 8085 fetches the next instruction pointed to by its PC register. However, depending upon the program logic, it sometimes becomes necessary to change that sequence. Control needs to be transferred to a different region of the program code. The program branching supported by the 8085 can be of several types. One such classification is *conditional* vs. *unconditional* branching. In conditional branching, control branches to the specified target address based on some condition around the status flags becoming true, whereas, for unconditional branching, control transfer happens always. Another classification is *jump* vs. *subroutine call/return*. In normal jump, control simply branches to the target address. For subroutine call, before branching to the target address, the current PC content is saved in the stack, so that at the end of subroutine execution, a return instruction may be executed that restores the PC content from the stack to resume the operation of the main program, from where the original transfer took place. Various branch instructions are enumerated next.

1. **JMP.** This is the unconditional branch instruction. The 16-bit target address is specified with the instruction itself. Thus, it is a 3-byte instruction. The PC register is loaded with the specified target address, so that the next instruction to be executed is fetched from the target location. An example of the instruction is as follows.

```
JMP 2056H ; Branch to memory address location 2056H
```

2. **J*x*.** This is the set of conditional branching instructions. The condition code framed using the status flags is captured in the character *x*. The 3-byte instruction contains the 16-bit target address. The branching to the target address takes place if the condition is true. Table 3.2 shows the usage of condition codes in framing the conditional jump instructions.

   Some examples are as follows.

```
JZ 2300H   ; Branch to 2300H if Z = 1
JNC 4500H  ; Branch to 4500H if CY = 0
```

3. **CALL.** This is a 3-byte instruction with a target address specified for the subroutine/procedure to be executed. Compared to JMP instruction, in case of CALL, the return address (current PC value) is saved onto the stack, SP is decremented and the PC is loaded with the target address. For example,

TABLE 3.2 Conditional Jump instructions

Mnemonic	Description	Status Flag
JC	Jump if Carry	CY = 1
JNC	Jump if no Carry	CY = 0
JP	Jump if Positive	S = 0
JM	Jump if Minus (Negative)	S = 1
JZ	Jump if Zero	Z = 1
JNZ	Jump if not Zero	Z = 0
JPE	Jump if Parity Even	P = 1
JPO	Jump if Parity Odd	P = 0

consider the following CALL instruction located at memory address 4000H and the current SP value being 2000H.

```
4000H   CALL 3000H; Memory[199F] gets 30H, Memory[199E] gets 00H
                  ; SP becomes 199EH, PC gets 3000H
```

4. **C*x*.** This is the 3-byte conditional CALL instruction. The meaning of *x* and the instruction variants are similar to the conditional jump shown in Table 3.2. The control branches to the subroutine/procedure only if the specified condition is met by the status flags. Some examples of conditional call are as follows.

```
CZ 2000H   ; Call procedure at 2000H if Z = 1
CNC 2500H  ; Call procedure at 2500H if CY = 0
```

5. **RET.** This is a single-byte instruction. No other operand needs to be specified. In the execution of this instruction, the 8085 takes off the two top most entries from the stack and loads them into the PC. The SP is incremented by two. The RET instruction is used in conjunction with subroutines. It is placed at the end point(s) of the subroutine so that the PC can be restored with the return address saved at the time of calling the subroutine. For example, if the SP is currently 2000H and the stack top locations corresponding to memory addresses 2000H and 2001H contain 00H and 25H, respectively, on executing the RET instruction, PC will be loaded with 2500H and the SP value will become 2002H. Thus, the next instruction to be executed will be fetched from the memory address 2500H. The instruction is as follows.

```
RET        ; Load PC from top of the stack, increment SP by 2
```

6. **R*x*.** This single-byte instruction is the conditional return statement. It does not have any other arguments in it. The content of top two locations of the stack is copied to PC if the specific condition built around the status flags is met. The instructions can be formulated similar to conditional jump, shown in Table 3.2. Some specific examples are as follows.

```
RZ         ; Return if Z = 1, PC updated from stack, SP
             incremented by 2
RC         ; Return if CY = 1, PC updated from stack, SP
             incremented by 2
```

7. **RST.** This is restart instruction, 3-byte long and is used as software interrupts. The format of the instruction is "RST *n*" where *n* is a value between 0 and 7. The branch address for the processor is given by, $8 \times n$. Table 3.3 shows the branching addresses for different values of *n*. In the execution

TABLE 3.3 RST branch addresses

RST Instruction	Branch Address
RST 0	0000H
RST 1	0008H
RST 2	0010H
RST 3	0018H
RST 4	0020H
RST 5	0028H
RST 6	0030H
RST 7	0038H

of the RST instruction, the current PC value is saved onto the stack and PC is loaded with the branch address. These instructions are more commonly known as *software interrupts* and will be discussed in more detail in the context of interrupts.

3.3.5 Other Miscellaneous Instructions

LO 6

Some more instructions of 8085 are noted in the following which serve many useful purposes. Two of the instructions are related to stack operations while the others are related to machine control.

1. **PUSH.** This is a single-byte instruction that stores the specified 16-bit register value into the stack. The SP is decremented accordingly. The operand may be B, D, H and PSW. The instruction "PUSH B" saves the registers B and C onto stack. Similarly, the instructions "PUSH D" and "PUSH H" save the register pair DE and HL onto the stack. "PUSH PSW" saves the A register and status flags onto the stack. Assuming the SP to be equal to 2000H, the following are a few examples of PUSH instruction.

```
PUSH B    ; Saves B at Memory[199FH], C at Memory[199EH], SP
            becomes 199EH
PUSH PSW  ; Saves A at Memory[199FH], flags at Memory[199EH], SP
            becomes 199EH
```

   It may be noted that "PUSH C", "PUSH E", "PUSH L" etc. are not valid.

2. **POP.** This is the reverse of PUSH instruction. The single-byte instruction can be used to retrieve the two topmost entries from the stack and put into the specified register pair. Assuming that the SP value is 2000H, the following are a few examples of POP instruction.

```
POP B     ; C gets Memory[2000H], B gets Memory[2001H], SP becomes
            2002H
POP H     ; L gets Memory[2000H], H gets Memory[2001H], SP becomes
            2002H
```

3. **EI, DI.** These two instructions are related to the interrupts in the 8085. The EI (Enable Interrupt) instruction enables all the maskable interrupts, however does not have any effect on the non-maskable interrupt line TRAP. The DI (Disable Interrupt) instruction disables all the maskable interrupts. None of the status flags are affected by these instructions. The instructions will be discussed in more detail while enumerating the interrupts.

   Programmer has to use EI and DI instructions judiciously and ensure that the interrupts do not get disabled perpetually in some execution sequence of the program.

4. **NOP.** This is the no-operation instruction. The instruction is fetched and decoded, but no operation is undertaken. The instruction simply introduces a delay of four clock cycles in program execution.

5. **HLT.** The instruction halts the operation of the 8085 CPU. This is a single-byte instruction to halt the 8085. A reset or interrupt is necessary to bring the 8085 back into operation.

6. **PCHL.** This is a single-byte instruction that copies the content of HL-pair into the PC register. The higher order byte of PC gets the H register content, while the lower order byte of PC gets the content of the L register. Thus, this may be regarded as an indirect branch instruction where the branch target is specified through the HL-pair.

7. **SPHL.** This instruction is similar to PCHL; however, instead of PC register, the SP is loaded. This is an indirect way of loading the SP register (direct one being the "LXI SP" instruction).
8. **XTHL.** It stands for "exchange stack top with HL". The content of location Memory[SP] is exchanged with L register while the location Memory[SP+1] is exchanged with the H register. The SP is not affected.
9. **RIM, SIM.** These are two special multipurpose instructions. The single-byte instruction RIM stands for *Read Interrupt Mask* while SIM stands for *Set Interrupt Mask*. The RIM instruction loads accumulator with the data available in the SID pin and some status information about the interrupts. Similarly, the SIM instruction can be used to send data serially through the SOD pin and set masks for the interrupts. The accumulator bit patterns for RIM and SIM instructions have been shown in Figure 3.2. The instructions will be discussed in more detail while dealing with interrupts and serial data transfer.

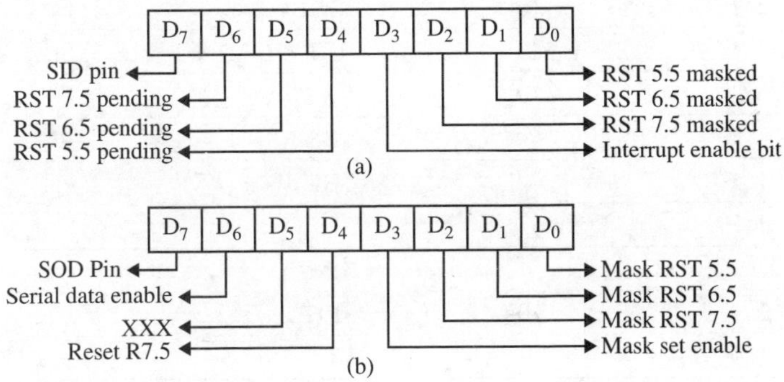

FIGURE 3.2 Accumulator bits for (a) RIM, (b) SIM

Depending upon the type of operands, the instructions take different number of machine cycles and T-states. Table 3.4 summarizes the instructions in terms of those parameters. Each instruction has a machine code format which may be obtained by consulting the manual or from an *assembler*.

TABLE 3.4 Machine cycles and T-states for 8085 instructions

Instruction	Machine Cycles (No.: Types)	T-States
ACI <data>	2: F, R	7
ADC <reg>	1: F	4
ADC M	2: F, R	7
ADD <reg>	1: F	4
ADD M	2: F, R	7
ADI <data>	2: F, R	7
ANA <reg>	1: F	4
ANA M	2: F, R	7
MVI <reg>, <data>	2: F, R	7
MVI M, <data>	3: F, R, W	10
NOP	1: F	4
ORA <reg>	1: F	4
ORA M	2: F, R	7
ORI <data>	2: F, R	7
OUT <port>	3: F, R, O	10
PCHL	1: S	6

(Continued)

TABLE 3.4 *(Continued)*

Instruction	Machine Cycles (No.: Types)	T-States
ANI <data>	2: F, R	7
CALL <addr>	5: S, R, R, W, W	18
C<*x*> <addr>	5: S, R, R, W, W	9–18
CMA	1: F	4
CMC	1: F	4
CMP <reg>	1: F	4
CMP M	2: F, R	7
DAA	1: F	4
DAD <rp>	3: F, B, B	10
DCR <reg>	1: F	4
DCR M	3: F, R, W	10
DCX <rp>	1: S	6
DI	1: F	4
EI	1: F	4
HLT	2: F, B	5
IN <port>	3: F, R, I	10
INR <reg>	1: F	4
INR M	3: F, R, W	10
INX <rp>	1: S	6
J<*x*> <addr>	3: F, R, R	7–10
JMP <addr>	3: F, R, R	10
LDA <addr>	4: F, R, R, R	13
LDAX <rp>	2: F, R	7
LHLD <addr>	5: F, R, R, R, R	16
LXI <rp>, <data>	3: F, R, R	10
MOV <reg>, <reg>	1: F	4
MOV M, <reg>	2: F, W	7
MOV <reg>, M	2: F, R	7

Instruction	Machine Cycles (No.: Types)	T-States
POP <rp>	3: F, R, R	10
PUSH <rp>	3: S, W, W	12
RAL	1: F	4
RAR	1: F	4
R<*x*>	3: S, R, R	6–12
RET	1: F	4
RIM	1: F	4
RLC	1: F	4
RRC	1: F	4
RST <n>	3: S, W, W	12
SBB <reg>	1: F	4
SBB M	2: F, R	7
SBI <data>	2: F, R	7
SHLD <addr>	5: F, R, R, W, W	16
SIM	1: F	4
SPHL	1: S	6
STA <addr>	4: F, R, R, W	13
STAX <rp>	2: F, W	7
STC	1: F	4
SUB <reg>	1: F	4
SUB M	2: F, R	7
SUI <data>	2: F, R	7
XCHG	1: F	4
XRA <reg>	1: F	4
XRA M	2: F, R	7
XRI <data>	2: F, R	7
XTHL	5: F, R, R,W,W	16

F: Fetch (4 cycle), S: Fetch (6 cycle), R: Read, W: Write, I: Input, O: Output, B: Bus idle

3.3.6 Instruction Timing Diagrams (LO 7)

The detailed timing diagrams for some representative instructions have been presented in the following.

1. MVI. Consider the following instruction at location 1000H.

```
1000H    MVI A, 56H
```

The instruction when assembled produces code 3EH (for MVI) at location 1000H and 56H at location 1001H. Behaviour of the 8085 has been shown in Figure 3.3.

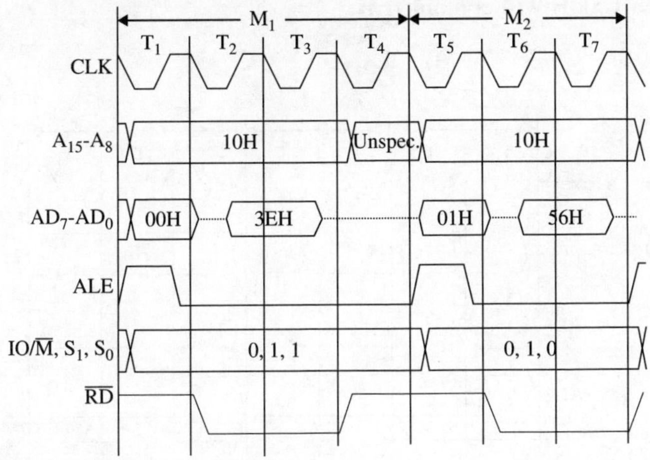

FIGURE 3.3 Timing diagram for "MVI A, 56H"

2. **LXI.** Consider the following instruction with mnemonic LXI.

```
1000H        LXI H, 4056H
```

The instruction when assembled produces code 21H (for LXI H) at location 1000H, 56H at location 1001H and 40H at location 1002H. While executing the instruction, the behaviour of the 8085 has been shown in Figure 3.4.

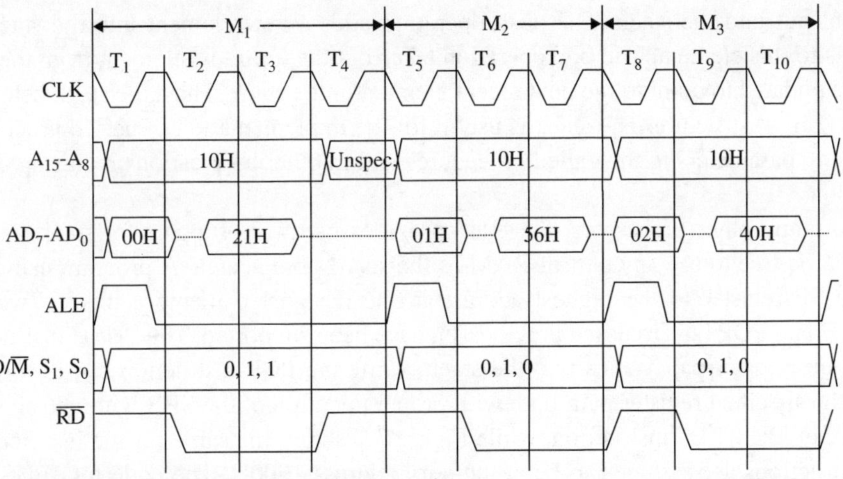

FIGURE 3.4 Timing diagram for "LXI H, 4056H"

3. **MOV.** Next consider the instruction "MOV A, M" in which the content of HL-pair is used as the address of the memory location whose content will be moved to A. The timing diagram for the execution of this instruction has been shown in Figure 3.5. The opcode for the instruction is 7EH. Thus, the memory location 1000H will contain 7EH.

```
1000H      MOV A, M
```

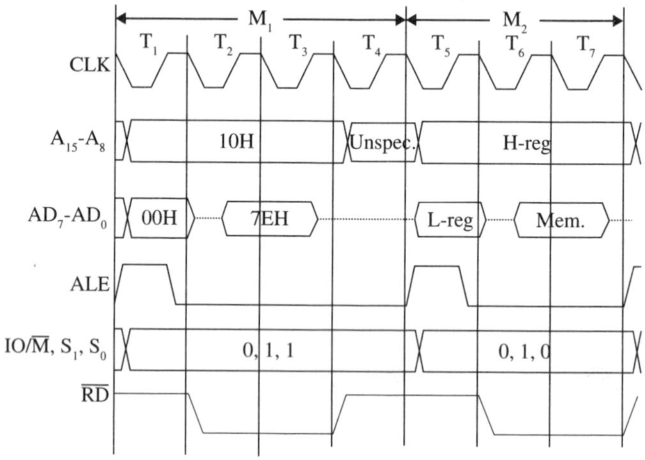

FIGURE 3.5 Timing diagram for "MOV A, M"

3.4 The Stack

LO 6

For the programming community, a stack is an abstract data structure that supports two operations – *push* and *pop*. The stack as got associated with it a pointer, conveniently called *top* that points to the latest element put into the stack. The push operation adds a new element into the stack and the top pointer points to this element. The pop operation takes out the top-most element from the stack and the top pointer is updated to point to the next recent element in the stack. Thus, the stack behaves as a *Last-In-First-Out* (LIFO) structure. The stack is useful for the implementation of nested procedure calls and returns, passing parameters to the called procedures and in other applications requiring the LIFO style of computation.

Similar to many other processors, the 8085 provides hardware support for stack implementation. A programmer is free to use any part of RAM as the stack storage area. A program using stack should initialize the SP register to the highest address in the designated memory block. Two instructions, namely PUSH and POP (as discussed previously), have been supported. The PUSH instruction can save specified 16-bit register pair values into the stack, while the POP instruction can restore values from the stack to the specified register pair, both with proper updation of the SP. Figure 3.6 shows the timing diagram for the "PUSH D" instruction, while Figure 3.7 shows the same for the instruction "POP D". Both the instructions are assumed to be at memory address 1000H. The code for "PUSH D" is D5H while for "POP D" it is D1H.

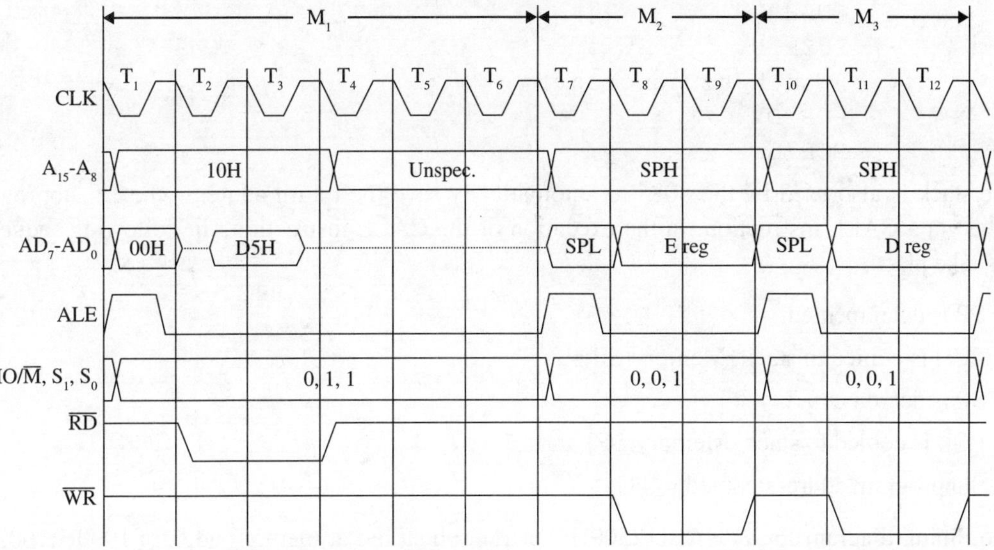

FIGURE 3.6 Timing diagram for "PUSH D"

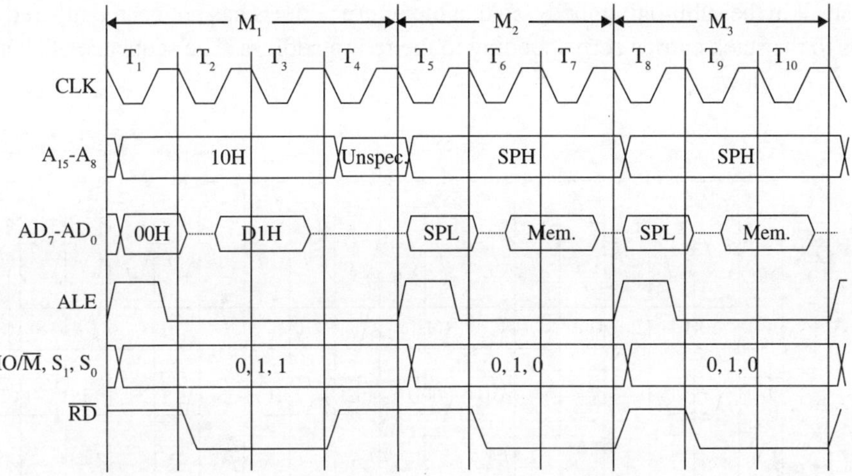

FIGURE 3.7 Timing diagram for "POP D"

The strategy of saving the register contents via PUSH, into the stack, may be used to pass parameters to the called subprogram. In the subprogram, the passed parameters could be retrieved via POP. Similarly, the calling program may wish that the contents of its B, C, D and E registers should not be affected by the instructions in the subprogram. Thus, it may save those registers into the stack before branching to the subprogram, as shown next. It should be noted that the registers be popped out in the reverse order of their pushing into the stack.

```
PUSH B
PUSH D
Branch to subprogram
POP D
POP B
```

The stack is also used by the 8085 to automatically save the return address when a subprogram is invoked via a CALL instruction. In the execution of the CALL instruction, the following sequence of events take place.

a. SP is decremented.

b. PCH is copied to stack, Memory[SP].

c. SP is decremented.

d. PCL is copied to stack, Memory[SP].

e. Subprogram address copied to PC

The timing diagram for a "CALL 2500H" instruction stored at memory address 1000H:1002H has been shown in Figure 3.8, where CDH is the code for CALL.

To return from a subprogram, the RET instruction uses the stack to restore the PC register. It is assumed that the top of the stack contains the return address. Hence, the programmer needs to be careful while using stack in the subprogram body, so that the return address has not been tampered with and that the SP points to the stack entries corresponding to the return address. The sequence of events for a RET instruction are as follows.

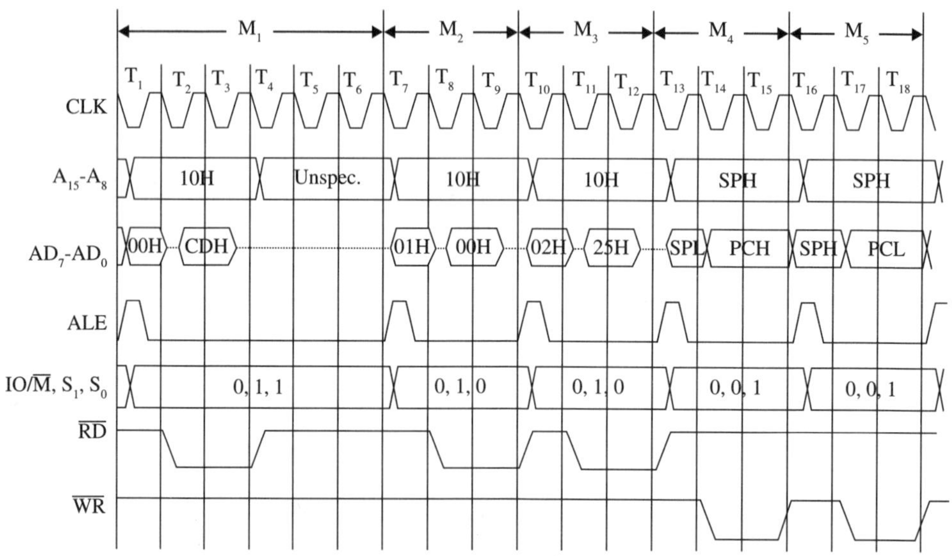

FIGURE 3.8 Timing diagram for "CALL 2500H"

a. Stack top (that is, Memory[SP]) copied to PCL.
b. SP incremented.
c. Stack top (that is, Memory[SP]) copied to PCH.
d. SP incremented.

The timing diagram for "RET" instruction, assumed to be available at memory address 1000H, has been shown in Figure 3.9. The code for RET is C9H.

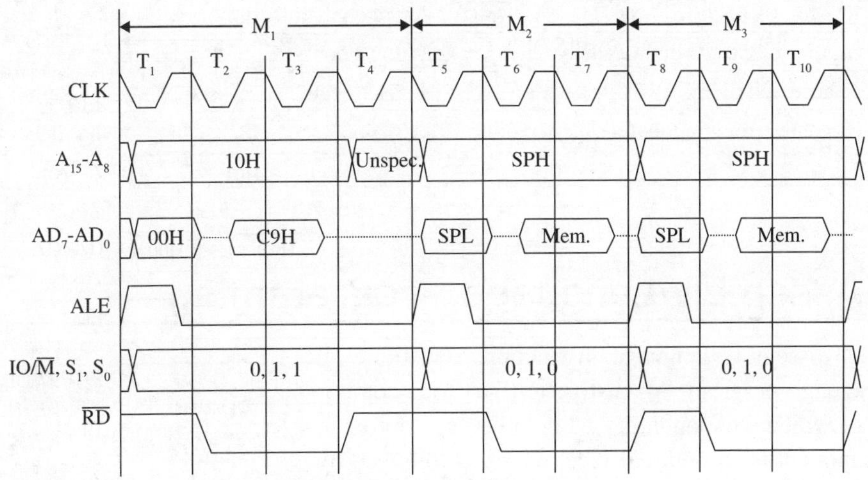

FIGURE 3.9 Timing diagram for "RET"

3.5 Memory-Mapped-I/O vs. I/O-Mapped-I/O

LO 8

> Memory-mapped I/O treats I/O devices as memory locations only.

The I/O devices in the 8085 can be treated either as memory locations or as I/O locations. From the discussion in Section 2.5, the decoding logic can be configured in such a way the memory and I/O devices are driven by the same decoder. In that case, the I/O devices occupy some memory addresses out of the total range of 0000H-FFFFH. If separate decoding logic are used for memory and I/O with the $IO/\overline{M}$ line enabling individual decoders, the memory will have address range of 64 kB while I/O devices will have address range of 256 bytes. When some memory locations are reserved for I/O devices, the mode of I/O operation is called *memory-mapped-I/O*. On the other hand, if the I/O devices use their own I/O address range, the mode is called *I/O-mapped-I/O*. It should be noted that the IN and OUT instructions are available only in the I/O-mapped-I/O mode. On the other hand, the memory-mapped-I/O can use all memory access instructions to access the I/O devices as well. A comparison between the two methods has been shown in Table 3.5.

TABLE 3.5 I/O-mapped-I/O vs. memory-mapped-I/O

I/O-Mapped-I/O	Memory-Mapped-I/O
I/O device treated as I/O device and is given I/O address	I/O device treated like a memory device and given a memory address
8-bit I/O address	16-bit memory address
I/O devices given IOR/IOW control signals	I/O devices given MEMR and MEMW control signals
Decoding is easier due to lesser address lines	Decoding is more complex due to more address lines
Decoding is cheaper	Decoding is more expensive
Works faster due to less delay	More delay due to more logic gates
Allows a maximum of 256 I/O devices	Allows up to 65536 I/O devices
Devices can be accessed by IN, OUT instructions	Devices can be accessed using any memory instruction
Only A register can be used to transfer data with the I/O device	Any register can be used to transfer data with the I/O device

3.6 Assembly Language Programming

LO 9

> Format of assembly language program may differ across the assemblers for same processor.

A sequence of assembly language instructions constitute an assembly language program. An instruction has four components, some of which are mandatory while others are optional. The first component is an optional label which is followed by a mandatory mnemonic. Depending upon the instruction, a number of operands follow the mnemonic. Finally, an optional comment (often demarcated by a ';' character) can be put to include additional comments. The overall structure of an instruction is as follows.

[label:] mnemonic [operands] [; comment]

For example, consider the following instruction.

```
HERE: MOV A, C    ; A gets content of C
```

HERE is a label, MOV is the mnemonic, A and C are two operands. The portion after the ';' character is a comment and is ignored by the assembler. The label HERE may be used in some other instruction to refer to this "MOV A, C" instruction as jump target etc.

An assembly language program may have some additional components that help in better organization of the program. One such component is called *assembler directives* while another one is used for storage declarations. It may be noted that the exact format of these components may vary across the assemblers. In the following, a generic discussion has been presented, which may differ to some extent in any individual assembler.

3.6.1 Assembler Directives

Most of the 8085 assemblers support the three directives – ORG (origin), EQU (equate) and END (end of source code).

1. **ORG:** This directive is used to instruct the assembler the address of first instruction in the assembly language instruction block. It is of the form,

```
ORG < address >
```

For the instruction "ORG <*address*>", the target code for the next instruction is generated at memory location *address*. For example, consider the following code sequence.

```
ORG 2000H
MOV A, B
```

Here, the machine code for the instruction "MOV A, B" is to be placed at memory location 2000H. There may be several ORG statements in an assembly language program. Each ORG statement indicates start address of its immediately following code fragment.

2. **EQU:** This directive is used to define a constant. It may be noted that the assembler does not allocate any storage space for the constant. A typical syntax for EQU directive is as follows.

```
NUM     EQU     100
```

Here, NUM is a label. All occurrences of the label NUM in the source code are replaced by the constant 100, before translating to the machine code. For example, with the previous definition of NUM, the assembly language statement "MVI B, NUM" is same as "MVI B, 100".

It may be noted that the availability of the EQU directive helps in writing more structured programs, with the ease of modification. For example, the constant NUM may correspond to the total number of numbers (100 in this case) in an array. In the program, it may be necessary to load several registers/memory locations with this value of 100. Later on, if the number of numbers needs to change to 1000, all such program instructions need to change from 100 to 1000. This may introduce errors in the program code. However, in the presence of definition "NUM EQU 100" and using the constant NUM in all those instructions, it is very easy to introduce the change from 100 to 1000 by just changing the constant definition to "NUM EQU 1000". The modification gets reflected to all the relevant instructions.

3. **END:** The END directive instructs the assembler not to consider anything beyond this point in the source code to generate the machine code. Thus, it marks the end of the assembly process.

3.6.2 Storage Declaration

The source program needs some variables to operate upon. The variables are defined via storage declarations. Most of the assemblers provide a pseudo-opcode DB (Define Byte). It defines 1 byte of storage, which may/may not be initialized with some value. The storage declarations are also called pseudo-opcodes. The code generated corresponding to a pseudo-opcode is not for execution by the CPU, hence the name pseudo-opcode. Consider the following code fragment.

```
    ORG   1000H
X   DB    20H
Y   DB    30H
```

The code defines two 1-byte variables X and Y. Since there is an ORG statement before the storage declarations, the variable X is allotted memory address 1000H, and Y is allotted the address 1001H. The two locations are also initialized. The location 1000H is initialized for X with 20H and 1001H for Y with 30H. In the following, a few definitions of multibyte structures have been shown as examples, utilizing the DB directive.

```
STR1        DB    "Microprocessors and Microcontrollers"
ARR         DB    10, 20, 30, 40
```

Here, STR1 defines and initializes the storage of 36 bytes for the string "Microprocessors and Microcontrollers". ARR defines an array of 4 bytes with the successive bytes initialized to 10, 20, 30 and 40, respectively.

3.6.3 Important Files

The files involved in the assembly process are as follows. Figure 3.10 defines the relations between the files.

- *Source file*: This is a text file in which the assembly language program is written. A typical extension of such files is ".asm". The assembler translates (assembles) this file to produce a binary file containing the machine code.
- *Hex file*: This is a binary file containing the program code. Typically, the file has the extension ".hex". Since the content is binary in nature, the hex file is not human understandable.
- *List file*: Since the .hex file is not readable, in order to aid in the debugging and understanding process, another text file is produced that lists side-by-side the machine code generated and the instruction translated. Aptly, this file is called a list file and often has the extension ".lst".

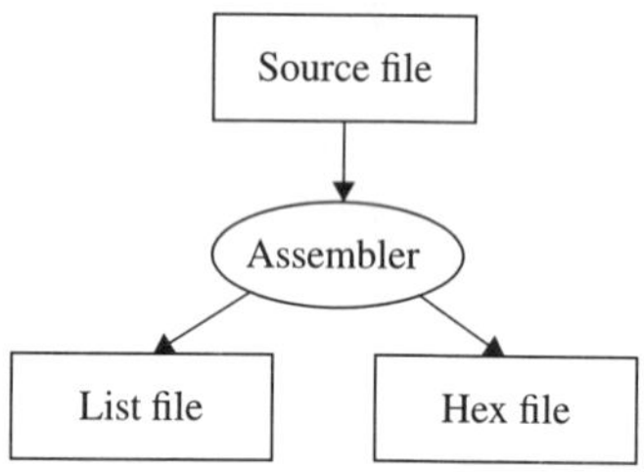

FIGURE 3.10 Assembler input and output files

3.6.4 Program Examples

In this section, we shall look into a number of example programs. For the first example, the list file has also been noted.

1. **Statement:** Count number of 1s in an 8-bit number available in register C. The count should be available in register B.

```
          ORG 1000H
          MVI B, 00H     ; Initialize B register to 0 (no 1 seen so far)
          MVI D, 08H     ; D register holds number of bits to check
          MOV A, C       ; Get C register content into A
BACK:     RAR            ; Rotate A to move D0 to Carry (CY)
          JNC SKIP       ; If CY not set, bit was 0, do not increment B
          INR B          ; Else bit was 1, increment B
SKIP:     DCR D          ; Decrement number of bits yet to check
          JNZ BACK       ; If 8 bits not seen yet, go back
          HLT            ; Else halt
          END
```

The program, when passed through an 8085 assembler, generates the machine code. The list file produced as a by-product of the assembly process is as follows. The program is assumed to start at 1000H, specified by the ORG statement. Each line of the list file contains the address, the machine code and the program text.

```
                          ORG 1000H
1000: 06 00               MVI B, 00H ; Initialize B register to 0 (no 1
                                       seen so far)
1002: 16 08               MVI D, 08H ; D register holds number of bits to
                                       check
1004: 79                  MOV A, C   ; Get C register content into A
1005: 1F         BACK:    RAR        ; Rotate A to move D0 to Carry (CY)
1006: D2 0A 10            JNC SKIP   ; If CY not set, bit was 0, do not
                                       increment B
1009: 04                  INR B      ; Else bit was 1, increment B
100A: 15         SKIP:    DCR D      ; Decrement number of bits yet to check
100B: C2 05 10            JNZ BACK   ; If 8 bits not seen yet, go back
100E: 76                  HLT        ; Else halt
                          END
```

2. Statement: Sort in ascending order 10 numbers stored from memory location 2300H.

```
          ORG 2300H
          X DB 20,25,10,45,34,12,78,87,65,32
          ORG 1000H
          MVI B, 09        ; Use B register as a counter
START:    LXI H, X         ; Initialize memory pointer
          MVI C, 09H       ; Use register C as another counter
BACK:     MOV A, M         ; Get the number in A
```

```
          INX H           ; Increment memory pointer
          CMP M           ; Compare number with next number
          JC SKIP         ; If less, don't interchange
          JZ SKIP         ; If equal, don't interchange
          MOV D, M        ; Exchange - get next number in D
          MOV M, A        ; Save A at next number slot
          DCX H           ; Decrement pointer H
          MOV M, D        ; Store next number in that slot
          INX H           ; Restore H
SKIP:     DCR C           ; Decrement counter 2
          JNZ BACK        ; If not zero, repeat
          DCR B           ; Decrement counter 1
          JNZ START       ; If not zero try next number
          HLT             ; Terminate program execution
          END
```

3. **Statement:** Assume that a set of 10 numbers are stored from memory location 2000H. A number available in C register is to be searched in this array of numbers. If the number is present, the corresponding address will be saved at address 2200H:2201H. If the number is not present, 0 will be stored at both the locations 2200H and 2201H.

```
          ORG 2000H
          X DB 20,25,10,45,34,12,78,87,65,32
          ORG 1000H
          LXI H, X        ; Initialize memory pointer HL
          MVI B, 10       ; Initialize counter B
BACK:     MOV A, M        ; Get the next number
          CMP C           ; Compare with the given byte in C
          JZ LAST         ; Go last if match occurs
          INX H           ; Increment memory pointer
          DCR B           ; Decrement counter
          JNZ BACK        ; If not zero, repeat
          LXI H, 0000H    ; Load 0 to HL pair
LAST:     SHLD 2200H      ; Store matched address at 2200H:2201H
          HLT
          END
```

4. **Statement:** In this problem, we write a subprogram called "ADDSUB" that adds or subtracts the operands specified in B and C registers, the result being available in the B register. The choice is

specified in D register. D = 0 means addition, D = 1 means subtraction operation. This subprogram is to be used to compute the expression "(*x* + *y*) – (*z* + *w*)", where *x*, *y*, *z* and *w* are the memory locations 2000H, 2001H, 2002H and 2003H. The final result should be available at 2004H. The most important thing to note here is the initialization of the SP. Since the subprogram call and return will use stack for saving and restoring the return addresses, it is advisable to initialize the SP to a free region of the memory that could be used as stack without overwriting any program code or data area.

```
ORG 2000H
X DB 45
Y DB 29
Z DB 26
W DB 15
RES DB
ORG 1000H
LXI SP, 5000H ; Initialize SP so that stack works
                properly
LXI H, X
MOV B, M      ; Get X in B
INX H
MOV C, M      ; Get Y in C
MVI D, 0      ; Reset D to mean addition
CALL ADDSUB   ; Call subprogram to have B = B + C
MOV E, B      ; Save X + Y in E
INX H
MOV B, M      ; Get Z in B
INX H
MOV C, M      ; Get W in C
MVI D, 0      ; Reset D to mean addition
CALL ADDSUB   ; Call subprogram to have B = B + C
MOV C, B      ; Save Z + W in C
MOV B, E      ; Get X + Y in B
MVI D, 1      ; Set D to mean subtraction
CALL ADDSUB   ; Call subprogram to have B = (X + Y) -
                (Z +W)
INX H         ; HL points to RES
MOV M, B      ; Save (X + Y) - (Z+W) at RES
HLT           ; Halt execution
```

```
ADDSUB: MOV A, D        ; Subprogram performs B+C if D=0 and
                          B-C if D=1
        CPI 0
        JZ SUM          ; Do summation if D=0
        MOV A, B
        SUB C           ; Else subtraction
        JMP OVER
SUM:    MOV A, B
        ADD C           ; Summation
OVER:   MOV B, A        ; Get result in B
        RET             ; Return to main program
        END
```

5. **Statement:** Next, we shall look into code fragment to insert delay between two important instructions in the program. One such example may be displaying a sequence of patterns to some output port, particularly if it is a visual pattern for human eyes. Consider the following code fragment.

```
        MVI B, 0FFH
LOOP:   DCR B
        JNZ LOOP
```

Here, the MVI instruction needs 7 T-states, DCR 4 T-states and JNZ 10 T-states. Delay introduced by the code segment is given by,

$T_{delay} = T_O + T_L$, where T_O is the delay outside the loop and T_L is the delay of the loop.

In this case, T_O = 7 T-states and $T_L = (14 \times 255) - 3 = 3567$ T-states. For the last cycle, JNZ requires 7 T-states only, instead of 10. Thus, total delay T_{delay} = 3574 T-states. Assuming the clock frequency to be 3 MHz, the delay introduced is 1191.33 µS. In case we need more delay, two loops can be nested, a 16-bit register pair may be used or even 16-bit register pairs may be used in a nested fashion. The following code fragment uses the 16-bit BC register pair to generate a higher delay. It may be noted that unlike the DCR instruction, DCX does not affect the status flags. Thus, to check if the BC pair has reached zero, the contents of B and C registers are to be OR-ed and the zero flag be checked.

```
        LXI B, 1000H
LOOP:   DCX B
        MOV A, C
        ORA B
        JNZ LOOP
```

The instruction LXI takes 10 T-states, DCX 6 T-states, MOV 4 T-states, ORA 4 T-states and JNZ 10/7 T-states. Thus, total delay introduced by the code fragment is $(10 + (24 \times 4096) - 3) = 98301$ T-states.

SUMMARY

LO 9

The 8085 has a rich set of instructions that can be used by the programmers to code their algorithm into assembly language program. Each instruction has a mnemonic and a maximum of two operands. The operands may be specified using different addressing modes. The operands may be registers, memory locations, immediate operands or port addresses. The major arithmetic operations supported are addition and subtraction. On the other hand, the logical operands like AND, OR, XOR and NOT are available. Branching instructions perform conditional/unconditional control transfers. Control transfer may also take care of the subprogram calls. Full support is provided for implementation of stack. The chapter has provided a detail discussion on each of the topics. Furthermore, a number of example programs have been included to illustrate the assembly language programming for the 8085. The next chapter will deal with the interrupt facilities supported by the 8085.

EXERCISES

Multiple Choice Questions

1. The operand present in all arithmetic/logic operations in 8085 is
   (a) A (b) Immediate value
   (c) B (d) HL
2. NOP instruction introduces
   (a) Address (b) Delay
   (c) Memory location (d) None of the above
3. In 8085, the numbers 45H and 7EH are added. The resulting flag settings will be as follows.
   (a) Both CY and AC are set
   (b) CY Set AC reset
   (c) CY reset AC set
   (d) Both CY and AC reset
4. In the 8085 processor, suppose the DE register pair contains value 2. In the following code, how many times is the JNZ instruction executed?

```
Loop:   DCX D
        JNZ Loop
```

   (a) 1 or infinity
   (b) 2 or infinity
   (c) Infinity
   (d) Cannot be determined
5. Which of the following is NOT a valid 8085 instruction?
   (a) PUSH PSW (b) PUSH B
   (c) PUSH C (d) PUSH D
6. Suppose that the Stack Pointer (SP) of 8085 contains 2004H. After executing a CALL instruction, the value of SP will be
   (a) 2002H (b) 2000H
   (c) 2006H (d) 2008H
7. In 8085, auxiliary carry bit is set if there is a carry during addition
   (a) Bits 3 to 4 (b) Bits 4 to 5
   (c) Bits 2 to 3 (d) All of the above
8. In an 8085 processor, how to clear the carry bit?
   (a) CLC
   (b) STC followed by CMC
   (c) CMC followed by STC
   (d) CAL
9. An 8085 processor has executed a HLT instruction. How to make it resume its normal activity?
   (a) By interrupt
   (b) By Reset
   (c) Either by interrupt or reset
   (d) None of the above

10. Which signal line in 8085 is always activated in the first clock period of each machine cycle during normal program execution (that is without interrupt)?
(a) ALE (b) RD
(c) WR (d) IO/M

11. 8085 instruction that can be used to load program counter is
(a) LXI PC, <immediate value>
(b) PCHL
(c) HLPC
(d) PCDE

12. In an 8085 processor, suppose the accumulator content is FFH and the carry flag is 0. What will be the content of the accumulator in RAR and RRC instructions?
(a) Both FFH
(b) Both EFH
(c) RAR - FFH, RRC - EFH
(d) RAR - EFH, RRC - FFH

13. Consider the following sequence of 8085 instructions.

```
        MVI A, 82H
LOOP:   ORA A
        JP DSPLY
        XRA A
DSPLY:  OUT PORT1
        HLT
```

The output at PORT1 is
(a) 00H (b) FFH
(c) 92H (d) 11H

14. In 8085, which instruction can set the zero flag?
(a) XRA A
(b) MVI A, 0
(c) Both the XRA and MVI options
(d) None of the above

15. Which instruction(s) has/have similar effect as NOP?
(a) MOV A, A (b) HLT
(c) ADI 0 (d) None of the above

16. In 8085, which instruction can be used to clear A register using least code size?
(a) MVI A, 0
(b) XRA A
(c) ANI A, 0
(d) All of the above

17. In a 8085 processor, which of the following cannot be done by a single instruction?
(a) Adding two registers
(b) Adding two memory locations
(c) Adding a memory location to a register
(d) None of the above

18. In 8085, which instruction can set the parity flag?
(a) XRA A
(b) MVI A, 0
(c) Both the XRA and MVI options
(d) None of the above

19. Between DCR and DCX instructions, the status flags are affected by
(a) Only DCR
(b) Only DCX
(c) Both DCR and DCX
(d) Neither DCR nor DCX

20. Which addressing mode is not used in 8085?
(a) Register
(b) Immediate
(c) Register indirect
(d) Relative

21. Register Indirect addressing mode can be used with the register pair
(a) BC
(b) DE
(c) HL
(d) All of the above

22. The register pair that cannot be used in LDAX instruction is
(a) BC
(b) DE
(c) HL
(d) None of the above

23. After executing a CMA instruction, the carry flag will be
    (a) 0 (b) 1
    (c) Inverted (d) Not affected
24. The CMC instruction will affect flag(s)
    (a) Carry (b) Carry, Zero
    (c) Carry, Parity (d) Carry, AC
25. Maximum number of I/O devices that could be connected in I/O-mapped-I/O is
    (a) 48
    (b) 128
    (c) 256
    (d) 65536

Answers to Multiple Choice Questions

1. a	2. b	3. c	4. a	5. c	6. a	7. a	8. b	9. c	10. a
11. b	12. d	13. a	14. a	15. a	16. b	17. b	18. a	19. a	20. d
21. d	22. c	23. d	24. a	25. c					

Long Answer Type Questions

1. What are the different addressing modes in 8085?
2. Discuss about the data transfer instructions in 8085.
3. Enumerate the arithmetic instructions in 8085.
4. Explain the operation of logical instructions in 8085.
5. Enumerate different types of branch instructions in 8085.
6. Explain how PUSH and POP instructions work in 8085.
7. Distinguish between conditional and unconditional return instructions in 8085.
8. Distinguish between the instructions PCHL, SPHL and XTHL in 8085.
9. Explain the operation of RIM and SIM instructions of 8085.
10. Enumerate the assembler directives commonly available in 8085 assemblers.
11. Draw the timing diagram of "LXI D, 2030H" instruction.
12. Explain how the stack can be useful in implementing nested subprograms.
13. Write an 8085 assembly language program to sort a set of numbers in descending order.
14. Write an assembly language program to decide whether a string stored at memory is a palindrome or not.
15. Write an assembly language program to reverse a string stored at memory.

4

Interrupts in 8085

LEARNING OUTCOMES

After reading this chapter, students will be able to:

LO 1 Classify the interrupts of 8085

LO 2 State the interrupt triggers available for 8085

LO 3 Explain the role of RIM and SIM instructions for interrupt processing

LO 4 Write interrupt service routines for 8085

LO 5 Draw timing diagram for INTR processing of 8085

LO 6 Explain interrupt priorities of 8085

LO 7 Draw block diagram of 8259 interrupt controller

LO 8 Explain the operation of 8259 in the context of 8085

LO 9 Explain software interrupts in 8085

4.1 Introduction (LO 1)

Interrupt is a mechanism to inform the processor about the occurrence of an unusual event. In its normal course of action, a processor fetches successive instructions from memory and executes them. However, some special event may occur, for example, a key on the keyboard may be pressed. Now the processor needs to read the key code. For this, the processor needs to suspend the instruction sequence currently being executed and branch to the keyboard routine. Thus, the key press is an *interrupt* to the processor. This suspension of current execution sequence and carrying out the special operations pertaining to the interrupt event takes place in a well-defined manner in all processors, including the 8085. This chapter presents a detailed discussion on handling of interrupts in the 8085.

Interrupts can be classified into several categories. In the following, some such categorizations have been discussed, particularly in the context of 8085.

1. *Hardware vs. Software Interrupts*: The 8085 possesses a number of pins that can be used by an external device to send an interrupt to the 8085. The pins are TRAP, RST5.5, RST6.5, RST7.5 and INTR. These are called hardware interrupts. On the other hand, a program in execution can also give an interrupt to the 8085 by executing an RST instruction. It may be recalled that the 8085 supports eight different RST instructions, namely RST 0 to RST 7. The software interrupts are generally used by the Operating System (OS) designers to provide some special OS services to the application program. A typical such service may be to run a new program, to do some I/O operation etc.

> Software interrupts are machine instructions only.

2. *Maskable vs. Nonmaskable Interrupts*: Among the interrupts of 8085, the interrupt TRAP is nonmaskable, while all others are maskable in nature. A programmer can inhibit the sensing of maskable interrupts via program instructions. The 8085 will not respond to any interrupt line that has been masked off. On the other hand, the nonmaskable interrupt TRAP cannot be inhibited via programming. The TRAP interrupt is generally reserved for critical operations in the system.
3. *Vectored vs. Non-vectored Interrupts*: A vectored interrupt is one for which the service routine address is fixed and on occurrence of such an interrupt, the processor branches to that routine after saving the return address into stack. In 8085, all the hardware and software interrupts, excepting INTR, are vectored interrupts. For the software interrupt "RST *n*" ($n = 0$ to 7), the value of n is multiplied by 8 to determine the service routine address. For the hardware interrupts, TRAP (equivalent to RST4.5), RST5.5, RST6.5 and RST7.5, a similar policy of multiplication by 8 is followed to determine the service routine address. Thus, the service routine address for TRAP is $8 \times 4.5 = 36_{10}$ (24H). For RST5.5, 6.5 and 7.5, the corresponding service routine addresses are 2CH, 34H and 3CH, respectively. For the non-vectored interrupt INTR, the address for the service routine is provided by the interrupting device in *interrupt acknowledge* machine cycles executed by the 8085, on occurrence of the interrupt.

> Vector address for "RST 4" is 20H and TRAP (RST4.5) is 22H. The two byte memory space 20H-21H is sufficient for a JMP instruction to take control to the actual service routine of TRAP. Similarly between "RST 1" and "RST 2" there are only 8 bytes. If the space is not sufficient to hold the service routine, a JMP instruction may be used to reach the actual service routine address.

The 8085 contains a flip-flop controlling the enabled/disabled status of interrupts. The flip-flop, known as *Interrupt Enable* (*IE*) flip-flop, can be set and reset by using the Enable Interrupt (EI) and Disable Interrupt (DI) instructions in the program, respectively. Upon reset or the acknowledgement of an interrupt, the IE flip-flop is reset, that is, all interrupts (excepting TRAP) get disabled. The EI instruction has to be used explicitly to make the 8085 sense the interrupts. Moreover, on the occurrence of an interrupt, the IE flip-flop gets reset. Thus, it is required that inside the service routine of that interrupt, the EI instruction is used at an

> A common mistake while writing Service routine is to forget the EI Instruction.

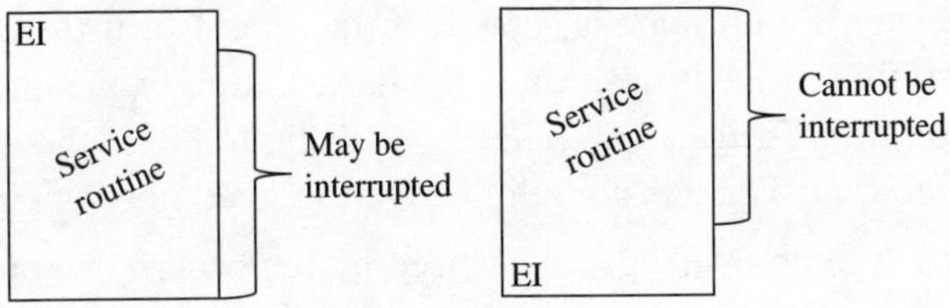

FIGURE 4.1 Placement of EI in service routine

appropriate place to enable the interrupts. The 8085 will notice the occurrence of interrupts only after the execution of the EI instruction. Thus, if the EI instruction is put at the beginning of the service routine, the 8085 may be interrupted while executing the service routine itself. On the other hand, if EI is the last instruction of the service routine, the 8085 cannot be interrupted before finishing the service routine. The concept has been illustrated in Figure 4.1.

Similarly, the DI instruction can be used in the program code to reset the IE flip-flop. This will inhibit the interrupts affecting the 8085 till the IE flip-flop is set again via the EI instruction. Thus, if the programmer wishes that the execution of some part of the code should not be interrupted, the DI-EI instruction pair can be used for guarding that code segment. This is typically used by the OS code. The situation has been shown in Figure 4.2.

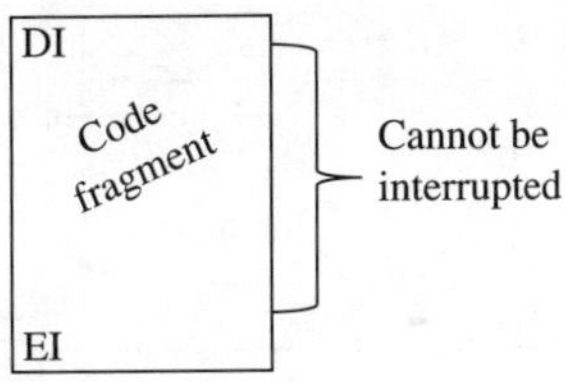

FIGURE 4.2 DI-EI to inhibit interrupt

4.2 Triggering of Interrupts

LO 2

Triggering refers to the required condition of the signal at an interrupt pin for it to be sensed by the CPU. As noted earlier, the 8085 has five hardware interrupts – TRAP, RST7.5, RST6.5, RST5.5 and INTR. From the point of view of triggering, these pins can be classified into three categories – *level triggered*, *edge triggered* and *level-and-edge triggered*. In the following, these triggering methods have been enumerated.

> External interrupting device must ensure to give sufficient input for triggering to take place.

1. *Level triggered*: A level triggered interrupt is accepted by the processor if the pin has got a signal of appropriate level applied to it. As a digital signal has two logic levels, level triggered interrupts may be either *low-level triggered* or *high-level triggered*. In 8085, the interrupt lines INTR, RST5.5 and RST6.5 are high-level triggered interrupts. The 8085 gets interrupt on these lines, if the corresponding pins are kept high for sufficient time. There is no low-level triggered interrupt in 8085.

2. *Edge triggered*: These interrupt lines are accepted by the processor if there is a low-to-high (rising edge) or a high-to-low (falling edge) transition in them. Accordingly, the interrupt line is called to be *rising-edge triggered* or *falling-edge triggered*, respectively. In 8085, the interrupt pin RST7.5 is a rising-edge triggered interrupt. There is no falling-edge triggered interrupt in 8085.
3. *Edge-and-Level triggered*: Such interrupts are sensed by the occurrence of a proper edge followed by the level. Thus, an interrupt may be *rising-edge-high-level triggered* or *falling-edge-low-level triggered.* The TRAP interrupt in 8085 is *rising-edge-high-level triggered.* There is no interrupt in 8085 belonging to the other category.

Figure 4.3 shows different types of triggering in the 8085. It also shows the IE flip-flop. From the diagram, it can be observed that if the IE flip-flop is enabled, then only the interrupts can reach the processor. The interrupts RST7.5, RST6.5 and RST5.5 have mask flip-flops also. If any of these interrupts is masked, the corresponding flip-flop gets set to 1, inhibiting that interrupt. Further, to remember the occurrence of RST7.5 edge, another flip-flop is available. This flip-flop has been shown in the figure, as well.

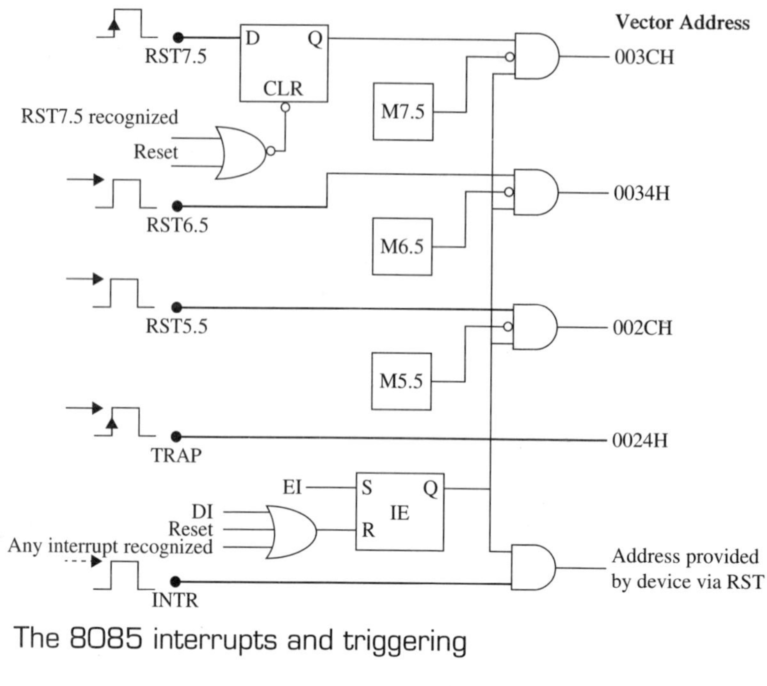

FIGURE 4.3 The 8085 interrupts and triggering

4.3 Instruction SIM – Set Interrupt Mask (LO 3)

The individual masks for the hardware interrupt pins RST7.5, RST6.5 and RST5.5 can be controlled via the instruction SIM. The instruction uses the accumulator bit pattern to specify the masking. The bit pattern for SIM has been shown in Figure 4.4. The bit 0, 1 or 2 is to be set to 1 if the interrupt line RST5.5, RST6.5 or RST7.5 is to be masked off, respectively. Any of these bits put to 0 will unmask

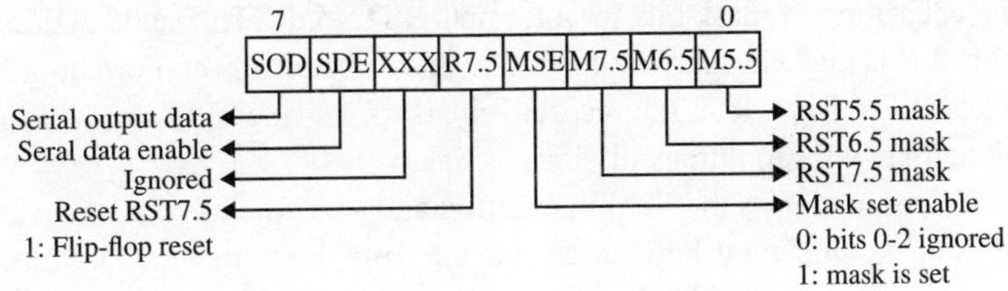

FIGURE 4.4 SIM instruction and interrupt masking

the corresponding interrupt pin. Bit 3, named *Mask Set Enable* (*MSE*), needs to be set to 1 if the SIM instruction is intended to control masking via bits 2 to 0. Bit 4, named R7.5 (*Reset RST7.5*), can be used to reset the flip-flop corresponding to RST7.5. If this bit is set to 1, the RST7.5 flip-flop is reset. It may be used to override (that is, ignore) the RST7.5 interrupt without servicing it. The last two bits, bit 6 and 7, are used for serial transmission and will be discussed in conjunction with serial communication. The following code fragment enables all the three RST interrupt pins.

```
EI              ; Enable interrupts
MVI A, 08H      ; Bit pattern to enable RST7.5, RST6.5 and RST5.5
SIM             ; Set masking
```

The following code fragment can be used to reset RST7.5 interrupt flip-flop, without servicing it.

```
MVI A, 18H      ; Set R7.5 to 1
SIM             ; Reset RST7.5
```

4.4 Instruction RIM – Read Interrupt Mask

LO 3

The SIM instruction, discussed in the last section, can be used to mask out the interrupt lines. In reference to that, the RIM (Read Instruction Mask) instruction can be used to get the current mask setting read into the accumulator bits. The instruction also gives information about the pending interrupts. Apart from interrupts, the RIM instruction also takes part in receiving serial data via SID pin. The accumulator bit pattern after a RIM instruction has been shown in Figure 4.5. The least significant three bits, M5.5, M6.5 and M7.5,

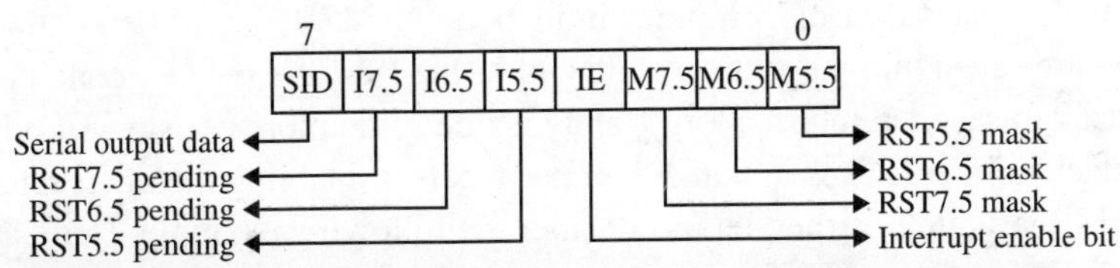

FIGURE 4.5 RIM instruction and pending interrupts

correspond to the current setting of mask bits for interrupts RST5.5, RST6.5 and RST7.5, respectively. The current value of IE flip-flop is made available at bit 3. The bits 4, 5 and 6 give information about any pending interrupt at pin RST5.5, RST6.5 and RST7.5, respectively. Bit 7 gets the value available at pin SID and will be discussed in conjunction with serial transmission.

The following code fragment uses the RIM instruction and subsequently checks whether the interrupt RST5.5 is pending or not. This may happen as the interrupt lines in the 8085 are having different priorities (discussed later); thus, an RST5.5 interrupt line may not be serviced immediately.

```
RIM                  ; Read interrupt mask
MOV D, A             ; Save mask information
ANI 10H              ; Check if RST5.5 pending
JZ L1                ; If zero, not pending
EI                   ; Enable interrupt
MOV A, D             ; Get mask bit pattern
ANI 0EH              ; Enable RST5.5 by setting mask M5.5 to 0
ORI 08H              ; Enable SIM by setting MSE
SIM                  ; Set interrupt mask
CALL SRV_RTN         ; Jump to service routine
L1:                  ; No pending interrupt
```

4.5 Interrupt Service Routine

LO 4

On occurrence of an interrupt, the 8085 first completes the current instruction, saves the return address (present PC value) onto stack and then branches to the *Interrupt Service Routine* (*ISR*) address. As shown in Figure 4.3, the ISR addresses are fixed for the RST interrupts, and between two such ISR addresses, only eight bytes of memory locations are free. For any non-trivial service, these eight bytes are not sufficient for the ISR. Thus, typically a JMP instruction is put at this ISR address to branch to a free chunk of memory area where the ISR could be put. The second important point to be noted while writing an ISR is that the ISR may be using a number of CPU registers, in its body. However, at the time of occurrence of an interrupt, the processor was executing some other program. An ISR developer cannot presume the set of registers that the interrupted program might have been using. Thus, to ensure that those register contents are not inadvertently modified by the ISR, it is advisable that the ISR first saves all the registers that it is going to use in its body and restore them before returning to the interrupted program. For example, if the ISR is using the registers A, B, C, H and L in its body, the ISR should be as shown in Figure 4.6. First, those registers are saved into the stack. At the end of the ISR, the register contents are restored. As the interrupts use stack to save the return address and may be some more registers (used by the ISR), it is necessary that the stack pointer (SP) is initialized properly before enabling the interrupts.

To illustrate the ISR structure further, let us consider the following example. Here, a program continually displays the patterns 00H to FFH in increasing sequence on port with address 05H. However, the

```
PUSH PSW
PUSH B
PUSH H
 [Body]
POP H
POP B
POP PSW
EI
RET
```

FIGURE 4.6 Saving and restoring registers in ISR

system has got a switch connected to RST6.5 pin, which if pressed will interrupt the 8085 to blink the output port bits once (that is, output 00H followed by FFH once). The ISR is assumed to be loaded from address 4000H. The program code has been shown next.

```
            ORG 0034H
            JMP 4000H         ; Put a jump at address for RST6.5
            ORG 3000H         ; Assume the main routine to be at 3000H
            LXI SP, 7000H     ; Initialize SP
            EI                ; Enable interrupts
            RIM               ; Read current interrupt mask
            ORI A, 08H        ; Enable RST6.5
            SIM               ; Set mask
            MVI C, 00H        ; C register holds next pattern
L1:         MOV A, C          ; Get pattern in A
            OUT 05H           ; Output pattern
            CALL DELAY_RTN    ; Put delay
            INR C             ; Increment C
            JMP L1            ; Repeat with next pattern
            ORG 4000H         ; Interrupt routine at 4000H
RST6.5_RTN: PUSH PSW          ; Save used registers
            MVI A, 00H        ; Get 0 in A
            OUT 05H           ; Output 0 to port
            CALL DELAY_RTN    ; Put delay
            CMA               ; Complement accumulator
            OUT 05H           ; Output FFH
            CALL DELAY_RTN    ; Put delay
```

```
            POP PSW        ; Restore A
            EI             ; Enable interrupts
            RET            ; Return from ISR
DELAY_RTN:  PUSH B         ; Save B
            PUSH PSW       ; Save A
            LXI B, 1000H   ; Initialize BC pair
L2:         DCX B          ; Decrement BC pair
            MOV A, C       ; Get C in A
            ORA B          ; A gets (A OR B)
            JNZ L2         ; Check if A is zero
            POP PSW        ; Restore A
            POP B          ; Restore B
            RET            ; Return from DELAY_RTN
```

4.6 Interrupt Line INTR

LO 5

> It is the responsibility of the interrupting device to provide code of relevant RST instruction in the interrupt acknowledge machine cycle.

Unlike other interrupts of 8085, the interrupt INTR is non-vectored. Thus, the ISR address for INTR is not fixed. The address is provided by the interrupting device. On sensing the INTR interrupt, the 8085, after finishing the current instruction, enters into an *interrupt acknowledge* machine cycle. In this cycle, the $\overline{\text{INTA}}$ line is pulled low to acknowledge the interrupting device. The content on the data bus is read by the 8085. The content is expected to be one of the eight instructions – "RST 0" through "RST 7". The structure of RST instructions has been shown in Figure 4.7. Various addresses for the RST instructions have been shown in Table 4.1. In the interrupt acknowledge cycle, one of the RST codes is available to the 8085 from the interrupting device. Control branches to the ISR address mentioned in Table 4.1. Figure 4.8 shows a mechanism by which the interrupting device may provide an RST code to the 8085. The structure shown in the figure generates the code EFH (that is, "RST 5"). The corresponding binary pattern "11101111" can be generated via an octal tristate buffer. The buffer is enabled by the interrupt acknowledge line $\overline{\text{INTA}}$. The individual inputs of the tristate buffers are tied to either 5V or GND depending upon the binary pattern to be generated to provide the RST code.

RST *n*:	1	1	←	*n*	→	1	1	1

FIGURE 4.7 Structure of RST instruction

TABLE 4.1 RST instruction details

Instruction	Binary Code	Hex Code	ISR Address (Hex)
RST 0	11000111	C7	0000
RST 1	11001111	CF	0008
RST 2	11010111	D7	0010
RST 3	11011111	DF	0018
RST 4	11100111	E7	0020
RST 5	11101111	EF	0028
RST 6	11110111	F7	0030
RST 7	11111111	FF	0038

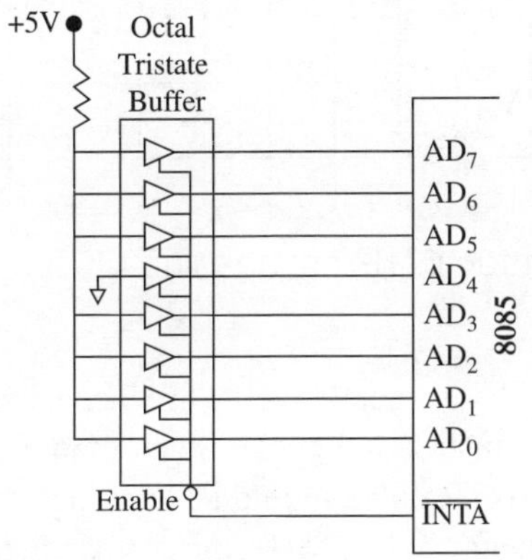

FIGURE 4.8 Generating RST 5 for ISR address

4.6.1 Timing Diagram for INTR Processing

The processing of the INTR interrupt spans over three machine cycles. The first one is a 6 T-state interrupt acknowledge cycle, while the second and the third machine cycles correspond to memory write. In the interrupt acknowledge machine cycle, the 8085 reads an RST instruction on the data bus and the corresponding ISR address is determined, as shown in Table 4.1. In the next two memory write cycles, the PC register is saved onto stack. In the second machine cycle, the SP is decremented and the content of higher order byte of Program Counter (PCH) is written onto the stack. The SP register provides the address bits A_0 to A_{15}. In the third machine cycle, the SP is decremented further and put onto the address bus, while the content of the lower order byte of Program Counter (PCL) is written onto the stack. The control then branches to the ISR by loading the PC with the RST address. A timing diagram depicting these three machine cycles has been shown in Figure 4.9.

From the timing diagram, it may be noted that the INTR interrupt is sensed if the pin is in high state. To answer the question about the minimum and the maximum duration for which the pin should be high

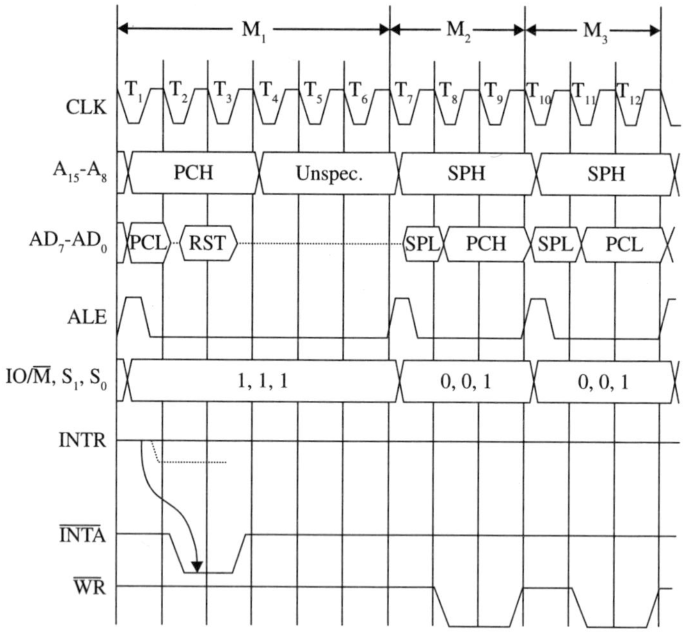

FIGURE 4.9 Timing diagram for INTR processing

for proper operation of the interrupt, let us recall that the 8085 checks for the occurrence of any interrupt at the last but one clock cycle of an instruction cycle. Since the maximum number of T-states in 8085 is 18 (for a CALL instruction), the INTR pulse should be kept high for at least 17.5 T-states for reliable operation. Assuming the clock frequency to be 3MHz, the corresponding duration turns out to be 5.8 μS. The INTR pulse must go low before the EI instruction is executed in the service routine; otherwise, the high level of INTR will be interpreted as another new interrupt on the line. This will lead to the ISR being invoked repeatedly.

4.6.2 Interrupt-Driven Data Transfer

Interrupts can be used for transferring data between the 8085 and an I/O device. In a normal data transfer process, the 8085 needs to check continually whether an input device is ready to send data to the 8085. Similarly, for an output device, the 8085 has to check whether that device is ready to receive data or not. The check has to be done periodically, perhaps in a busy loop. With interrupt facilities being available, it may be advisable that the I/O device interrupts the processor when it is ready to transmit/receive data, instead of the processor checking device status continually. The actual data transfer may take place as part of the ISR.

Figure 4.10 shows the structure of an input device and its interfacing with the 8085. The device has a number of two-position switches S1–S8, which the user can set to input a particular 8-bit data through the device. After a pattern has been set in S1–S8, another push-button switch Ready is to be pushed for a while to activate the data transfer. In interfacing to the 8085, the device has been provided an address FFH.

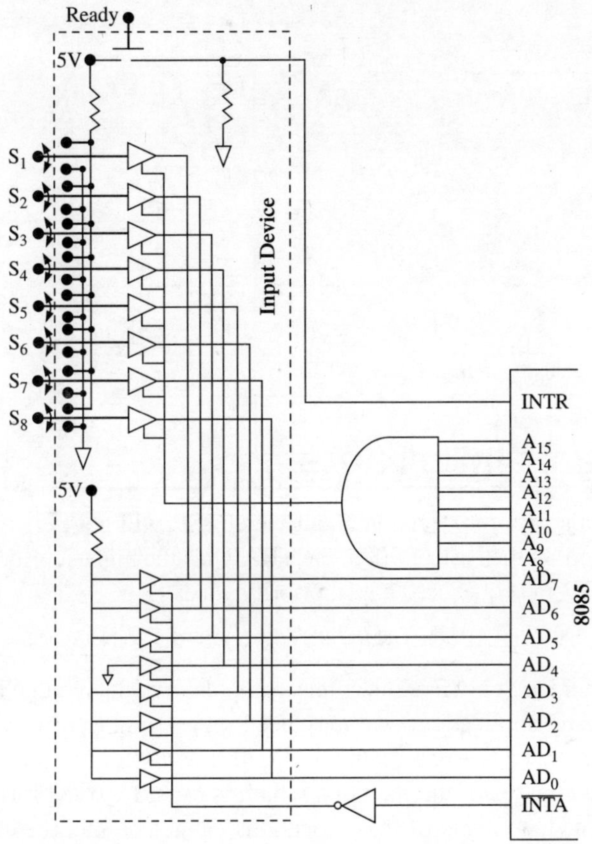

FIGURE 4.10 Interfacing an input device via interrupt

Pressing the Ready switch will interrupt the 8085 on its INTR line. The processor will enter into an interrupt acknowledge machine cycle. The device has been configured to provide the code for "RST 5" onto the data bus. As a result, the 8085 will branch to the service routine at 0028H. The actual ISR has been written from the memory address 5000H which transfers the byte read at port address FFH to the memory location 2000H. The overall program structure has been shown in the following. It is further assumed that on reset and power on, processor branches to address 3000H where the main routine starts.

```
ORG 0
JMP 3000H           ; Jump to main routine, on reset
ORG 0028H
JMP 5000H           ; Jump to actual ISR on "RST 5"
ORG 3000H
LXI SP, 7000H       ; Initialize Stack Pointer
```

```
EI              ; Enable interrupts
...             ; Other parts of main routine
ORG 5000H       ; ISR starts at 5000H
PUSH PSW        ; Save A register
IN 0FFH         ; Read port FFH
STA 2000H       ; Store read value at 2000H
POP PSW         ; Restore A register
EI              ; Enable interrupts
RET             ; Return from ISR
```

4.7 Priority of Interrupts

LO 6

The interrupt priority deals with the decision about which ISR will get executed first if more than one interrupt lines are activated. In the case of 8085, the TRAP has the highest priority, followed by RST7.5, RST6.5, RST5.5 and INTR in sequence. That is, priority wise,

$$\text{TRAP} > \text{RST7.5} > \text{RST6.5} > \text{RST5.5} > \text{INTR}$$

It may be noted that there are only five interrupts available in 8085 and that they have a fixed priority ordering. However, in many applications, we may need more number of interrupt lines, with varying priority requirements. Another very important issue with the 8085 interrupts is that all interrupts are vectored to the starting addresses of the memory, which is often reserved for ROM or EPROM. Once a system has been designed, contents of these locations cannot be modified easily. Thus, the system designers have to take enough precaution to route the interrupts to service routines loaded at different parts of the memory. To resolve these difficulties, a *Programmable Interrupt Controller* (*PIC*) chip, 8259A, has been introduced by Intel. The following section presents an overview of the chip.

4.7.1 Programmable Interrupt Controller 8259A

LO 7, 8

The programmable interrupt controller chip, 8259A, has been specifically designed to work with the INTR pin of 8085. However, it can be interfaced with advanced processors, like 8086. The chip provides many advantages. It can manage eight interrupt requests. The corresponding ISRs can be mapped to anywhere in the memory map – no additional RST hardware is necessary. However, the eight ISRs are spaced in an interval of four or eight locations. Different priority modes can be supported with eight different priority levels. The status of pending interrupts, in-service interrupts and masked interrupts can be read. The individual interrupt lines could be configured as level-triggered or edge-triggered. Additional 8259A chips can be cascaded to have upto 64 levels. Figure 4.11 shows a block diagram of the 8259A chip. The modules inside it are as follows.

1. *Read/Write Logic*: This works as the typical read/write control logic to write a command or read a status.

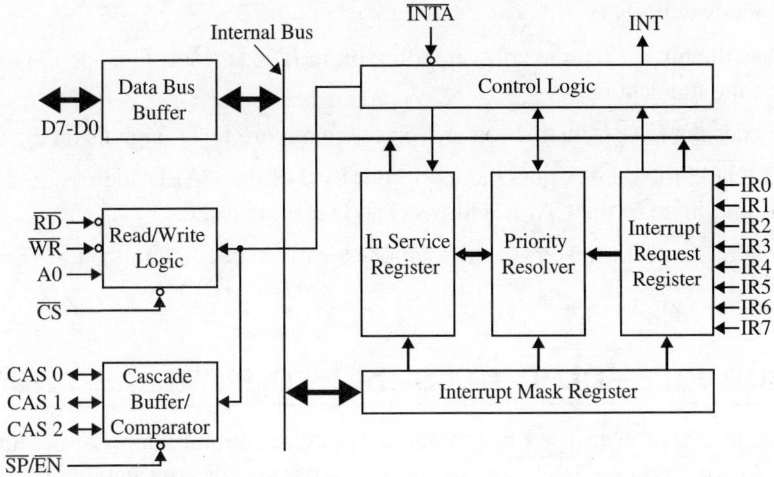

FIGURE 4.11 The 8259A block diagram

2. *Control Logic*: It controls all other modules in the chip. It has got two pins – INT (Interrupt) as output and $\overline{\text{INTA}}$ (Interrupt Acknowledge) as input. The INT pin is connected to the INTR line of 8085, while $\overline{\text{INTA}}$ is connected to the $\overline{\text{INTA}}$ line of 8085.
3. *Interrupt Request Register (IRR)*: It has eight input lines IR0 to IR7 for interrupts. When these lines go high, the requests are stored in the request register.
4. *In-Service Register (ISR)*: It stores the levels of all the interrupts currently being served.
5. *Interrupt Mask Register (IMR)*: It stores the masking bits of all the interrupt lines.
6. *Priority Resolver*: It consults the three registers – IRR, ISR and IMR to determine the priorities of the interrupt lines. Only one of the eight interrupts is sent to the processor. Three different interrupt modes are supported – Fully Nested, Automatic Rotation and Specific Rotation. In the Fully Nested mode, the IRs are arranged from the highest to the lowest priority with IR0 as the highest priority. Of course any other interrupt can be designated as the highest priority one and nesting starts from there. For example, the default order IR0 – IR1 – IR2 – IR3 – IR4 – IR5 – IR6 – IR7 can be modified as IR4 – IR5 – IR6 – IR7 – IR0 – IR1 – IR2 – IR3. In Automatic Rotation mode, once an interrupt has been serviced, it receives the minimum priority. Assuming that IR2 has been serviced last, the priority sequence becomes IR3 – IR4 – IR5 – IR6 – IR7 – IR0 – IR1 – IR2. In Specific Rotation mode, the user can select the interrupt to have the lowest priority. Other interrupt priorities are set similar to the Automatic Rotation.
7. *Cascade Buffer/Comparator*: The block can be used to increase the number of interrupt levels by cascading two or more number of 8259A chips.

The operating sequence of the 8259A chip is as follows.

1. The interrupt requests are stored in the IRR.
2. Priority Resolver determines the priority of the interrupting inputs and sets INT high, if required.

3. The 8085 responds with $\overline{INTA}$.
4. The 8259A sets the bit in ISR; the corresponding bit in IRR is reset. The opcode for CALL instruction is placed onto the data bus.
5. The 8085 decodes the CALL instruction and places two more $\overline{INTA}$ signals on the data bus.
6. In the second $\overline{INTA}$, the 8259A puts the low-order byte of the CALL address. At the third $\overline{INTA}$, it puts the high-order byte of the CALL address. The ISR bit is reset.
7. The 8085 program control transfers to the memory location specified in the CALL instruction.

4.8 Software Interrupts as Program Instructions (LO 9)

The RST instructions have been used to provide the service routine address in conjunction with the INTR interrupt. Apart from that, RST may be used as software interrupt in code fragments. Typically, the system services are provided to the application developers via RST software interrupts. One such support may be to provide a debugging facility to the application programmer. In the initial phase of program development, the user may need a facility to observe the contents of various registers intermittently and then again continue the program. In the following, we shall look into a mechanism where an ISR corresponding to "RST 6" will be used to output the selected register content to port 75H. The register to be selected is identified by the input port 70H. The registers get selected as per the input values specified in Table 4.2.

TABLE 4.2 Register selection

Input	Selection
0	A
1	B
2	C
3	D
4	E
5	H
6	L
7	Return from ISR

The program code for the ISR has been shown next.

```
        ORG 0030H
        JMP ISR_6
        ORG 5000H
ISR_6:  PUSH PSW          ; Save A register
```

```
L1:     IN 70H          ; Read register selection
        CPI 00H         ; A register?
        JNE L2          ; No, check other register
        POP PSW         ; Get original A value
        OUT 75H         ; Output A register
        PUSH PSW        ; Save original A register
        JMP L1          ; Repeat
L2:     CPI 01H         ; B register?
        JNE L3          ; No, check other register
        MOV A, B        ; Get B into A
        JMP L10         ; Jump to L10
L3:     CPI 02H         ; C register?
        JNE L4          ; No, check other register
        MOV A, C        ; Get C to A
        JMP L10         ; Jump to L10
L4:     CPI 03H         ; D register?
        JNE L5          ; No, check other register
        MOV A, D        ; Get D to A
        JMP L10         ; Jump to L10
L5:     CPI 04H         ; E register?
        JNE L6          ; No, check other register
        MOV A, E        ; Get E to A
        JMP L10         ; Jump to L10
L6:     CPI 05H         ; H register?
        JNE L7          ; No, check other register
        MOV A, H        ; Get H to A
        JMP L10         ; Jump to L10
L7:     CPI 06H         ; L register?
        JNE L8          ; No, check other register
        MOV A, L        ; Get L to A
L10:    OUT 75H         ; Output to port
        JMP L1          ; Repeat
L8:     CPI A, 07H      ; Return from ISR?
        JNE L1          ; No, repeat
        POP PSW         ; Restore original A
        RET             ; Return from ISR
```

```
      :
M1: RST 6
      :
M2: RST 6
      :
M3: RST 6
      :
```

FIGURE 4.12 Example program template

Figure 4.12 shows a program template in which the user wishes to check the register contents, whenever the control reaches the labels M1, M2 or M3. Accordingly, in the program, "RST 6" instruction has been put at positions M1, M2 and M3. It may be noted that "RST 6" is a single-byte instruction. Once the user is satisfied with the working of the program, the "RST 6" instructions can be replaced by single-byte NOP instructions, without affecting the program logic and branch target addresses. Thus, the mechanism can be used fruitfully to realize debuggers. The ISR can be very easily augmented to support modification of register contents/memory locations, to develop a full-fledged debugger.

SUMMARY

The 8085 processor has got a number of hardware and software interrupts supported. The hardware interrupt pins are TRAP (RST4.5), RST5.5, RST6.5, RST7.5 and INTR. Some of them are vectored, while INTR is non-vectored. Apart from the hardware interrupts, there are eight more software interrupts that may be used in conjunction with the INTR or as standalone program instructions. The interrupts are having defined priorities, which may be augmented further by means of programmable interrupt controller chip. The interrupts, apart from TRAP, can be masked selectively by using the SIM instruction. The current mask setting and the pending status of interrupts could be read via the RIM instruction. This chapter has presented a detailed discussion on all those issues. It has also shown the structure of an ISR and the issues related to its design. While the hardware interrupts can be used in different applications, including data transfer to/from I/O devices, the software interrupts are generally used for providing system support to the application developers.

EXERCISES

Multiple Choice Questions

1. With respect to 8085, match Column X with Column Y.

Column X	Column Y	
1. INTR	1. Non-maskable	(a) X1-Y2, X2-Y1, X3-Y4, X4-Y3
2. RST7.5	2. Level triggered	(b) X1-Y4, X2-Y2, X3-Y1, X4-Y3
3. RST4.5	3. Software	(c) X1-Y3, X2-Y2, X3-Y4, X4-Y1
4. RST 1	4. Non-vectored	(d) X1-Y1, X2-Y4, X3-Y2, X4-Y3

2. With respect to 8085, match Column X with Column Y.

Column X	Column Y
1. RST5.5	1. Edge triggered
2. RST 6	2. Level triggered
3. RST7.5	3. Edge and level triggered
4. RST 4.5	4. Software

(a) X1-Y3, X2-Y1, X3-Y4, X4-Y2
(b) X1-Y4, X2-Y2, X3-Y3, X4-Y1
(c) X1-Y2, X2-Y4, X3-Y1, X4-Y3
(d) X1-Y3, X2-Y4, X3-Y2, X4-Y1

3. Interrupt lines in 8085 are checked
(a) First T-state of an instruction
(b) Last T-state of an instruction
(c) Last but one T-state of an instruction
(d) Last but one T-state of a machine cycle

4. In 8085 if INTR and HOLD are activated simultaneously, in the middle of an instruction execution, which one is sensed?
(a) HOLD
(b) INTR
(c) Cannot be said deterministically
(d) None of them

5. Which interrupt of 8085 is maskable?
(a) RST7.5 (b) RST6.5
(c) RST5.5 (d) All of the above

6. The low-level triggered interrupt in 8085 is
(a) RST7.5 (b) RST6.5
(c) INTR (d) None of the above

7. Pending status of interrupts can be checked for
(a) TRAP (b) RST7.5
(c) INTR (d) None of the above

8. The address of the ISR for RST 6 is
(a) 0030H
(b) 0028H
(c) 0038H
(d) None of the above

9. The number of clock cycles in the interrupt acknowledge machine cycle is
(a) 4 (b) 6
(c) 8 (d) None of the above

10. The number of interrupts supported by the 8259A is
(a) 4 (b) 6
(c) 8 (d) 16

11. SIM stands for
(a) Set Interrupt Mask
(b) Start Interrupt Mask
(c) Soft Interrupt Mask
(d) Settle Interrupt Mask

12. RIM stands for
(a) Refer Interrupt Mask
(b) Read Interrupt Mask
(c) Rotate Interrupt Mask
(d) Resolve Interrupt Mask

13. In ISR, register restoration is done
(a) In the order of saving
(b) Reverse of the order of saving
(c) Any order
(d) Fixed order decided by the designers

14. Number of bytes for RST instruction is
(a) 1 (b) 2
(c) 3 (d) 4

15. EI stands for
(a) Eight Interrupts
(b) Enable Interrupts
(c) Enforce Interrupts
(d) None of the above

16. For proper operation, number of T-states for which INTR pulse should be high is
(a) 16 (b) 17.5
(c) 18.2 (d) None of the above

17. The highest priority interrupt in 8085 is
(a) INTR
(b) RST 5.5
(c) RST 7.5
(d) TRAP

18. Operating System services are supported by
(a) Software interrupts
(b) Hardware interrupts
(c) Both hardware and software interrupts
(d) None of the above

19. Which triggering type is not available for the 8085 interrupts?
(a) High-level triggered
(b) Low-level triggered
(c) Rising-edge triggered
(d) None of the above

20. The second machine cycle during an interrupt acknowledge cycle saves
(a) PCH (b) PCL
(c) SPH (d) SPL

21. Current interrupt enable status can be known by executing the instruction
(a) EI (b) DI
(c) RIM (d) SIM

22. The lowest priority interrupt of 8085 is
(a) INTR (b) RST 5.5
(b) RST 7.5 (d) TRAP

23. The machine code in binary for "RST 3" instruction is
(a) 11011111
(b) 01111111
(c) 10111111
(d) None of the above

24. The falling-edge-triggered interrupt in 8085 is
(a) RST 5.5
(b) INTR
(c) RST 7.5
(d) None of the above

25. The interrupt mode(s) in 8259A is/are
(a) Fully nested
(b) Automatic rotation
(c) Specific rotation
(d) All of the above

Answers to Multiple Choice Questions

1. b	2. c	3. c	4. a	5. d	6. d	7. b	8. a	9. b	10. c
11. a	12. b	13. b	14. a	15. b	16. b	17. d	18. a	19. b	20. a
21. c	22. a	23. a	24. d	25. d					

Long Answer Type Questions

1. Discuss about the classification of 8085 interrupts into different categories.
2. Enumerate different triggering schemes for interrupts in 8085.
3. Explain the utility of RIM and SIM instructions of 8085 in interrupt processing.
4. How is the INTR interrupt handled in 8085?
5. Explain how the software interrupts can aid in program debugging.
6. Show and explain the generic structure of an ISR.
7. Explain how RST 7.5 interrupt can be resolved without servicing.
8. State the advantages of interrupt-driven data transfer over programmed data transfer.
9. Draw the timing diagram of interrupt acknowledge cycle.
10. Explain how the RST instructions are converted into machine code.
11. Enumerate the priority setting of 8085.
12. Draw and explain the block diagram of 8259A.
13. Explain the priority resolving strategy in 8259A.
14. Show how 8259A can be interfaced with 8085.
15. Enumerate the operating sequence of 8259A.

5

I/O Interfacing with 8085

LEARNING OUTCOMES

After reading this chapter, students will be able to:

LO 1 Explain a generic I/O port structure

LO 2 Interface switches to a port as input to 8085

LO 3 Interface LED with 8085 port

LO 4 Interface a hexadecimal matrix keyboard with 8085

LO 5 Interface seven-segment display modules with 8085

LO 6 Interface 8279 chip with 8085

LO 7 Interface LCD module with 8085

LO 8 Interface programmable peripheral interface chip 8255 with 8085

LO 9 Interface ADC and DAC with 8085

LO 10 Interface 8254 – programmable interval timer with 8085

LO 11 Perform serial input-output operations via SID/SOD pins

5.1 Introduction

Microprocessors are used frequently in control applications that contain a number of devices connected to the system. These devices are either to sense the environment/plant parameters or control those values. The sensed physical quantity gets converted into a digital value and fed as input to the microprocessor. Thus, these devices act as input devices for the processor. The other set of devices, which are driven by digital values, output from the processor, are called output devices. The simplest possible input device is a switch with two positions – ON and OFF, which may correspond to digital values 1 and 0, respectively (or vice versa). On the other hand, a very simple output device may be a Light-Emitting Diode (LED) which can be turned ON and OFF to represent digital outputs 1 and 0, respectively (or vice versa). The LED may represent the output of a digital control routine being executed by the 8085. However, many of the input-output devices are electromechanical in nature, having special requirements in terms of voltage, current, timing etc. for their operation. For systems used in our day-to-day life, most of the display patterns output by the processor are for human understanding. A basic requirement for such displays is to be flicker-free. On the other hand, if the input switch is jittery in nature, one physical press of the switch may get interpreted as multiple presses. Similarly, a motor rotating at certain speed and direction will need a finite amount of time to increase/decrease speed, stop or reverse its direction. Thus, interfacing an I/O device to the 8085 requires two subtasks to be carried out carefully. The first task is to design the circuitry to connect to the device. The hardware has to be designed such that the digital outputs given at port bits are converted to appropriate voltage or current levels/pulses for the operation of the device. The second task is of course designing the software to be executed by the processor. The program should read the values at input port(s), process it through the desired control algorithm and finally send data to the output port(s). It may be noted that some of the I/O operations require special type of processing which may be available as separate chips. Those chips need to be interfaced with the 8085 for system development.

5.2 Generic I/O Port Structure

LO 1

The 8085 does not have dedicated ports available with the processor. Rather, the ports are identified by the address put on the address lines during the execution of IN and OUT instructions. It is assumed that the ports have capabilities to get enabled, depending upon the address bits put on the address lines. The I/O decoder may generate a chip which select signal to be used for enabling the connection between the input/output lines of the device and the CPU data bus. Figure 5.1(a) shows the structure of a generic input port, while Figure 5.1(b) shows a generic output port. Figure 5.1(c) shows a port structure that can be used either for input or for output purposes. All ports are assumed to be 8-bit wide, as the IN and OUT instructions work on 8-bit data. All ports are having *Enable* lines that need to be set high by the address decoding logic to enable the port. For a bidirectional port, the Enable line is ANDed with IOW and IOR lines to configure the port as an output port or as an input port.

> Microprocessors do not possess built-in I/O ports. Hence, additional hardware has to be connected to create such ports. On the other hand, microcontrollers have integrated I/O ports, in general.

5.2.1 Connecting Switch as Input

LO 2

The simplest input device is a switch that can be turned ON or OFF. Figure 5.2 shows different ways in which a switch can be connected to an input port. Here it is assumed that the switch is being interfaced with a generic input port bit line (I_0 through I_7) of Figure 5.1(a). For the structure shown in Figure 5.2(a), a logic high is read in if the switch is open. If the switch is closed, a logic low is read. The other two combinations read a logic high when the switch is closed. However, these two combinations have the problem of high current flow when the switch is

> Current flow through port be limited to avoid damage to the microprocessor data bus lines.

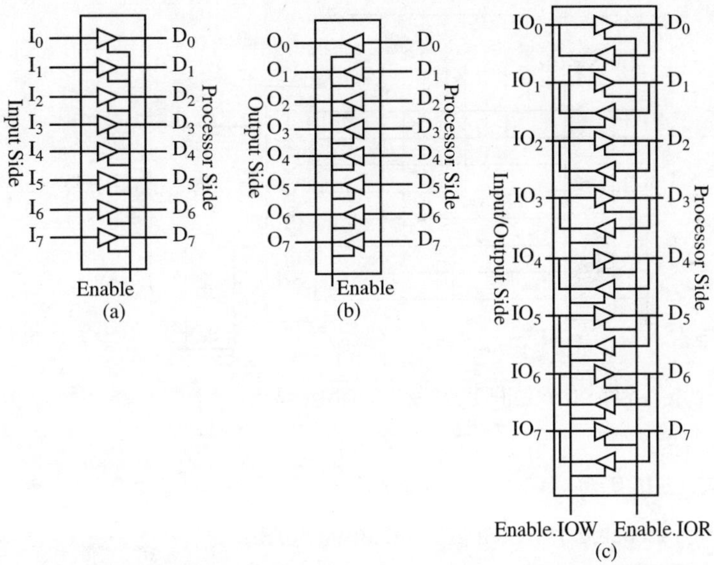

FIGURE 5.1 Generic port structure: (a) input, (b) output and (c) bidirectional

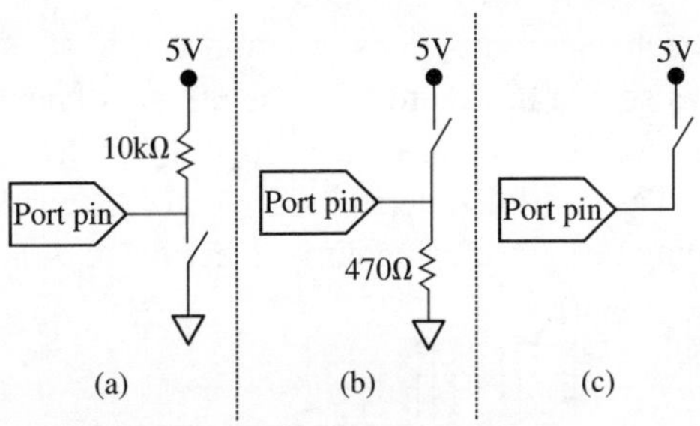

FIGURE 5.2 Connecting switch to a port pin

closed – this may damage the interface circuitry. Thus, the configuration shown in Figure 5.2(a) is the preferred one for connecting switches to the 8085.

It may be noted that a single switch can drive a single bit of the data bus. As the data bus in 8085 is 8-bit wide, eight such switches can be connected and using a single IN instruction, status of all the eight switches could be read. Such an assembly of eight switches is called a Dual in-line package (DIP) switch. Figure 5.3 shows how a DIP switch can be connected to the 8085 through a generic input port. Assuming that the I/O decoder enables the generic input port for the address DFH, the code fragment to read the DIP switch status into memory location 2000H is as follows.

```
IN 0DFH
STA 2000H
```

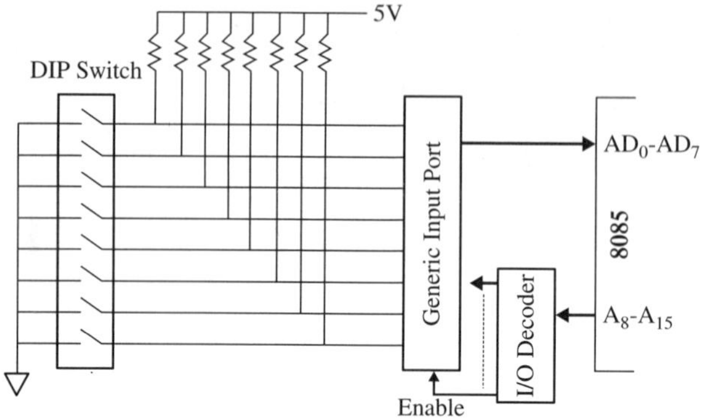

FIGURE 5.3 Schematic for DIP switch connection

Contact Debouncing

> When a mechanical switch is closed, it creates several makes-and-brakes. Contact debouncing is required to ensure that the switch is read once only.

A major problem in mechanical switch interfacing with the 8085 is that these switches work on the principle of make-and-break of contacts. A switch may be of different types, such as push button, toggle and electromechanical relay. The switches are normally open, leading to logic 1 read by the 8085. When the switch is pressed, due to mechanical vibrations, a number of small pulses are generated, as shown in Figure 5.4. The effect is known as *contact bounce* or *switch bounce*. As a result, the 8085 may read in a wrong data. Also, if the switch is connected to an interrupt line, a single press may get manifested as multiple interrupts. Special care has to be taken to avoid such undesirable situation. The problem may be solved using either a hardware or a software-oriented approach.

FIGURE 5.4 Bouncing effect in contacts/switches

> Works by discharging a capacitor quickly when a switch is closed. Charging time is more than the bouncing time of the switch.

- *Hardware Solution.* In this, an RC network is used to solve the bouncing problem. The structure has been shown in Figure 5.5. When the switch is open, the capacitor charges to logic high level, giving a '1' input to the port bit. If the switch is closed momentarily, the capacitor discharges quickly, making the port bit '0'. By choosing the R and C values carefully, it can be ensured that the RC time-constant is larger than the switch bounce time. This prevents the possibility of the capacitor getting charged quickly.

> Works by reading the switch status at least twice with an intervening delay more than the bouncing time of the switch.

- *Software Solution.* In software-based solution, the switch is read up to *N* times. Normally, the switch is open and the

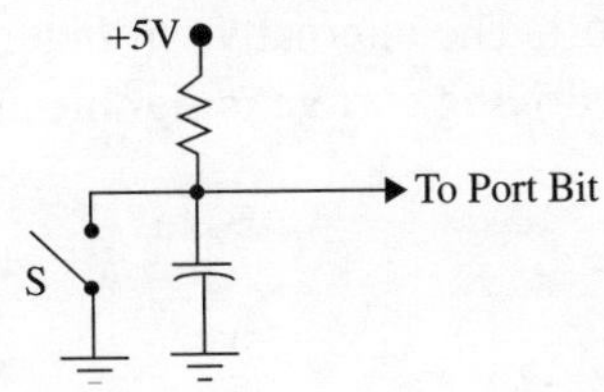

FIGURE 5.5 Hardware debouncing

port bit value read is 1. If the value read is 0, the key must have been pressed. However, this value is not taken as the final one, immediately. Rather, the reading process waits for an appropriate amount of time to overcome the bouncing period. The time is typically 10 ms. After this period, the port is read again to check whether the value read is still 0 or not. This ensures that the key press is registered only once. The whole process may be repeated for a number of times to have more confidence on the value read. A flowchart for the software debouncing process has been shown in Figure 5.6.

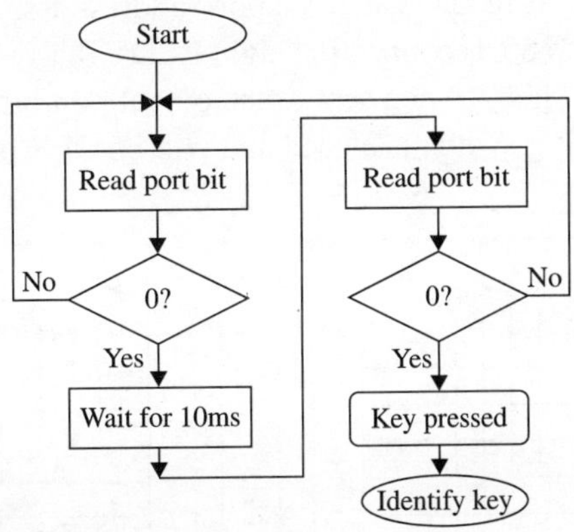

FIGURE 5.6 Software debouncing

5.2.2 Interfacing Light-Emitting Diode (LED)

LO 3

> Current must be sufficient to glow LED brightly and limited so that the LED is not damaged.

The simplest output device that could be interfaced to the 8085 is perhaps a single LED. The LED may be turned ON or OFF under program control to indicate two different logic levels, say 1 and 0, respectively, of the port bit. Figure 5.7 shows three different ways in which an LED can be connected to a generic output port bit. Out of the three configurations, the structure shown in Figure 5.7(a) is the safest one with good brightness of the LED. The LED will be turned ON when the port bit is 0. Most of the LEDs drop a voltage of about 1.7 V across it. Thus, to limit the current to about 10 mA, a resistance of value around $(5 - 1.7)/10 = 330\ \Omega$ or higher has to be connected. The case in Figure 5.7(b) is risky, as large current will flow through the LED with a possibility of damaging it and the port. The alternative in Figure 5.7(c) may have similar problem if the current from the port is high. If the driving current is significantly low, then also this alternative may result in poor brightness of the LED.

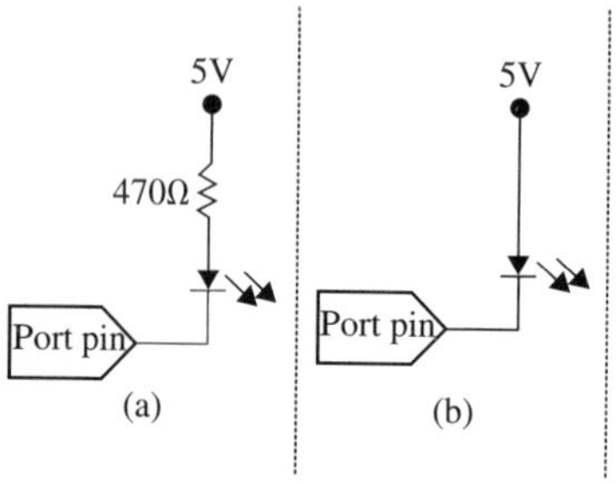

FIGURE 5.7 Connecting LED to a port pin

In the following, an interface has been shown that connects an eight-input DIP switch and eight LEDs to the 8085. The hardware connection has been shown in Figure 5.8. The user will set an 8-bit pattern in the DIP switch. The switch setting is read via an input port (address assumed to be 75H) and is sent to the output port (address assumed to be 85H) connecting eight LEDs. If a switch is closed, the corresponding bit is read as 0 by the processor. This 0, when sent to the output port bit, will turn on the LED. Thus, an LED will glow if the corresponding switch is closed and will be OFF, otherwise.

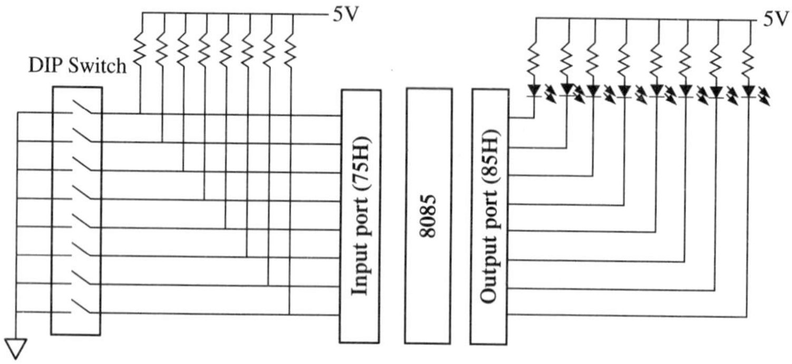

FIGURE 5.8 Interfacing eight switches and eight LEDs

The 8085 assembly language program for the problem has been noted next. In this case, debouncing of keys has not been taken into consideration as it is assumed that the switches are kept in a set position for quite some time. As a result, the display will also become stable. The program has been assumed to have started from address 5000H.

```
    ORG 5000H
L1: IN 75H
    OUT 85H
    JMP L1
```

5.3 Hex Keyboard Interfacing

LO 4

> Normal keyboards connected to personal computers contain a processor to interrupt the main processor in the CPU on pressing a key.

Figure 5.9 shows the schematic to interface a hexadecimal matrix keyboard to the 8085 using two I/O ports. A 4 × 4 hex keyboard has 16 keys corresponding to the hexadecimal digits 0 to F. One side of each row and column are tied high. The other side of rows and columns are connected to the ports of 8085. The port 75H is an input port, while the other one, 85H is an output port. The output port is used for feeding the row lines, while the input port reads the columns. Since there are only four rows and four columns, the four least significant bits of the ports have been used for interfacing. A particular key can be identified by the row and the column it belongs to. For example, the key 6 is in row 1, column 2. Now, if the control program has made the bit 1 of the output port 85H 0 (with all other bits of 85H as 1) and the key 6 has been pressed, the bit 2 of the input port 75H also becomes 0. Assuming that none of the other keys 7, 5 or 4 has been pressed, the least significant 4 bits of the port 75H will be 1011. To determine a key press uniquely, the row lines are set to 0, one by one, while reading the corresponding column inputs. It is also necessary to ensure that initially all keys are in open state. This can be checked easily by outputting an all-zero pattern to the output port 85H and checking that the least significant 4 bits of the port 75H are all 1's. Also, debouncing

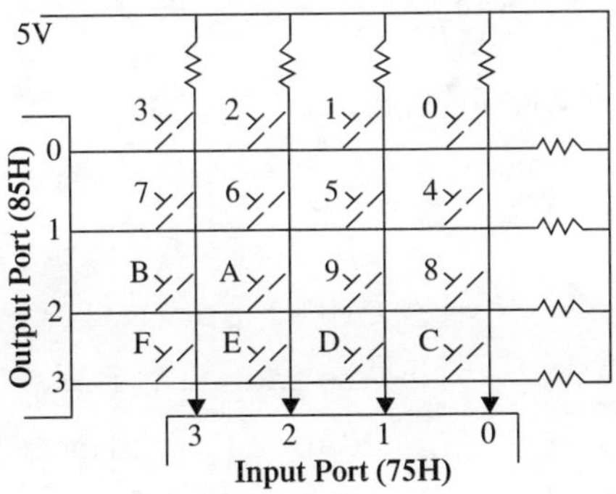

FIGURE 5.9 Interfacing a 4 × 4 hex keyboard

needs to be taken care of. As there is no additional capacitive circuitry connected in Figure 5.9, software debouncing is the ideal solution, in this case. The keyboard routine logic has been elaborated in Figure 5.10. After debouncing, if a key has been pressed, to detect it, one row is set to 0, by turn. This is followed by reading the columns. The column, read as 0, identifies the key. To get the key code, a code table needs to be consulted. For example, the co-ordinate of row 2, column 1, corresponds to the key code 9. The following is the assembly language program to scan the keyboard and get the code of the key pressed (if any) at RAM location 2000H.

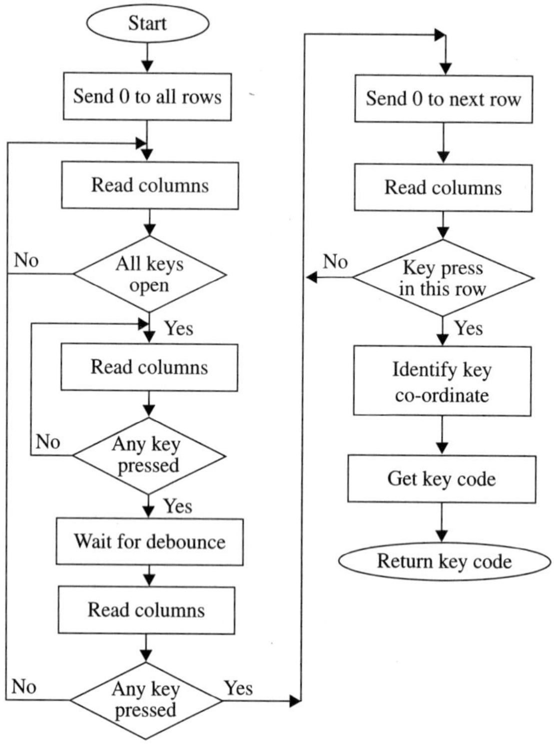

FIGURE 5.10 Flowchart for keyboard scan routine

```
    ORG 1000H
L1: XRA A                ; Clear accumulator
    OUT 85H              ; Ground all rows
L2: IN 75H               ; Read all columns
    ANI A, 0FH           ; Mask unused MSBs
    CPI 0FH              ; Check all keys released
    JNZ L2               ; If not released, wait in loop
    CALL DELAY           ; Some key may be pressed, wait for 30 ms
    IN 75H               ; Read all columns
    ANI A, 0FH           ; Mask unused MSBs
    CPI 0FH              ; Check for key press
    JZ L2                ; No key pressed, go back
    CALL DELAY           ; Key pressed, wait for 30 ms
```

```
        MVI A, 0FEH          ; Ground row 0
        OUT 85H              ; Ground row 0
        IN 75H               ; Read all columns
        ANI A, 0FH           ; Mask unused MSBs
        CPI 0FH              ; Check if key pressed in row 0
        JNZ ROW_0            ; Some key in row 0 pressed
        MVI A, 0FDH          ; Ground row 1
        OUT 85H              ; Ground row 1
        IN 75H               ; Read all columns
        ANI A, 0FH           ; Mask unused MSBs
        CPI 0FH              ; Check if key pressed in row 1
        JNZ ROW_1            ; Some key in row 1 pressed
        MVI A, 0FBH          ; Ground row 2
        OUT 85H              ; Ground row 2
        IN 75H               ; Read all columns
        ANI A, #0FH          ; Mask unused MSBs
        CPI 0FH              ; Check if key pressed in row 2
        JNZ ROW_2            ; Some key in row 2 pressed
        MVI A, 0F7H          ; Ground row 3
        OUT 85H              ; Ground row 3
        IN 75H               ; Read all columns
        ANL A, #0FH          ; Mask unused MSBs
        CPI 0FH              ; Check if key pressed in row 3
        JNZ ROW_3            ; Some key in row 3 pressed
        JMP L1               ; No key pressed, repeat scanning
ROW_0:  LXI H, KEY_ROW_0     ; HL = Start of Row 0
        JMP FIND_KEY         ; Find column for the key
ROW_1:  LXI H, KEY_ROW_1     ; HL = Start of Row 1
        JMP FIND_KEY         ; Find column for the key
ROW_2:  LXI H, KEY_ROW_2     ; HL = Start of Row 2
        JMP FIND_KEY         ; Find column for the key
```

```
     ROW_3: LXI H, KEY_ROW_3      ; HL = Start of Row 3
  FIND_KEY: RAR                   ; Rotate A and get column status to CY
            JNC MATCHED           ; Column found, if CY=0
            INX H                 ; HL = Next column address
            JMP FIND_KEY          ; Check for next column
   MATCHED: MOV A, M              ; Get code from the table
            STA 2000H             ; Get key code to RAM[2000]
            JMP L1                ; Repeat scanning

     DELAY: PUSH PSW              ; Save A register
            LXI B, 1000H          ; Load BC = 1000H
        L4: DCX B                 ; Decrement BC
            MOV A, C              ; Get C into A
            ORA B                 ; A = A or B
            JNZ L4                ; Loop if not zero
            POP PSW               ; Restore A
            RET

            ORG 3000H
KEY_ROW_0: DB '0', '1', '2', '3' ; Row 0
KEY_ROW_1: DB '4', '5', '6', '7' ; Row 1
KEY_ROW_2: DB '8', '9', 'A', 'B' ; Row 2
KEY_ROW_3: DB 'C', 'D', 'E', 'F' ; Row 3
           END
```

5.4 Seven-Segment Display Interfacing LO 5

After the keyboard interface, we shall next look into display modules typically used in microprocessor based small applications. While a single LED can display the status of a bit, a number of them are often organized in the shape of digit 8, so that numeric outputs can be shown there. This has been detailed in Figure 5.11. The display has seven segments, each segment being illuminated by an LED. The LED segments can be turned ON or OFF individually to display patterns 0–9. Apart from the seven LEDs forming the seven segments, one more LED is used to show a dot. This actually makes the display

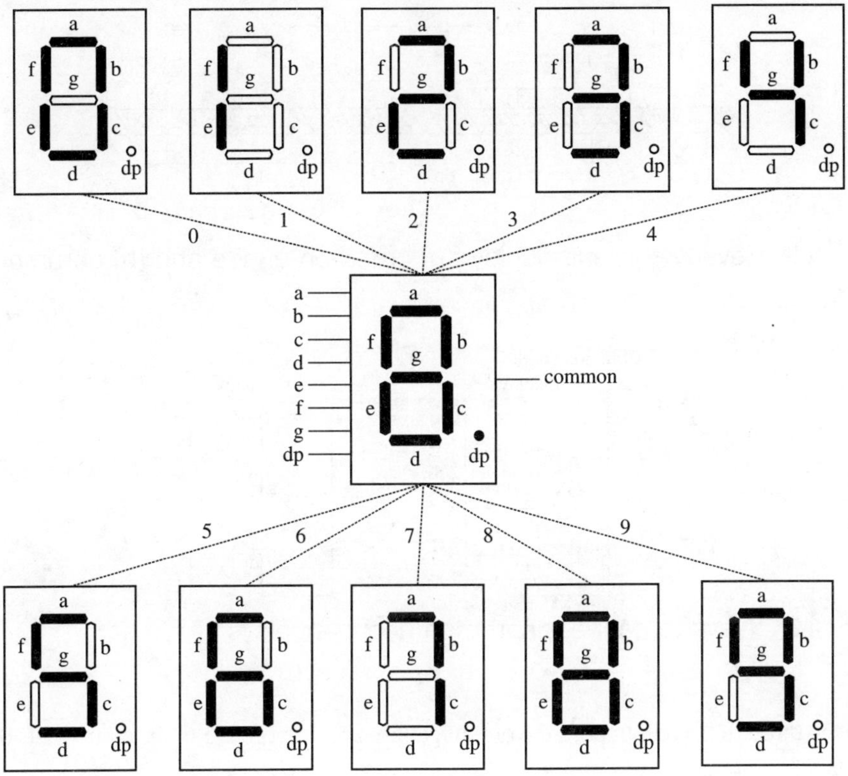

FIGURE 5.11 Seven-segment module and digits

eight-segment one. The segments are named a, b, c, d, e, f, g and dp. One such module can show a single digit. A number of such modules can be used together for displaying more number of digits. Such multi-digit displays are common in usages like Counters, Token Numbers, Date and Time.

There are two types of seven-segment displays – *common anode* and *common cathode*. In common anode module, the anode terminals of the LEDs are joined together into a terminal called common. To glow any segment in this module, the common terminal should have a logic high (+5 V) while the input corresponding to LED segment be put at logic low (0 V). Such a module has been shown in Figure 5.12(a).

On the other hand, a common cathode module (shown in Figure 5.12(b)) has all the cathode terminals joined into a common terminal. To glow a segment, the input corresponding to that segment be set to high (+5 V) and the common terminal grounded. As the current sinking capacities of TTL devices are generally higher than their sourcing capacity, the common anode configuration is preferred by many of the designers.

Figure 5.13 shows two ways in which the current limiting resistances can be connected in seven-segment display modules. Out of these two, the approach shown in Figure 5.13(b) is better. In Figure 5.13(a), brightness will vary depending upon the number of LEDs turned on. For a common anode display module with

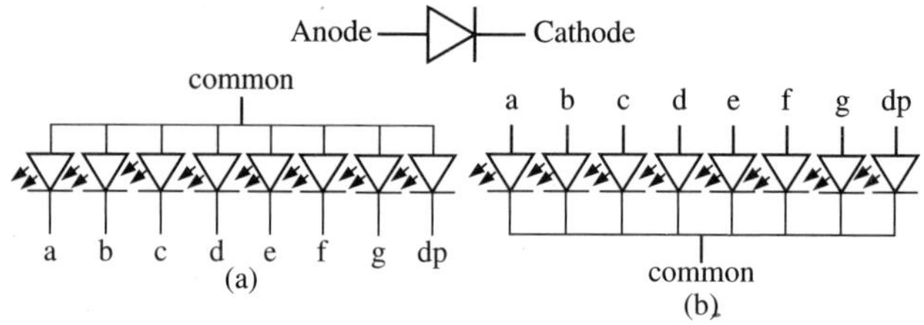

FIGURE 5.12 Seven-segment display: (a) common anode and (b) common cathode

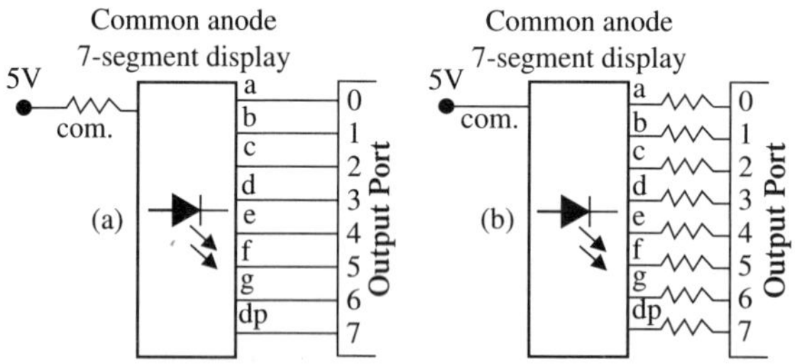

FIGURE 5.13 Ways to connect common anode displays

inputs a to dp fed through the output port, the output patterns corresponding to different digits 0–9 have been shown in Table 5.1.

TABLE 5.1 Patterns corresponding to digits

	Pattern Bits								
	7	6	5	4	3	2	1	0	Pattern
Digit	dp	g	f	e	d	c	b	a	**Hex**
0	1	1	0	0	0	0	0	0	C0
1	1	1	0	0	1	1	1	1	CF
2	1	0	1	0	0	1	0	0	A4
3	1	0	1	1	0	0	0	0	B0
4	1	0	0	1	1	0	0	1	99
5	1	0	0	1	0	0	1	0	92
6	1	0	0	0	0	0	1	0	82
7	1	1	1	1	1	0	0	0	F8
8	1	0	0	0	0	0	0	0	80
9	1	0	0	1	0	0	0	0	90

5.4.1 Multiplexed Display

> Assuming 8-bit ports, two output ports are sufficient to connect upto eight seven-segment modules in a multiplexed fashion.

A major problem with the seven-segment display interface, shown in Figure 5.13, is that connecting multiple such modules require large number of I/O port pins. Each module requires one output port of width 8 bits to drive the LED segments. However, for many applications involving LED display, a number of such modules are to be integrated together. For example, at least six seven-segment displays (two each for the hour, minute and second parts) are needed in a real-time clock module. To sort out this problem, most of the interfacing solutions rely upon the *persistence* property of human vision. A visual impression is retained by human eye and brain for about 1/30th of a second. Hence, if a pattern is taken off after displaying it with sufficient brightness and redisplayed within the duration of the persistence window, human eyes will see a steady image only. In a multimodule multiplexed display, the pattern for the first module is output. The second module is selected next and the corresponding bit pattern is sent to the port, and so on. The whole process goes on continually ensuring that each module gets selected within 1/30th of a second with its pattern sent to the port. The port bits providing the pattern are connected parallel to all the display modules. The other port selects the module to display the pattern. The pattern affects the digit only in the selected module. The scheme has been shown in Figure 5.14.

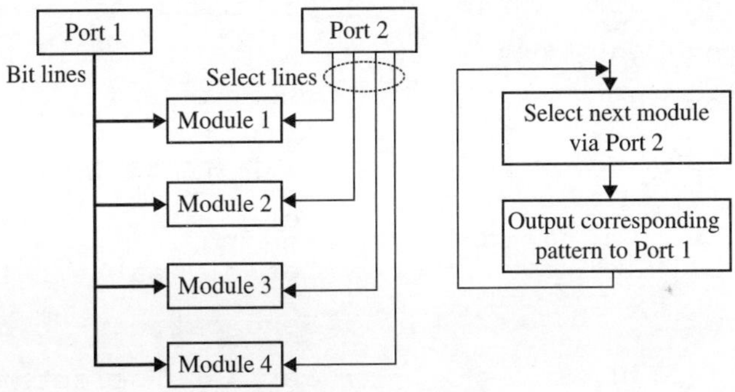

FIGURE 5.14 Multiplexing seven-segment displays

Figure 5.15 shows the hardware connection for four seven-segment display modules connected in a multiplexed fashion to the 8085 ports. The module selection is done via the least significant 4 bits of port 2, while bit patterns for digit display are sent by port 1. As module drives need more current, *p-n-p* transistors *BC557* have been used in the circuit. To display the string of four decimal digits "8085", continually into the four multiplexed segments, bit patterns are to be fed to the ports. The string is assumed to be stored at 4 bytes starting at RAM location 2000H. The program starts from memory location 1000H. The program uses a subprogram named BIT_PATT that gets the bit pattern to be output to the port 1 for a particular digit of the string. Index of the digit is made available in register B before making the call. It returns the bit pattern to glow the LED segments in register A. The display program is in an infinite loop so that the pattern displayed is stable. The bit pattern to glow the required segments for a digit has been stored in the array DGT_PATT (as shown in Table 5.1).

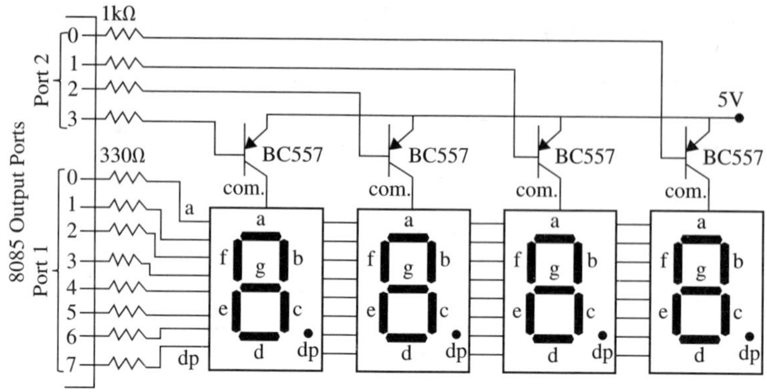

FIGURE 5.15 Four multiplexed seven-segment displays

```
          ORG 1000H
          PORT1 EQU 75H               ; Port 1 address
          PORT2 EQU 85H               ; Port 2 address

DGT_PATT: DB 0C0H, 0CFH, 0A4H,        ; Bit patterns for digits
             0B0H, 99H, 92H, 82H,
             0F8H, 80H, 90H

                                      ; Display most significant digit
      L1: MVI B, 0                    ; Set B
          CALL BIT_PATT               ; Get bit pattern in A
          OUT PORT1                   ; Send pattern to glow segments
          MVI A, 0F7H                 ; Set A to select module
          OUT PORT2                   ; Select module
          CALL DELAY_RTN              ; Put a small delay
                                      ; Display second digit
          MVI B, 1
          CALL BIT_PATT
          OUT PORT1
          MVI A, 0FBH
          OUT PORT2
          CALL DELAY_RTN

                                      ; Display third digit
```

```
        MVI B, 2
        CALL BIT_PATT
        OUT PORT1
        MVI A, 0FDH
        OUT PORT2
        CALL DELAY_RTN
                                ; Display fourth digit
        MVI B, 3
        CALL BIT_PATT
        OUT PORT1
        MVI A,0FEH
        OUT PORT2
        CALL DELAY_RTN
        JMP L1                  ; Repeat display
BIT_PATT: LXI H, STRING         ; Get start address of string
        MOV A, B                ; Get digit index in A
        ADD A, L                ; Update A to have A + L
        MOV L, A                ; Get updated L
        MOV A, M                ; Get the digit (character format)
                                  in A
        SUI A, '0'              ; Subtract ASCII code of '0'
        MOV H, DGT_PATT         ; HL = Start of digit display array
        ADD A, L                ; Update A to have A + L
        MOV L, A                ; Get updated L
        MOV A, M                ; Get bit pattern into A
        RET                     ; Return
DELAY_RTN: LXI B, 1000H         ; Load BC = 1000H
     L4: DCX B                  ; Decrement BC
        MOV A, C                ; Get C into A
        ORA B                   ; A = A or B
        JNZ L4                  ; Loop if not zero
        RET
```

```
        ORG 2000H
STRING: DB "8085"                   ; String to be displayed
        END
```

5.5 The 8279 – Programmable Keyboard/Display Interface LO 6

In the last two sections, we have seen how to interface a matrix keyboard with 16 keys and also to connect a number of seven-segment display modules in a multiplexed fashion. However, a major problem with the approaches is that the 8085 will be busy for a considerable amount of time in scanning the keyboard and refreshing the display. The chip 8279 has been introduced by the Intel to address this keyboard and display interfacing problem. The 8279 is a 40-pin device with two major parts in it – the keyboard segment and the display segment. The keyboard segment can connect to a 64-key matrix, with integrated debouncing and an internal First-In-First-Out (FIFO) memory to store the key entries. An interrupt is generated on every key press. The display section can interface 16-character scanned LED displays. It contains a 16 × 8 RAM that can be used to read/write display information. The display can be either in left-entry or in right-entry format. Figure 5.16 shows the grouping of pins of the 8279. Figure 5.17 shows the internal block diagram of the chip. Based on functionalities, the 8279 can be divided into four sections – keyboard section, scan section, display section and processor interface section. A brief description of each of the sections has been presented next.

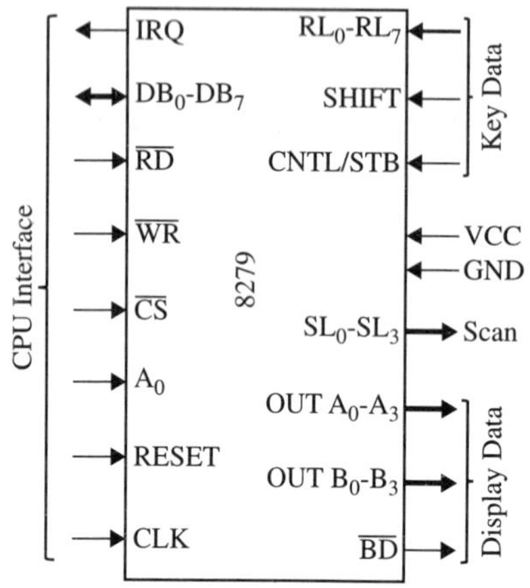

FIGURE 5.16 The 8279 pins

Keyboard Section: The 8279 can support keys distributed over eight columns – RL_0–RL_7. Apart from that, status of two additional keys, named Shift and CNTL/STB (Control/Strobe) can be captured and stored with each key closure event. Debouncing of keys happens automatically. There are two modes in which the keyboard can operate – *two-key lockout* and *N-key rollover*. In the two-key lockout mode, if

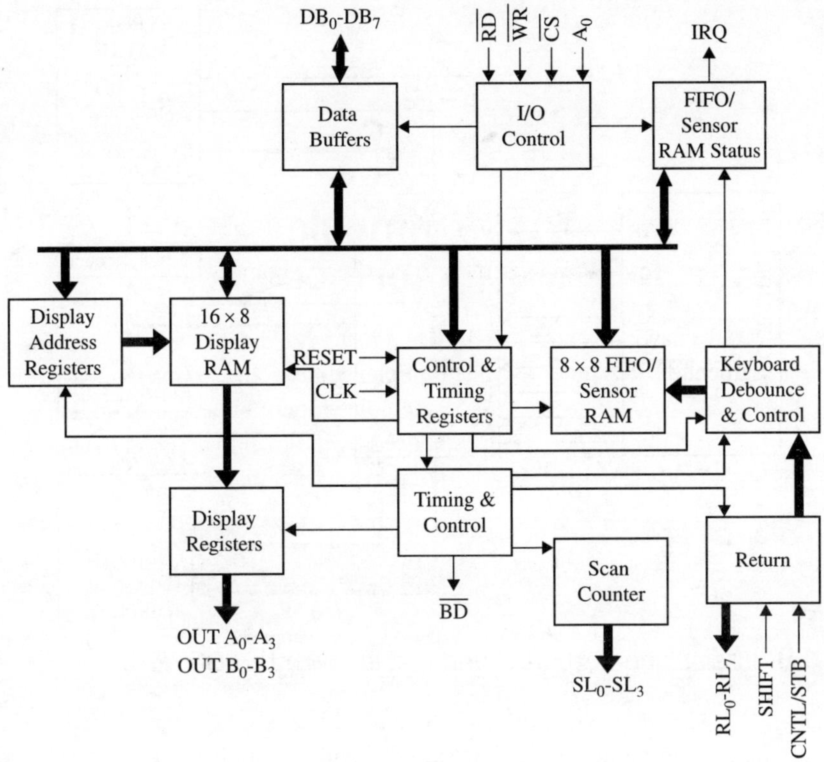

FIGURE 5.17 The 8279 block diagram

two keys are closed almost simultaneously, only the first key gets recognized. In the N-key rollover mode, all the keys pressed simultaneously are recognized and their codes are stored in the buffer. The chip can also be programmed so that no key is recognized until only a single key remains pressed. The 8 × 8 FIFO RAM stores the eight keyboard entries. An IRQ interrupt is generated when the FIFO is not empty.

Scan Section: The scan section contains a scan counter with four scan lines SL_0–SL_3. The lines can be decoded using a 4-to-16 decoder to generate 16 scan lines. The scan lines (or their decoded version) can be connected to the rows of a keyboard matrix and/or to the digit drivers of multiplexed display.

Display Section: There are eight output lines in this section, A_0–A_3 and B_0–B_3. The lines can be used either as one of eight lines or two groups of four lines for a multiplexed display, in conjunction with the scan lines. The section contains 16 × 8 display RAM that can be read/written by the processor. The display can be turned blank by using the $\overline{BD}$ line.

Processor Interface Section: This section is responsible for interfacing the 8279 with the 8085 or any other processor. It includes eight bidirectional data lines DB_0–DB_7, one interrupt request line IRQ and other six control lines. Out of the control lines, when the line A_0 is high, the signals are considered control words, while a low A_0 means the lines contain data. Whenever data entries are stored in the FIFO, the IRQ line goes high, which may be used to interrupt the processor.

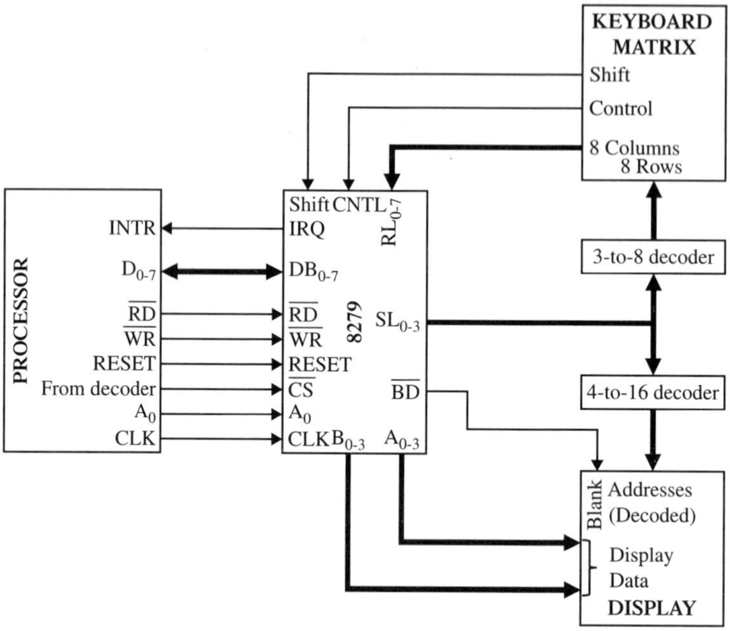

FIGURE 5.18 Interfacing keyboard and display with the 8279

Figure 5.18 shows an interfacing of keyboard and display modules with the 8279 connected to a processor. The keyboard is assumed to be a matrix keyboard with 64 keys arranged in a 8 × 8 fashion. The rows are scanned through the scan lines SL_0–SL_2, which is decoded by a 3-to-8 decoder to feed the eight rows of the keyboard. The columns are read by the return lines RL_0–RL_7. The 8279 interrupts the processor whenever a key is pressed. The display part is assumed to have 16 characters. The address of a character is identified by the 4-to-16 decoder outputs connected to the scan lines. There are several control words to configure the 8279 to operate in various keyboard and display modes. The details for the same could be found in the Intel manual pages for the 8279. The processor needs to send those control words to the 8279 for displaying characters and reading the code of key pressed.

5.6 Liquid Crystal Display (LCD) Interfacing (LO 7)

Liquid Crystal Display (LCD) is an electronic display device working on the principle of light modulating properties of liquid crystals, combined with polarizers. LCD can display arbitrary images, preset words, digits etc. In the recent time, LCDs have become much more preferable to the seven-segment LED displays due to the following reasons. The price of such devices has reduced significantly. LCDs can display digits, characters and graphics, while the seven-segment LED displays are useful for digits and possibly a limited set of characters. Most of the LCD devices have built-in refreshing circuitry. Thus, the processor need not do anything special to keep the display stable. Programming of such devices is also relatively simple.

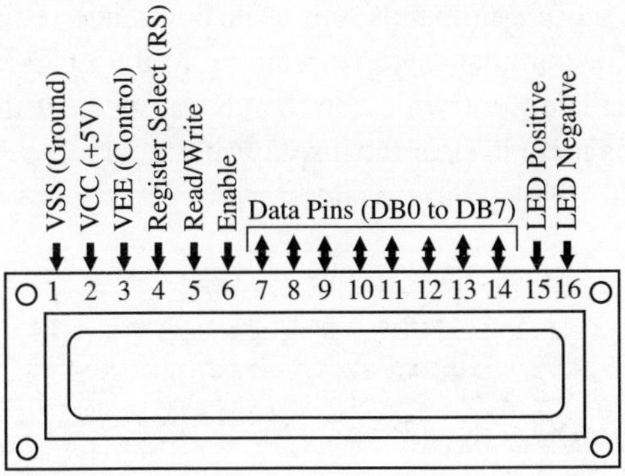

FIGURE 5.19 An LCD module

An LCD module with its pinout has been shown in Figure 5.19. The display consists of two rows, with each row being capable of printing several characters. One line in an LCD can accommodate 16, 20 or 40 characters. A typical LCD contains one, two or four lines. A 40 × 2 LCD has two lines with 40 characters in each line. The individual character positions are specified by the bits DB7 through DB0. The bit DB7 is always 1, DB6 is 0 for line 1 and 1 for line 2. The 40 positions (0–39) are identified by the lines DB5 through DB0. These bits can take up the values in the range 00000 (decimal 0) to 100111 (decimal 39). Hence, the addresses 80H to A7H belong to line 1, while line 2 contains the addresses C0H–E7H. In the following, the pins of the display have been listed and grouped according to their functionality.

- **VCC, VSS, VEE:** The display works with 5 V power supply. The pin VCC corresponds to the +5 V supply while VSS provides the ground. Contrast of the LCD is controlled by the pin VEE through a controllable voltage input.
- **RS:** *Register Select.* The LCD module has two important registers in it. First one, named the command register, holds the instruction command, while the second one, called the data register, holds the data. For selecting the instruction command code register, RS should be set to 0. To select the data register, RS should be set to 1.
- **Read/Write:** To perform a writing operation to the LCD registers, the pin should be set to 0. To read contents of the registers, the pin should be set to 1.
- **Enable:** A high-to-low pulse of a minimum duration of 450 ns applied to this pin is used to latch the data into the LCD registers.
- **Data Pins (0–7):** The DB0–DB7 data lines transfer data to or from the LCD registers. To display a character, the corresponding ASCII code is sent to the LCD with RS = 1 and R/W = 0. The data pin 7 has an additional function of status information. Setting RS = 0 and R/W = 1, the value read at data pin 7 gives the status of the display. If the bit is 1, it indicates the LCD to be busy with its internal operations. No new data is accepted till the bit is found to be 0, at which the LCD is ready to accept further data. Before initiating the data transfer, it is advisable to check the pin 7 status.
- **LED+, LED–:** Anode and cathode for LED backlight.

The LCD display accepts a set of commands sent to its command register for proper display of patterns. Table 5.2 shows the major commands corresponding to the LCD module – the command codes and the corresponding actions. The 8-bit wide commands can be sent through the pins DB0 through DB7. Some example commands include clear the display, move cursor to different positions on the LCD screen etc. It is assumed that the display has two lines.

TABLE 5.2 Codes for LCD commands

Code (in Hex)	Command details
01	Clear screen
02	Return to home
04	Decrement cursor (Shift cursor left)
06	Increment cursor (Shift cursor right)
05	Shift display right
07	Shift display left
08	Display off, cursor off
0A	Display off, cursor on
0C	Display on, cursor off
0E	Display on, cursor blinking off
0F	Display on, cursor blinking on
10	Shift cursor left
14	Shift cursor right
80	Force cursor to the beginning of 1st line
C0	Force cursor to the beginning of 2nd line
38	Use 2 lines and 5 × 7 matrix

Figure 5.20 shows the timing diagram for the behaviour of interfacing signals for the read operation of the LCD command register. Similarly, Figure 5.21 corresponds to the write operation. In both the cases, RS is set to 0. The R/W line is set to 1 for the read operation (Figure 5.20) and 0 for the write operation (Figure 5.21). The read operation is effected by a low-to-high pulse on the E line (Figure 5.20), while the write operation requires a high-to-low pulse (Figure 5.21). Typical values of Setup Time, Hold Time, Enable Pulse Width and Data Setup Time are 140 ns, 10 ns, 450 ns and 195 ns, respectively. These minimum delays must be provided by the interface routine. However, a better method is to determine whether the LCD module is ready to accept the next command. This can be done by monitoring the DB7 line with RS = 0 and R/W = 1, after sending a command.

Figure 5.22 shows an interfacing of an LCD module with the 8085 ports. The data pins DB0 through DB7 are driven by the port 2, which is a bidirectional one. Other important pins RS, R/W and E are connected via the port 1. It is assumed that the LCD module is a 40 × 2 one with backlit LEDs present.

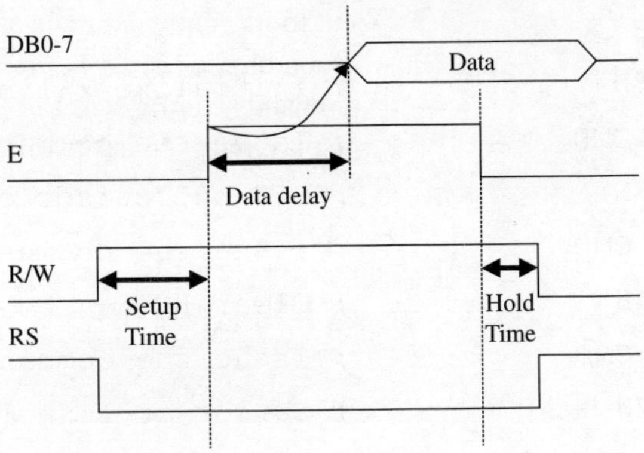

FIGURE 5.20 LCD read operation

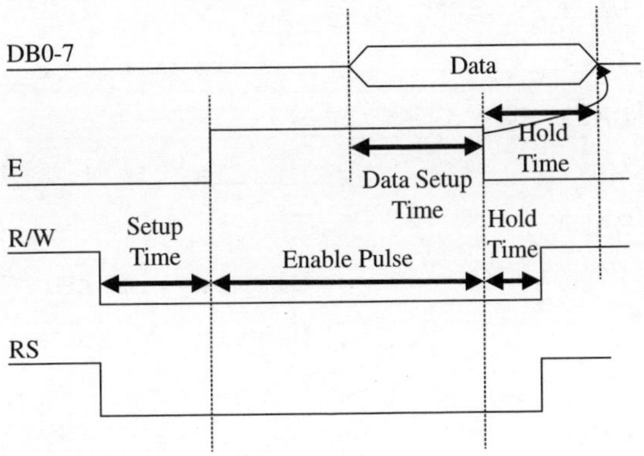

FIGURE 5.21 LCD write operation

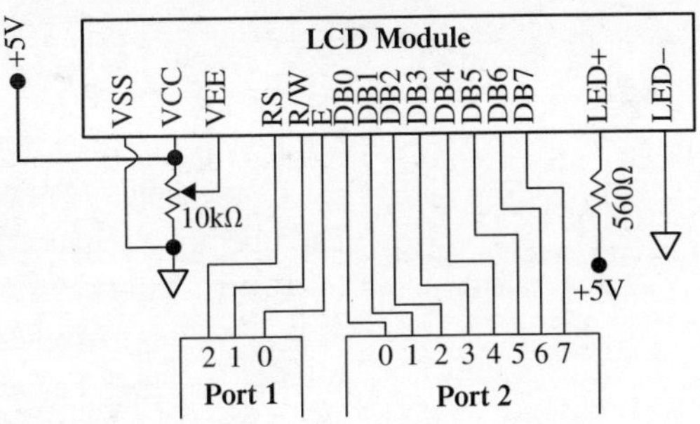

FIGURE 5.22 Interfacing LCD module with the 8085

The 8085 assembly language program to display the string "8085" starting at line 1, position 18 has been noted in the following. The program checks the busy flag, DB7 pin status, before issuing any command or data to the LCD.

```
          ORG 1000H
          MVI A, 38H              ; Initialize 2 line LCD, 5 × 7 matrix
          CALL ISS_CMD            ; Issue the command
          MVI A, 0EH              ; LCD on, cursor on
          CALL ISS_CMD            ; Issue the command
          MVI A, 01H              ; Clear LCD
          CALL ISS_CMD            ; Issue the command
          MVI A, 92H              ; Cursor at line 1, position 18
          CALL ISS_CMD            ; Issue the command
          MVI A, '8'              ; Display letter '8'
          CALL DSP_DATA
          MVI A, '0'              ; Display letter '0'
          CALL DSP_DATA
          MVI A, '8'              ; Display letter '5'
          CALL DSP_DATA
          MVI A, '5'              ; Display letter '1'
      L1: JMP L1                  ; Remain at this position

 ISS_CMD: CALL CHK_READY          ; Check if LCD ready
          OUT PORT2               ; Issue the command code
          MVI A, 01H              ; Make RS = 0, R/W = 0, E = 1
          OUT PORT1               ;
          CALL DELAY_RTN          ; Give a delay
          MVI A, 00H              ; Make E = 0 to latch command
          OUT PORT1               ; Set E = 0 to latch the command
          RET                     ; Return

DSP_DATA: CALL CHK_READY          ; Check if LCD ready
          OUT PORT2               ; Send data
          MVI A, 05H              ; Make RS = 1, R/W = 0, E = 1
          OUT PORT1               ;
          CALL DELAY_RTN          ; Give a delay
```

```
           MVI A, 00H           ; Make E = 0 to latch data
           OUT PORT1            ; Set E = 0 to latch data
           RET                  ; Return

CHK_READY: MVI A, 02H           ; Make RS = 0, R/W = 1, E = 0 for read
           OUT PORT1            ;
     BACK: MVI A, 00H           ; Set E = 0 for low-to-high pulse
           OUT PORT1
           CALL DELAY_RTN       ; Put some delay
           MVI A, 01H           ; Set E = 1 for low-to-high pulse
           OUT PORT1
           IN PORT2             ; Read Port 2 for DB7
           ANI 80H              ; AND with 80H to keep only DB7 relevant
           JNZ BACK             ; Loop if DB7 is clear
           RET                  ; Return

DELAY_RTN: LXI B, 1000H         ; Load BC = 1000H
       L2: DCX B                ; Decrement BC
           MOV A, C             ; Get C into A
           ORA B                ; A = A or B
           JNZ L2               ; Loop if not zero
           RET
           END
```

5.7 Programmable Peripheral Interface – The 8255

LO 8

The chip 8255 is called a *Programmable Peripheral Interface* (*PPI*). This is one of the most widely used programmable, parallel I/O devices that can be used to do data transfers in different modes. It can be interfaced with almost any processor. Figure 5.23 shows the pin diagram of the chip. The 8255 is a 40-pin chip having three independent 8-bit ports, A, B and C. Each of the three ports can be configured as input or output. Apart from the 24 port pins (PA0–PA7, PB0–PB7, PC0–PC7), 8255 has 16 additional

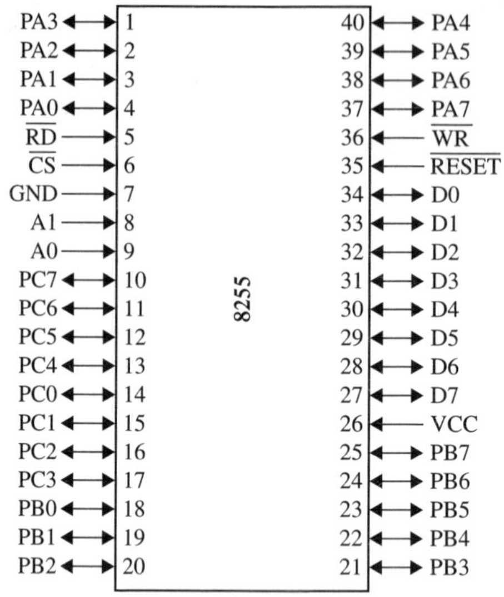

FIGURE 5.23 8255 pin layout

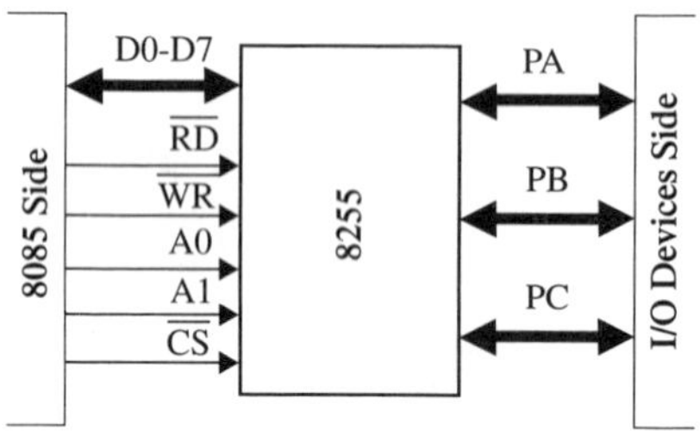

FIGURE 5.24 Grouping of pins of 8255

lines used for interfacing it with the 8085 (or any other processor). Figure 5.24 shows the grouping of pins as per their functionality and recommended connection pattern. It has been detailed in the following.

- **D0–D7:** These eight pins are used to connect with the data bus lines of the 8085 to send data to the 8255 or back.
- **$\overline{\text{RD}}$, $\overline{\text{WR}}$:** These control signals are provided by the 8085 to indicate a read or a write operation onto the registers/ports of the 8255.
- **PA, PB, PC:** These are the three 8-bit ports. The ports can be programmed independently as input or output ports. Moreover, only the port A bits can be configured for simultaneous bidirectional operation (both input and output) also. Port C can be configured into two parts – *Port C Upper CU* (PC4–PC7) and *Port C Lower CL* (PC3–PC0). The upper and the lower parts of port C can be programmed to behave as input or output, independently.

- **$\overline{CS}$, A1, A0:** These lines are used to select the 8255 chip and individual ports/registers within it. For any data transfer to the 8255, the $\overline{CS}$ must be set to low. The values of A1 and A0 will decide the port/register selection as noted in Table 5.3.

TABLE 5.3 8255 port selection

$\overline{CS}$	A1	A0	Port/Register
0	0	0	Port A
0	0	1	Port B
0	1	0	Port C
0	1	1	Control Register
1	x	x	Not selected

8255 Modes: The 8255 can be made to operate in one of the following modes. The mode can be selected by writing a code word to the control register. Figure 5.25 shows the bit setting needed for different modes of operation of the ports. The ports A, B and C are put into two groups. *Group A* corresponds to the port A and the upper 4 bits of port C (CU). *Group B* corresponds to the port B and the lower 4 bits of port C (CL). The mode settings are as follows.

1. **Mode 0, Simple I/O.** Mode 0 is used to program any of the ports A, B, CU and CL as input or output. It may be noted that individual port bits cannot be programmed separately. For ports A and B, all 8 bits are to be configured as either input or output. For ports CU and CL, entire 4-bits get programmed in the same way. Most of the 8255 based interfacings use Mode 0 programming. A typical example is to set all the three ports in output mode. For this, the control word is "10000000", that is, 80H. On the other hand, to program port A as input and all other ports (that is, port B, CU and CL) as output, the control setting is "10010000", that is, 90H.
2. **Mode 1, Handshaking.** This mode is used for handshaking interfaces. It enables the ports A and B to be used as input or output. The handshaking signals are provided by port C.
3. **Mode 2, Bidirectional Handshaking.** In Mode 2, port A is used as bidirectional I/O port with handshaking signals through port C. Port B bits can be programmed independently in either Mode 0 or Mode 1.

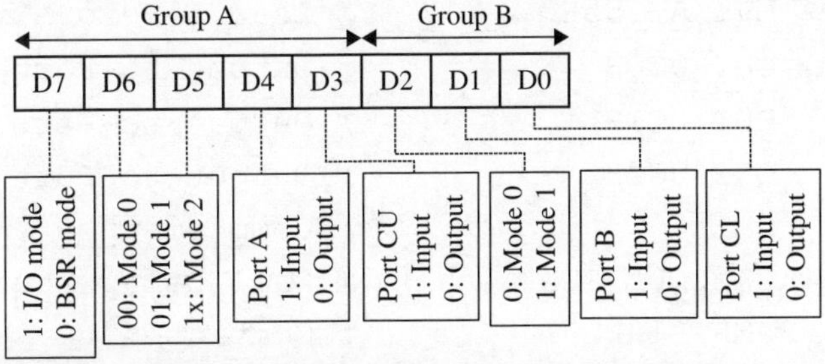

FIGURE 5.25 Control word format of 8255

4. **Bit Set/Reset (BSR) Mode.** This mode provides the facility to program the individual bits of port C independently.

Figure 5.26 shows the interfacing of 8255 with the 8085. It has been assumed that the 8255 will get selected only for addresses with A15 = A14 = 1. As shown in Table 5.3, port A gets selected for A1 = 0, A0 = 0. Thus, under memory-mapped-I/O, for data transfer through port A, the address should be "11xxxxxxxxxxxx00". The memory-mapped-I/O addresses for the ports B, C and control register are "11xxxxxxxxxxxx01", "11xxxxxxxxxxxx10" and "11xxxxxxxxxxxx11", respectively. One set of addresses for the ports A, B, C and the control register could be, C000H, C001H, C002H and C003H, respectively. Under I/O-mapped-I/O operation, the addresses A0–A7 are copied onto A8–A15. Hence, in I/O-mapped-I/O, the addresses of A, B, C and control ports are C0H, C1H, C2H and C3H, respectively.

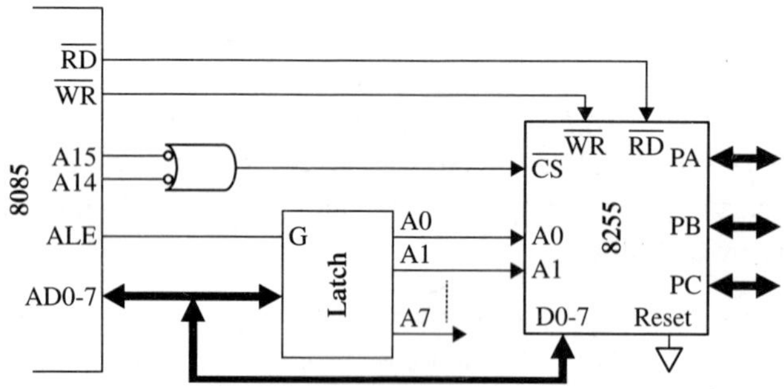

FIGURE 5.26 Interfacing 8255 with the 8085

The following assembly language program can be used to output patterns to the ports A, B and C in such a fashion that the 8-bit outputs in each port are "01010101" (55H) followed by "10101010" (AAH) continually. First, 80H is written to the control port to configure all ports in output mode. The program works in an infinite loop to output the patterns.

```
         ORG 2000H
         MVI A, 80H      ; Load control word
         OUT 0C3H        ; Issue control word
         MVI A, 55H      ; Load 55H to A
     L1: OUT 0C0H        ; Output to Port A
         OUT 0C1H        ; Output to Port B
         OUT 0C2H        ; Output to Port C
         CMA             ; Complement A
         CALL DL_RTN     ; Call Delay routine
         JMP L1          ; Continue
 DL_RTN: LXI B, 1000H    ; Load BC = 1000H
```

```
        PUSH PSW            ; Save A
    L2: DCX B               ; Decrement BC
        MOV A, C            ; Get C into A
        ORA B               ; A = A or B
        JNZ L2              ; Loop if not zero
        POP PSW
        RET
        END
```

5.8 Analog-to-Digital Converter (ADC) Interfacing

LO 9

Analog-to-Digital Converter (ADC) is a very important device that often needs to be interfaced with a processor interacting with the environment. The 8085 (or any other microprocessor/microcontroller) is a digital device that works on binary values. On the other hand, the signals of the physical world are analog in nature (for example, temperature, pressure, velocity, light intensity etc.). A sensor/transducer converts such signals into some equivalent electrical signals. To process this electrical signal by the 8085, it has to be converted into digital values. An ADC performs this function. The number of bits (n) used to represent the digital signal produced is called the *resolution* of the ADC, typical values of n being 8, 10, 12, 16 etc. The *step size* of an ADC is the smallest change in the input that can effect a change in the digital output. With a resolution of $n = 8$, there can be 256 steps. Assuming the range of a voltage signal that this ADC converts, to be 5 V, the step size is equal to 5 V/256 = 19.53 mV. To keep the step size to say 10 mV, the conversion range for this ADC will be 2.56 V.

Several ADCs are available in the market. A typical one is ADC0808/0809 from National Semiconductor (currently, Texas Instruments), an 8-bit ADC ($n = 8$). The chip contains eight analog channels that are handled in a multiplexed fashion. It may be noted that the two converters, the ADC0808 and ADC0809, are functionally identical except that the ADC0808 has a total unadjusted error of ±1/2 LSB while for ADC0809 the unadjusted error is ±1 LSB. Figure 5.27 shows a schematic for the chip with the following signal line distribution.

- **VCC, GND:** The chip operates on a 5 V power supply.
- **Clock:** The nominal clock frequency is 640 kHz. However, the chip can operate up to a maximum clock speed of 1280 kHz.
- **Vref(+), Vref(–):** These correspond to the reference voltages in the conversion process. With Vref(+) set at 5 V and Vref(–) grounded, the step size becomes 19.53 mV, as explained earlier. The step size can be set to other desired values by selecting proper reference voltages.
- **IN0–IN7:** The chip has eight analog input channels. The ADC can operate by multiplexing these channels.

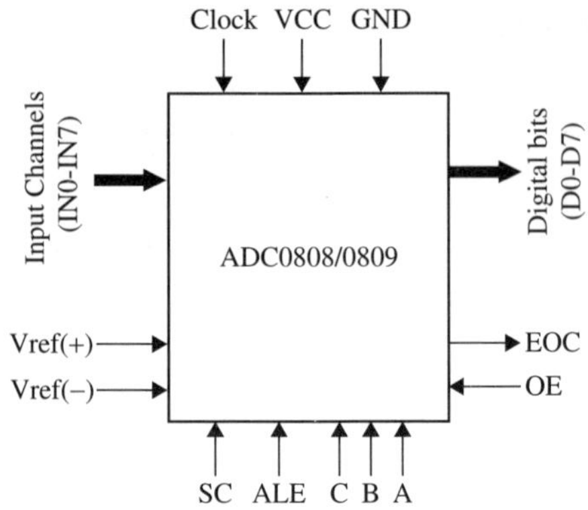

FIGURE 5.27 ADC0808/0809 schematic

- **C, B, A, ALE:** A three bit value in the lines A, B, C (with C as MSB) selects the analog input channel. The selection is registered via a low-to-high pulse given to the ALE input.
- **SC:** The *Start Conversion* input is used to dictate the ADC to start the conversion process. A low-to-high pulse in the SC line initiates the conversion process.
- **EOC:** The *End of Conversion* signal comes out of the ADC. It marks the completion of the conversion process. A high-to-low transition of this line marks the completion.
- **OE:** To read out the conversion value, the Output Enable (OE) input has to be activated. The digital value is made available at pins D0–D7 on applying a low-to-high pulse on the OE pin.

The sequence of steps in the analog-to-digital conversion process has been shown in Figure 5.28. It consists of the following steps.

1. The channel for the analog input is selected, identified by the inputs C, B, A.
2. A low-to-high transition is applied on the ALE input to the ADC to latch the selection.
3. A low-to-high pulse is applied on the SC line to start the analog-to-digital conversion process.
4. End of conversion is marked by the ADC by giving a high-to-low transition on EOC line.
5. If a low-to-high transition is applied on the OE line, the digital value is made available on D0–D7.

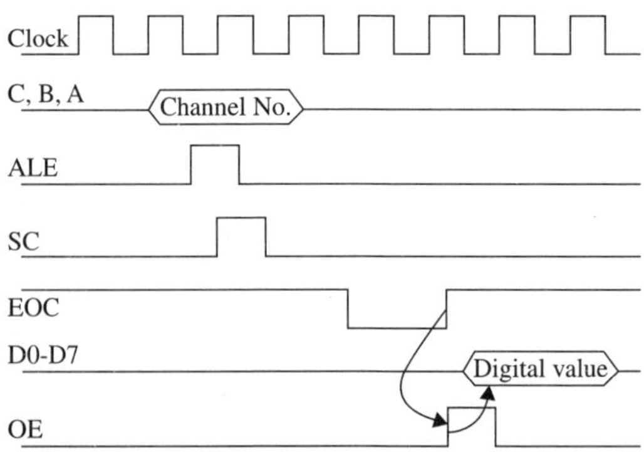

FIGURE 5.28 Timing diagram for ADC0808/0809

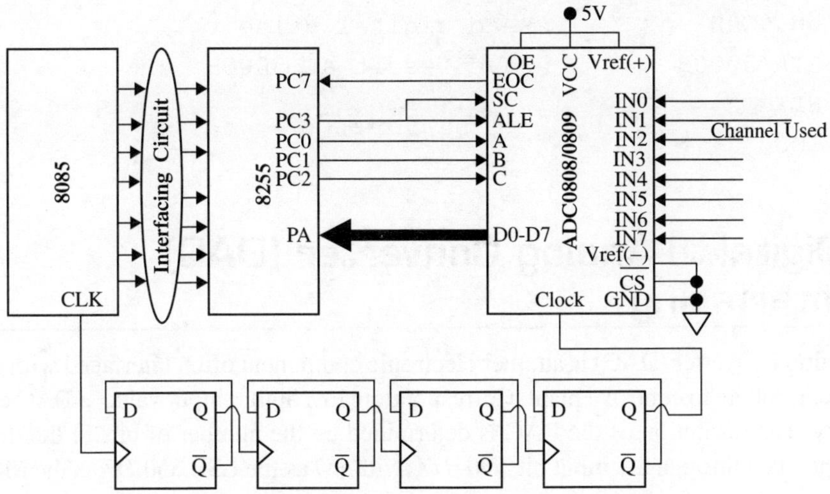

FIGURE 5.29 Interfacing ADC0808/0809 with the 8051

Figure 5.29 shows the interfacing of ADC0808/0809 with the 8085 through 8255. The port A of 8255 has been used to read the lines D0–D7. Lower C-port has been used to drive the pins A, B, C, SC and ALE. The upper C-port bit PC7 has been used to read the end-of-conversion pin EOC of the ADC. Since ADC0808/0809 runs at a clock frequency of about 640 kHz, the 8085 clock has been divided by employing a set of four D-flip-flops in cascade to achieve the required frequency. The interfacing between the 8085 and 8255 has not been shown separately, as it is similar to the circuit shown in Figure 5.26. The lines ALE and SC have been tied together. Similarly, the lines OE and Vref(+) have been tied to VCC. The program to do the analog-to-digital conversion for the analog input on Channel 1 has been noted in the following.

```
        ORG 2000H
        MVI A, 98H      ; Set port A and upper-C as input,
        OUT 0C3H        ; lower-C as output
        MVI A, 01H      ; Set to select IN1
        OUT 0C2H        ; Output to Port lower-C to select IN1
        MVI A, 09H      ; Make ALE, SC, A high
        OUT 0C2H        ; Output to Port lower-C
        MVI A, 01H      ; Make ALE, SC low
        OUT 0C2H        ; Output to Port lower-C
L1:     IN 0C2H         ; Read for EOC (PC7)
        RAL             ; Rotate left to check if C7 = 1
        JNC L1          ; If C7 ≠ 1, read EOC again
```

```
IN 0C0H          ;Read digital value
STA 5000H        ;Save result at 5000H
HLT
END
```

5.9 Digital-to-Analog Converter (DAC) Interfacing

LO 9

Digital-to-Analog Converter (DAC) is another electronic component often interfaced with the microprocessors/microcontrollers to control a plant. Corresponding to a digital input value, a DAC can equivalent analog quantity. The resolution of the DAC is determined by the number of bits in the digital input. A digital input specified through the input bits *D0–D7* (with *D0* as the LSB and *D7* as the MSB) gets converted into an output current I_{out} as a fraction of the reference output current I_{ref}.

$$I_{out} = I_{ref}\left(\frac{D7}{2}+\frac{D6}{4}+\frac{D5}{8}+\frac{D4}{16}+\frac{D3}{32}+\frac{D2}{64}+\frac{D1}{128}+\frac{D0}{256}\right)$$

The current I_{ref} is decided by the reference voltage V_{ref} and is generally set to about 2.0 mA. In the output by an 8-bit DAC, there can be 256 different current levels. It may be noted that with increasing resolution of the DAC, it becomes more expensive and may not be very much useful for microprocessor/microcontroller based applications. The I_{out} can be converted to voltage by simply passing the current through a resistor; however, in most of the cases, an operational amplifier is used to separate the DAC from other stages.

Figure 5.30 shows the pin diagram of a very popular digital-to-analog converter DAC0808. The converter has 8-bit resolution. The pins A1 to A8 work as digital inputs, through the pins numbered 5 to 12 (with A1 as MSB and A8 as LSB). Power supply VCC is via pin 13 and GND at pin 2. Pin 3 corresponds to VEE, the negative power supply. The positive and negative reference voltage inputs VREF+ and VREF– are applied via the pins 14 and 15. A compensation capacitor may be connected to pin 16. The output of IO is available at pin 4. The current output at pin 4 needs to be converted to voltage for analog output.

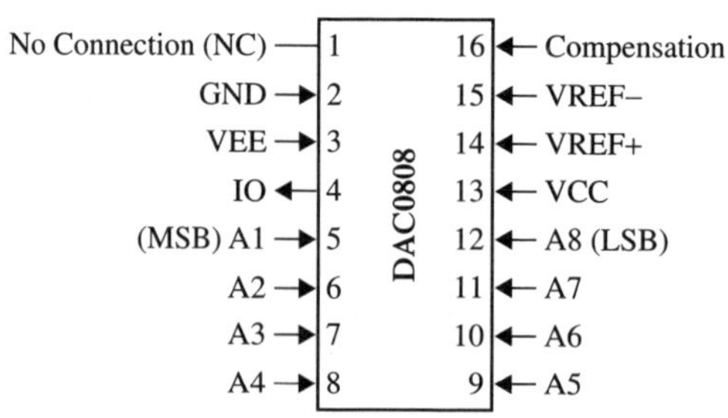

FIGURE 5.30 Pin layout of DAC0808

Figure 5.31 shows an interface of DAC0808 with the 8255, which is assumed to be further driven by the 8085. The digital input to be converted to analog current/voltage is provided to the DAC via port A of the 8255. The pins VEE and VCC are connected to –15V and +5V power supplies, respectively. In the interface, VREF+ has been connected to 10 V, VREF– to 0 V and the resistance as 5 kΩ. This makes the reference current I_{ref} to be 10 mA. To convert the output current to analog voltage, an operational amplifier (such as 741) can be used with feedback resistance 5 kΩ.

The interface shown in Figure 5.31 can be used to generate different types of analog waveforms. In the following, two examples have been noted – one to generate a triangular wave and the second one to generate a sine wave.

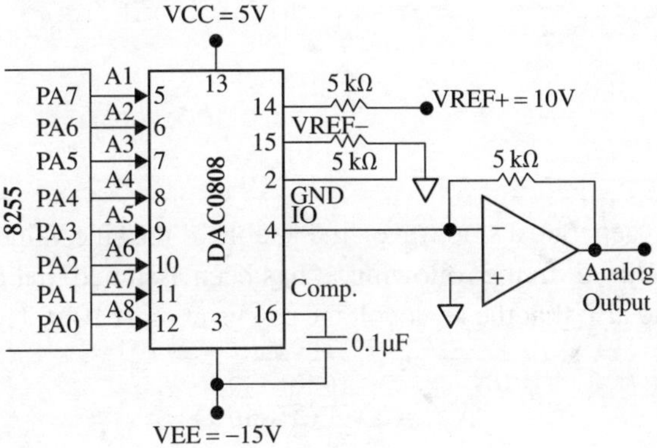

FIGURE 5.31 Interfacing DAC0808 with 8051

Triangular Wave Generation. To generate a triangular wave, the values 0 to 255 can be output to port PA followed by decrementing it up to 0. The digital values, when converted by the DAC will provide an analog output with triangular waveform. The 8085 source code for the same has been presented in the following. Between the successive digital value outputs, a small delay has been introduced to allow time to the DAC to convert the pattern. Increasing this delay value will convert the waveform to be more of staircase in shape. An oscilloscope may be used to see shape of the waveform, the delay value may be adjusted accordingly.

```
    ORG 2000H
    MVI A, 80H        ; Control word 8255 - all ports output
    OUT 0C3H          ; Configure 8255
    XRA A             ; Clear accumulator
L1: OUT 0C0H          ; Send data to DAC via 8255 port A
    CALL DLA          ; Insert delay
    CPI 0FFH          ; Check if 255 reached
    JZ L3             ; Jump to decrement if 255 reached
```

```
L2: INR A            ; Increment A
    JMP L1           ; Repeat increment
L3: DCR A            ; Continue
    OUT 0C0H         ; Send data to DAC via 8255 port A
    CALL DLA         ; Insert delay
    CPI 01H          ; Check if 1 reached
    JNZ L3           ; Repeat if not 1
    XRA A            ; Clear accumulator
    JMP L1           ; Repeat increment
DLA: NOP             ; Small delay routine
    NOP
    RET
    END
```

Sine Wave Generation. To generate a sine wave, first a table of the sine values of different angles in the range 0–360 degrees is produced. In the following, it has been assumed that corresponding to 0 degree, the output voltage is 5 V and also that the peak voltage can be at most 10 V. The equation governing such a sine wave is as follows.

$$V_{out} = 5\,\text{V} + (5 \times \sin\theta)$$

The 10 V range is covered by 256 digital levels of inputs. Table 5.4 shows the different angles and their corresponding digital input to be sent to the DAC for generating the sine wave.

TABLE 5.4 Table for Sine wave generation

Angle (θ)	sin θ	V_{out}	Digital Value
0	0	5	128
30	0.5	7.5	192
60	0.866	9.33	238
90	1.0	10	255
120	0.866	9.33	238
150	0.5	7.5	192
180	0	5	128
210	–0.5	2.5	64
240	–0.866	0.669	17
270	–1.0	0	0
300	–0.866	0.669	17
330	–0.5	2.5	64
360	0	5	128

The assembly language program for the sine wave generation is as follows. It stores the values noted in Table 5.4 and outputs those serially to produce the sine wave. The values noted in the table are 30 degree apart. As a result, the generated waveform will not be very smooth. To produce a better waveform, more intermediary values are required to be kept in the table to have a smoother wave.

```
        ORG 2000H
WAVE:   DB 128, 192, 238, 255,          ; The Table
           238, 192, 128, 64, 17,
           0, 17, 64, 128
        MVI A, 80H
        OUT 0C3H                        ; Configure 8255, all ports output
  L1:   LXI H, WAVE                     ; HL points to Table
        MVI C, 13                       ; C holds count
  L2:   MOV A, M                        ; Get next sine value
        OUT 0C0H                        ; Output to DAC via 8255 port A
        INX H                           ; HL points to next
        DCR C                           ; Decrement C
        JNZ L2                          ; Output next value
        JMP L1                          ; Generate next wave
        END
```

5.10 The 8254 – Programmable Interval Timer

LO 10

As noted in Chapter 3, the 8085 can be programmed to provide delay between two operations. This is primarily done by initializing one register or a pair of registers to some desired value. Next, the 8085 is made to work in an idle loop, till the register(s) content becomes zero. However, the method does not provide accurate delays, which may be needed in some precise control applications. Intel has designed the 8254 chip for incorporating hardware-controlled precise delays that can be useful for applications like real-time clock, event counter, waveform generators etc. The 8254 is a 24-pin chip, operating with 5 V power supply. The chip contains three identical 16-bit Counters. A counter, initialized with a 16-bit value, gets decremented with events in the corresponding CLK line. When the count value becomes 0, a pulse is generated in the OUT pin that can be used to interrupt the 8085. Apart from that, a count value can also be read by the 8085 when the counter is decrementing.

Figure 5.32 shows the block diagram of 8254. It consists of three Counters (0, 1 and 2), Data Bus Buffer, Read/Write Logic and a Control Word Register. Each counter has two input pins – CLK and GATE, one output pin – OUT. The Data Bus Buffer is an 8-bit bidirectional buffer that can be used to connect to the data bus of 8085. The Read/Write Logic block has inputs $\overline{RD}$, $\overline{WR}$, $\overline{CS}$, A_0 and A_1.

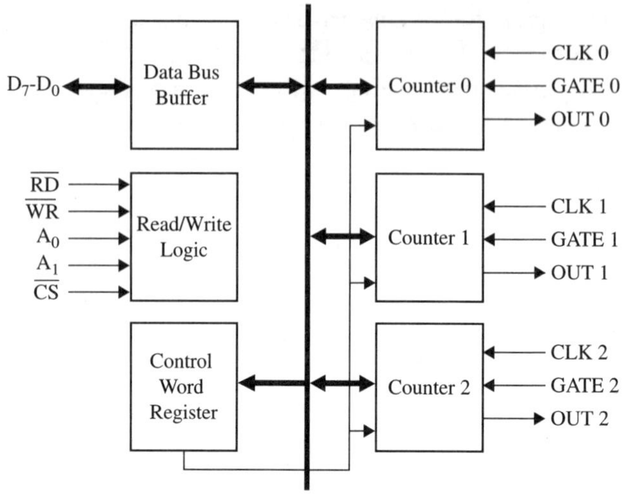

FIGURE 5.32 The 8254 block diagram

The chip gets selected if $\overline{CS}$ is low. Reading and writing operations on the counters are controlled by the $\overline{RD}$ and $\overline{WR}$ signals. Selection of counters is done by the A_0 and A_1 lines, as follows.

A_1	A_0	Selection
0	0	Counter 0
0	1	Counter 1
1	0	Counter 2
1	1	Control Word Register

The Control Word Register is accessed with A_1 and A_0 both equal to 1. This can be used to write a command word, specifying the ways in which the counters are to be used. The control word format has been shown in Figure 5.33. Parts of the control word are as follows.

- **SC1, SC0:** These bits are used to select the counter which is to be controlled by the specified command. Settings 00, 01 and 10 select Counters 0, 1 and 2, respectively. Setting 11 is for *read-back* operation, to read the present count of the specified counter.
- **RW1, RW0:** These bits are used to specify the read/write operation on the counters. The details have been specified in the figure.
- **M2, M1, M0:** These bits identify one of the six modes for a counter selected by the SC1, SC0 bits.
    - **Mode 0 – Interrupt on Terminal Count:** In this mode, the OUT is initially low. After the count value has been loaded into the counter, at every clock, the count gets decremented. Once the count becomes zero, the OUT goes high and it remains high till a new count or command word is loaded. The GATE input should be high for the entire decrement session. The decrement operation gets suspended temporarily whenever the GATE input is low. The OUT signal can be used to interrupt the 8085 on count becoming zero.

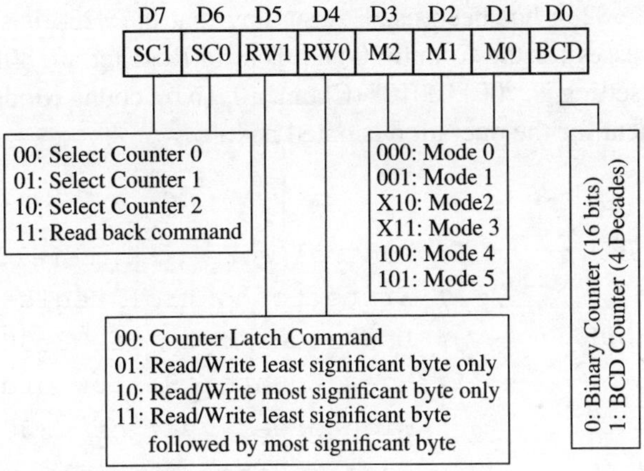

FIGURE 5.33 The 8254 control word

- **Mode 1 – Hardware-Retriggerable One-Shot:** In this mode, OUT is initially high. When GATE is triggered (becomes high), down count starts and OUT is made low. OUT returns to high once the count value reaches 0.
- **Mode 2 – Rate Generator:** The mode can be used to generate a pulse of duration equal to a given interval. When the count is loaded, OUT becomes high and it remains high till the count reaches 1. Then OUT becomes low for one clock period. The count is reloaded automatically and the pulse is generated continually. The counter cannot be loaded with the value 1, which is illegal in this case.
- **Mode 3 – Square-Wave Generator:** In this mode, OUT is initially high. The count value loaded into the counter gets decremented by two at every clock pulse. When the count reaches 0, the OUT changes its state, the count value is reloaded and the whole process is repeated. Thus, it generates a square wave of period equal to the period of the count.
- **Mode 4 – Software-Triggered Strobe:** In this mode, OUT is initially high and goes low for one clock period when count becomes zero. To control the subsequent values of OUT, the count value needs to be reloaded explicitly.
- **Mode 5 – Hardware-Triggered Strobe:** This is similar to the Mode 4, excepting that the counter triggering happens at the rising pulse of the GATE input. Decrementing the count value starts with the low-to-high rise of GATE. When count becomes zero, the OUT goes low for one clock pulse.

- **BCD:** This bit being 0 indicates the counter value to be a 16-bit binary number. If the bit is 1, the content is treated as a number with four BCD digits.

As an example, let us assume that we want to generate a delay of 25 ms using the Counter 0, in Mode 1. The CLK0 is assumed to be connected to a clock source of 2 MHz (time period 0.5 μs). Thus, the count value to be initialized with is 25 ms/0.5 μs = 50000 (C350H). The program first outputs the pattern 00H to a port (named PORT1) and after a delay of 25 ms, the output is changed to FFH.

To generate the delay, the 8254 has been used. Assuming that the 8254 is selected whenever the A_7 address bit is 1, the addresses for the Counter 0 and Control Register are 80H and 83H, respectively. The control word for the setting is, "00110010" (Counter 0, 16 bit count, Mode 1, Binary), that is, 32H. The program code fragment for the operation is noted next.

```
    ORG 2000H
    MVI A, 32H       ; Control word to initialize Counter 0
    OUT 83H          ; Write to control register
    MVI A, 50H       ; Low order byte of C350H
    OUT 80H          ; Load Counter 0, low order byte
    MVI A, 0C3H      ; High order byte of C350H
    OUT 80H          ; Load Counter 0, high order byte
    MVI A, 00H       ; Load 0 to A
    OUT PORT1        ; Output 0 to PORT1
L1: MVI A, 00H       ; Control word to latch a count
    OUT 83H          ; Write in Control Register
    IN 80H           ; Read low-order byte
    MOV D, A         ; Store low-order byte in D register
    IN 80H           ; Read high-order byte
    ORA D            ; OR with low-order byte
    JNZ L1           ; If Counter 0 ≠ 0, go to L1
    MVI A, 0FFH      ; Load FFH to A
    OUT PORT1        ; Output to PORT1
    END
```

5.11 Interfacing via SID/SOD Pins

LO 11

The 8085 provides two pins for serial input and output operations. The pin SID is used for serial input and SOD for serial output. Thus, the two pins act as single bit input and output ports. Data transfers through these ports can be done via RIM and SIM instructions, as noted in Chapter 3.

The RIM instruction transfers the bit at SID pin to the bit 7 of the accumulator. The following program fragment will clear the memory location 2000H if the input bit is 0, else the location is set to FFH.

```
ORG 3000H
LXI H, 2000H      ; Initialize the HL pair with 2000H
RIM               ; Read bit at SID pin
RLC               ; Get bit A7 into CY
```

```
        JNC CLEAR        ; If 0, clear 2000H
        MVI A, 0FFH      ; Load FFH into A
        MOV M, A         ; Store A into 2000H
        JMP OVER         ; Jump to OVER
CLEAR:  MVI A, 00H       ; Load 00H into A
        MOV M, A         ; Store A into 2000H
 OVER:
        END
```

Next, we shall look into a program code that continually blinks the SOD pin through the SIM instruction. The SIM instruction outputs the bit 7 of the accumulator onto SOD pin. However, it needs the bit 6 of the accumulator also set, for proper execution of the SIM instruction. The program has two subroutines OUT_0 and OUT_1 to output a 0 and 1 onto the SOD pin, respectively. Apart from that, the program assumes existence of a DELAY_RTN (not shown explicitly in the program) that inserts a delay between the successive values on the SOD pin.

```
        ORG 3000H
   L1:  CALL OUT_0        ; Output 0 on SIM line
        CALL DELAY_RTN    ; Call delay routine
        CALL OUT_1        ; Output 1 on SIM line
        CALL DELAY_RTN    ; Call delay routine
        JMP L1            ; Repeat continually
OUT_0:  STC               ; Set Carry
        RAR               ; Get 1 at bit 7 of A
        STC               ; Set Carry
        CMC               ; Complement carry
        RAR               ; Get 1 at bit 6 and 0 at bit 7 of A
        SIM               ; Output 0 on SOD pin
        RET               ; Return from subroutine
OUT_1:  STC               ; Set Carry
        RAR               ; Get 1 at bit 7 of A
        STC               ; Set Carry
        RAR               ; Get 1 at both bits 6 and 7 of A
        SIM               ; Output 1 on SOD pin
        RET               ; Return from subroutine
        END
```

SUMMARY

The 8085 can be interfaced with a number of I/O devices. The devices being mostly electromechanical in nature need special care to be taken. The simplest input device is a switch while the simplest output device may be an LED. This chapter has started with discussion about connecting such simple devices. Then, more complex devices have been taken up, such as matrix keyboard, multiplexed display, LCD interfacing and ADC/DAC interfacing. A programmable peripheral interfacing chip 8255 could be interfaced with the 8085 to provide general-purpose I/O ports with different functionalities. Many a times, generating accurate hardware-controlled delay between the steps of control algorithm is a necessity. The chapter has introduced the chip 8254 that can serve as general-purpose timer/counter. For single-bit input-output, the pins SID and SOD could be used. To summarize, this chapter has provided the techniques for I/O interfacing with the 8085 processor. The next chapter will discuss about the processor 8086, the successor of 8085.

EXERCISES

Multiple Choice Questions

1. For matrix keyboard, debouncing helps to
   (a) Read one key if multiple keys are pressed
   (b) Read a key only once avoiding vibrations
   (c) Sense one key multiple times
   (d) None of the above

2. Which type of LED displays need common side to be grounded?
   (a) Common anode
   (b) Common cathode
   (c) Both common anode and common cathode
   (d) None of the above

3. Single 8279 can interface keys in number up to
   (a) 8 **(b)** 16
   (c) 64 **(d)** 128

4. In LCD devices setting RS to 0 selects
   (a) Instruction command code register
   (b) Data register
   (c) Status register
   (d) Control register

5. In 8255, control word is written to port
   (a) A **(b)** B
   (c) C **(d)** None of the above

6. In 8255, bit addressability is available with port
   (a) A
   (b) B
   (c) C
   (d) D

7. The appropriate pin of a processor to which the End-of-Conversion (EOC) line from ADC be connected to is
   (a) Port bit
   (b) Interrupt
   (c) Either a port bit or interrupt
   (d) Neither port bit nor interrupt

8. Which of the following statements are true about DAC0808?
   (a) Parallel digital data to analog data conversion
   (b) It has current as an output
   (c) All of the mentioned
   (d) None of the above

9. Number of modes in 8254 is
   (a) 2 **(b)** 4
   (c) 6 **(d)** 8

10. Number of Counters in 8254 is
(a) 2 (b) 3
(c) 4 (d) 8

11. Essential components of debouncing hardware is/are
(a) Resistor
(b) Capacitor
(c) Both resistor and capacitor
(d) None of the above

12. Soft debouncing takes place via
(a) Delay routine (b) MOV
(c) STA (d) None of the above

13. Number of LEDs in a seven-segment display module is
(a) 7 (b) 8
(c) 9 (d) 16

14. Number of ports needed for a multiplexed display is
(a) 1 (b) 2
(c) 4 (d) None of the above

15. The IRQ line of 8279 is generally connected to the 8085 pin
(a) RD (b) ALE
(c) INTR (d) None of the above

16. In LCD, the pins LED+ and LED– control
(a) Backlight (b) Power
(c) Intensity (d) None of the above

17. Which device require refreshing?
(a) LED
(b) LCD
(c) Both LED and LCD
(d) Neither LED nor LCD

18. In 8255, port B gets selected for A_1 and A_0 setting
(a) 0, 0 (b) 0, 1
(c) 1, 0 (d) 1, 1

19. In 8255 Mode 1, handshaking signals are provided by the bits of port
(a) A (b) B
(c) C (d) None of the above

20. In a 10-bit ADC, the resolution is
(a) 10 (b) 16
(c) 20 (d) None of the above

21. Number of analog channels in ADC0808 is
(a) 2 (b) 4
(c) 8 (d) 16

22. Signal to be applied to start conversion line of ADC0808 is
(a) High (b) Low
(c) low-to-high (d) High-to-low

23. To select Counter 2 of 8254, the bits A_1 and A_0 are to be set to
(a) 0, 0 (b) 0, 1
(c) 1, 0 (d) 1, 1

24. Single bit input port of 8085 is
(a) SID (b) D0
(c) D7 (d) None of the above

25. Single bit output port of 8085 is
(a) SOD (b) D0
(c) D7 (d) None of the above

Answers to Multiple Choice Questions

1. b	2. b	3. d	4. a	5. d	6. c	7. c	8. a	9. c	10. c
11. c	12. a	13. b	14. b	15. c	16. a	17. a	18. b	19. c	20. a
21. c	22. c	23. c	24. a	25. a					

Long Answer Type Questions

1. Discuss how to interface a matrix keyboard with the 8085.
2. Enumerate the technique to interface 4, seven-segment displays in multiplexed fashion.

3. Explain how to interface keyboard and display units with 8279.
4. Write a program to use port A and port B of 8255 as input port and port C as output. Sum of values read at ports A and B is to be output to port C.
5. Explain how to utilize 8254 counting modes.
6. Explain interfacing of a DAC with the 8085.
7. Explain interfacing an ADC with the 8085.
8. Show the interface of 8279 with 8085.
9. Write an assembly language program in 8085 to generate a saw-tooth waveform.
10. Write an assembly language program that alternately converts the analog inputs on Channel 0 and Channel 1. The results of the two channels are to be added and stored at memory location 2000H.
11. Write an assembly language program to generate 50 ms delay using 8254.
12. Write an assembly language program to read SID pin 10 times, add those bits and store the result at location 2000H.
13. Write an assembly language program to generate a square wave on SOD pin with 75% duty cycle.
14. Write an assembly language program to output your name to an LCD module.
15. Compare and contrast between LED and LCD displays.

The 8086 Architecture

LEARNING OUTCOMES

After reading this chapter, students will be able to:

LO 1 Draw a pin diagram of 8086 and explain the operations of individual pins

LO 2 State physical memory interfacing to 8086 in minimum and maximum modes

LO 3 Draw the functional block diagram of 8086 and explain its components

LO 4 List the addressing modes of 8086

LO 5 Enumerate the instruction set of 8086

LO 6 Enumerate the interrupts of 8086

LO 7 Compare between 8086 and 8088 processors

6.1 Introduction

> Minimum mode: Uniprocessor
> Maximum mode: Multiprocessor

Microprocessor 8086 is the first 16-bit processor, with about 29000 transistors, released by Intel in the year 1978. It has ALU and internal registers that work with 16-bit words. The data bus is also 16-bit wide and it can read/write data to a memory location or port, either 16 bits or 8 bits at a time. Compared to the 16-bit address bus of 8085, 8086 has a 20-bit address bus, allowing it to address up to 2^{20} = 1 MB memory locations. The chip works at 6–10 MHz with a 5-V power supply. As far as the arithmetic operations are concerned, it supports fixed-point arithmetic, similar to the 8085. However, the 8086 can operate in two modes – the *minimum* mode and the *maximum* mode. In the minimum mode, the 8086 works as a single processor, whereas in the maximum mode, it can work with multiple bus masters including a coprocessor (8087) to support floating-point operations as well. To reduce the effective execution time, the 8086 can prefetch up to six instruction bytes from memory. This is a significant improvement over 8085 instruction execution since fetching of the next instructions proceed simultaneously with the execution of the current instruction. The phenomenon is commonly known as *fetch-execution overlap*. In the following, the architecture of the processor has been discussed in detail.

6.1.1 Pin Layout

LO 1

The 8086 is a 40-pin DIP chip. The pin layout of the chip has been shown in Figure 6.1. The important pins of the chip are as follows.

AD_0–AD_{15}: These are bidirectional *Address/Data* bus lines. The lines constitute low-order address bus and are multiplexed with the data bus lines.

A_{16}/S_3, A_{17}/S_4, A_{18}/S_5, A_{19}/S_6: These are output pins with multiplexed functionalities. While one function is to provide the higher order address lines, the other responsibility is to provide status signals from the 8086. During the clock cycle T_1, the lines provide the most significant address bits. For I/O operations, the lines are all-zero. For memory operations, the lines identify the operation being carried out over the clock periods T_2, T_3, T_4 and also wait cycles. The pin S_5 shows the status of the interrupt enable flag and is updated at each clock cycle. The status bit S_6 is always zero. The bits S_4 and S_3 identify the memory segment being accessed (explained later), as shown in Table 6.1.

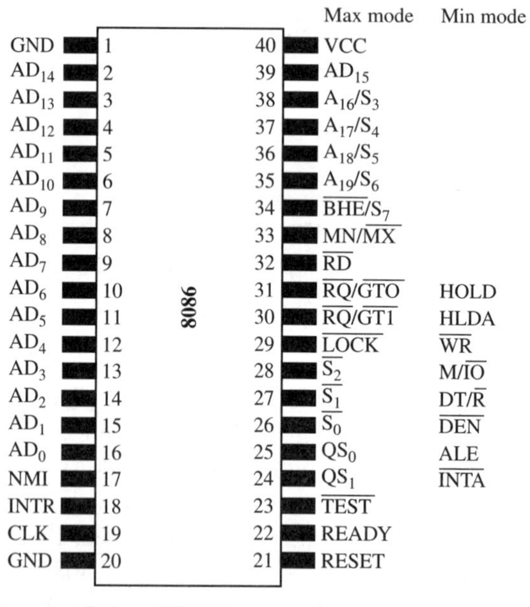

FIGURE 6.1 Pin diagram of the 8086

TABLE 6.1 Status lines

S4	S3	Segment Accessed
0	0	Extra
0	1	Stack
1	0	Code
1	1	Data

ALE signal can be used to separate out the address bits from the status bits, as in the 8085 processor.

$\overline{BHE}/S_7$: The BHE signal is used to enable data on the higher order data bus, D_8–D_{15}. It is multiplexed with the status signal S_7. The bit S_7 is always 1.

$MN/\overline{MX}$: As mentioned previously, the 8086 can operate in two modes – *maximum* and *minimum*. The mode can be selected by setting the input pin $MN/\overline{MX}$ to 0 (for maximum mode operation) or 1 (for minimum mode operation). In the minimum mode, the 8086 acts as a stand-alone processor while in the maximum mode, it can operate in a multiprocessor environment.

$\overline{RD}$: This active low output signal is used for read operation.

$\overline{TEST}$: This is an input pin. The status of this pin is tested by the 8086 while executing the *wait* instruction. During execution of this instruction, the 8086 enters into a wait state until some external hardware makes the pin low. It is often used in synchronizing the operation of 8086 with some external event.

READY: Used for interfacing slow I/O devices and memory chips. Its operation is similar to that in 8085.

RESET: An active high signal on this input pin will cause the 8086 processor to reset its operation. The signal should be kept high for at least four clock cycles.

CLK: An asymmetric clock with a 33% duty cycle is needed to be applied to this pin to provide clock signals to the internal modules of 8086.

INTR: This is a maskable interrupt line that is sampled during the last clock cycle in the execution of any instruction. If any interrupt is pending, the processor enters into an interrupt acknowledge cycle. A logic high level on the pin is taken as the occurrence of the interrupt.

NMI: This is a non-maskable interrupt line with priority higher than the INTR. On its occurrence, control branches to the fixed memory address 000A8H. The service routine is expected to be available at this address. The interrupt is sensed by a low-to-high transition on the pin.

The signals discussed so far, are common for both the minimum and the maximum mode of operation. For the pins noted next, the functionalities in the minimum and the maximum modes are different.

$DT/\overline{R}$: The *Data Transmit/Receive* is a minimum mode signal. It is an output from the 8086 controlling the direction of data flow through the data transceivers.

$\overline{DEN}$: The *Data Enable* output signal is a minimum mode signal used as output enable for the transceivers.

ALE: The *Address Latch Enable* signal. This minimum mode output signal is used to demultiplex address and data lines using external latches.

$M/\overline{IO}$: The minimum mode output signal is used to differentiate between memory and I/O accesses. For memory accesses, the pin is high, whereas for I/O operations (via IN and OUT instructions), the pin is low.

$\overline{WR}$: This minimum mode signal is for write control. It is asserted low by the 8086 whenever data has to be written to memory or I/O ports.

$\overline{INTA}$: This minimum mode output signal is made low by the 8086 to acknowledge the acceptance of interrupt request via the INTR pin.

HOLD: This minimum mode input signal to the 8086 can be used by other bus masters as a request to grant control of the bus. This is typically used by the DMA controller.

HLDA: This minimum mode output signal comes out of the 8086 to acknowledge the HOLD request. After relinquishing control of the bus, the 8086 asserts this pin to high.

$\overline{S_0}$, $\overline{S_1}$, $\overline{S_2}$: These are the status signals in the maximum mode of operation. The signals identify the type of operation that the 8086 is doing currently. It has been indicated in Table 6.2. Compared to 8085, memory access in the case of 8086 can be of three different types. It may be accessing program code, reading a data item from memory or writing a data item to memory. All these modes are indicated by the status bit settings.

$\overline{QS_0}$, $\overline{QS_1}$: These maximum mode signals are for the processor to indicate the *Queue Status*. External devices can track the status of the instruction queue via these lines. Interpretation of these signals has been indicated in Table 6.3. The various situations are as follows – the queue is empty, first byte of an instruction or its subsequent bytes are being accessed, the queue was not accessed in the previous clock cycle.

TABLE 6.2 Status signals in maximum mode

Status Signal			Machine Cycle
$\overline{S_2}$	$\overline{S_1}$	$\overline{S_0}$	
0	0	0	Interrupt acknowledge
0	0	1	I/O read
0	1	0	I/O write
0	1	1	Halt
1	0	0	Code access
1	0	1	Memory read
1	1	0	Memory write
1	1	1	Passive/Inactive

TABLE 6.3 Queue status signals in maximum mode

Queue Status		Queue Operation
$\overline{QS_1}$	$\overline{QS_0}$	
0	0	No operation. During the last clock cycle, nothing was taken from the queue
0	1	First byte. The byte taken was the first byte of the instruction
1	0	Queue empty. Queue reinitialized as a result of execution of a transfer instruction
1	1	Subsequent byte. The byte taken was a subsequent byte of the instruction

$\overline{RQ/GT_0}$, $\overline{RQ/GT_1}$: These signals, named *Bus Request/Bus Grant*, are the counterparts of HOLD signal, in the maximum mode of operation. The request lines could be used by other bus masters to request the 8086 to release the bus at the end of the current bus cycle. Request on $\overline{GT_0}$ has higher priority than $\overline{GT_1}$.

$\overline{LOCK}$: This is an output signal available in the maximum mode of operation. It is activated (made low) when the 8086 executes instructions with LOCK prefix and remains active until the completion of the instruction. It is to prevent other bus masters from gaining control of the system bus.

6.2 Physical Memory Organization

LO 2

Figure 6.2 shows the typical organization of physical memory for 8086. The one megabyte of physical memory of 8086 is organized into two banks – *odd bank* and *even bank*. Each bank has a capacity 512 KB. The banks are accessed in parallel by the 8086. Byte data with an even address is transferred through the data bus lines D7–D0, while the byte with an odd address is transferred over the lines D15–D8. The even bank is selected by the line A0 while the odd bank is selected by the line BHE. The selection is made as per Table 6.4. The division into banks help to get 16-bit data quickly.

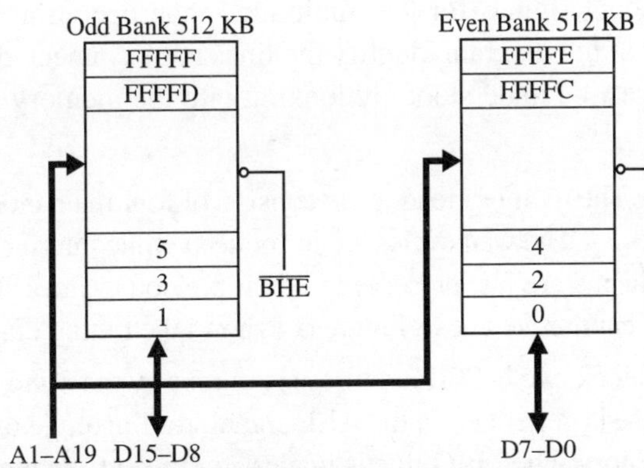

FIGURE 6.2 Physical memory organization

TABLE 6.4 Odd/Even memory access

BHE	A0	Meaning
0	0	Whole word (2 bytes)
0	1	Upper byte from/to odd address
1	0	Lower byte from/to even address
1	1	None

The processor utilizes different types of bus cycles to access the memory banks. The following are the alternatives.

- *Low bank byte access*: This corresponds to the byte access from the even bank. To access a location, its address X is put to both the banks over the address lines A19–A0. Now, X being even, the bit A0 of its address is 0, enabling the low/even bank. The BHE signal is kept high, disabling the high/odd bank. Byte data is written into or read from the even address X, transferred over the data lines D0–D7.
- *High bank byte access*: This gives byte from the odd addresses. An odd address (X + 1) is put onto the lines A19–A0. The bit A0 being 1, the low/even bank gets disabled. The line BHE is made low, enabling the high/odd bank. Byte data is written into or read from the odd address X + 1, transferred over the data lines D8–D15.
- *Even address word access*: The address X is put onto the lines A19–A0. The BHE signal is also low. As a result, word-wide data transfer takes place over D0–D15.
- *Odd addressed word access*: This is also called misaligned-word access, as a word being accessed spans over the addresses (X + 1) and (X + 2). The transfer requires two bus cycles. In the first cycle, byte at address (X + 1) is accessed from the high bank. The databus lines D8–D15 carry the data. The low bank remains disabled with A0 = 1. In the next cycle, byte at address (X + 2) is accessed from the low bank, data being available over D7–D0. The high bank remains disabled with BHE = 1.

6.2.1 Minimum Mode Memory Interfacing

Figure 6.3 shows the interfacing signals for the minimum mode operation of 8086. In particular, the signals at the right-hand side of the diagram identify the lines to be connected to the memory chips. The activities on these pins can best be understood by looking into the memory read and write operations, explained in the following.

- **Memory Read Cycle:** The memory read cycle consists of four main clock states – T_1, T_2, T_3, T_4 and one or more wait cycles T_w. The wait cycles are introduced while interfacing slow memory modules, as in 8085, different signals are just preserved to their previous values. The signals activated at different clock cycles are enumerated next. Figure 6.4 shows the timing diagram for the read cycle.
    - **Period T_1:** In this clock cycle, 20-bit address is output by the 8086. BHE signal is also output in this cycle. A pulse is generated in the ALE signal, trailing edge of which is used to latch the address for the memory. The $M/\overline{IO}$ line is made 1 for the entire memory read cycle. The $DT/\overline{R}$ line is also maintained at 0 for the entire read cycle.

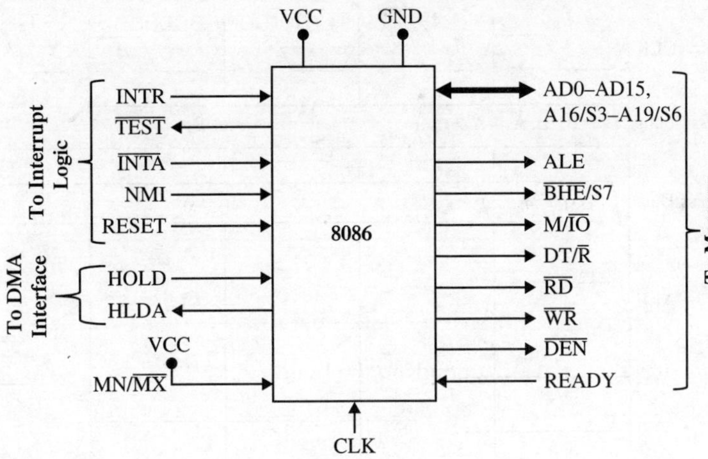

FIGURE 6.3 Minimum mode memory interfacing

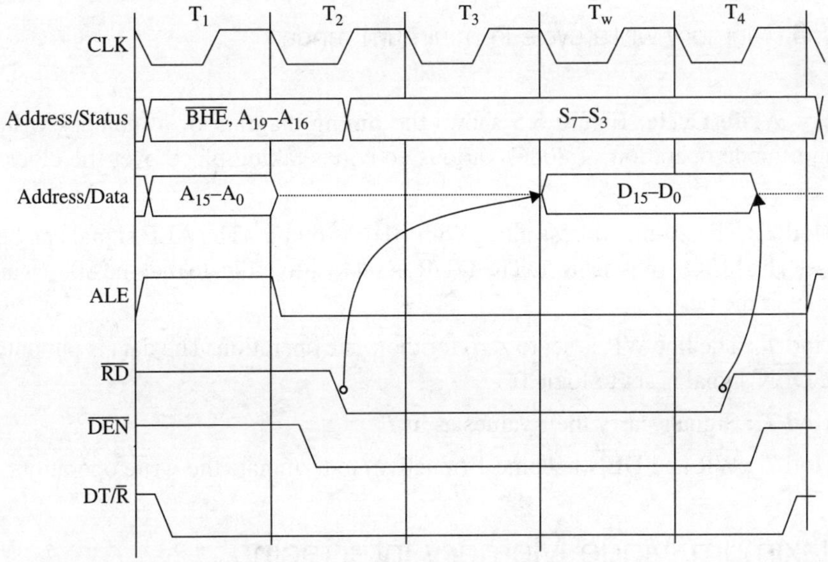

FIGURE 6.4 Memory read cycle in minimum mode

- **Period T_2:** Status bits S_3 through S_7 are made available in the upper four address lines and maintained over the periods T_3 and T_4. The lines AD0–AD7 are tristated. The $\overline{RD}$ line switches to 0, $\overline{DEN}$ also switches to 0 (to allow the external memory to transfer data to the data bus of 8086).
- **Period T_3:** Memory provides data to the 8086.
- **Period T_4:** $\overline{RD}$ and $\overline{DEN}$ become 1 (inactive) to terminate the read operation.

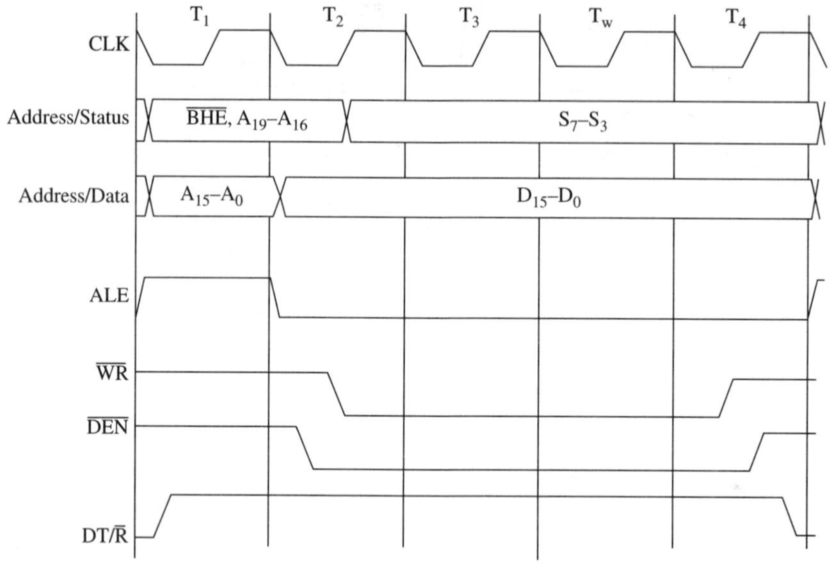

FIGURE 6.5 Memory write cycle in minimum mode

- **Memory Write Cycle:** Figure 6.5 shows the timing diagram of a memory write cycle in the minimum mode operation of 8086. Various activities taking place over the clock cycles are as follows.
  - **Period T_1:** The 20-bit address along with BHE is output. The ALE signal can be used to latch those. The $M/\overline{IO}$ pin is set to 1. The $DT/\overline{R}$ is set to 1 to indicate that the 8086 wants to transmit data over the bus.
  - **Period T_2:** The line $\overline{WR}$ is set to zero for the write operation. The data is put onto the data bus. The $\overline{DEN}$ signal is set to logic 0.
  - **Period T_3:** Signals carry their values as in T_2.
  - **Period T_4:** $\overline{WR}$ and $\overline{DEN}$ become 1 (inactive) to terminate the write operation.

6.2.2 Maximum Mode Memory Interfacing

In the maximum mode, 8086 provides signals to operate in a multiprocessor/coprocessor environment. A bus controller (8288) is generally employed to produce the bus control signals for the entire system. Figure 6.6 shows the connection pattern between 8086 and 8288. Based on the status signal lines $\overline{S}_0$, $\overline{S}_1$ and $\overline{S}_2$ from 8086, 8288 generates the read-write control signals for the interface. The signals like *DEN*, *DT/$\overline{R}$* and *ALE* are also produced by the 8288.

The timing diagrams for memory read and write operations have been shown in Figures 6.7 and 6.8, respectively. The major difference here, compared to the minimum mode of operation is the usage of status signals to produce control and advanced signals, such as, $\overline{AMWC}$. The operations are as follows.

- $\overline{S}_0$, $\overline{S}_1$, $\overline{S}_2$ are set at the beginning of the bus cycle.
- In T_1, 8288 controller outputs a pulse on the ALE line. The DT/$\overline{\text{R}}$ signal is set to low for read operation and high for a write operation.
- In T_2, DEN is set to 1, enabling the transceivers. For memory read operation, $\overline{\text{MRDC}}$ is generated. For write operation, $\overline{\text{MWTC}}$, $\overline{\text{AMWC}}$ are activated. The signals continue till T_4. The status lines $\overline{S}_0$, $\overline{S}_1$, $\overline{S}_2$ remain active till T_3.

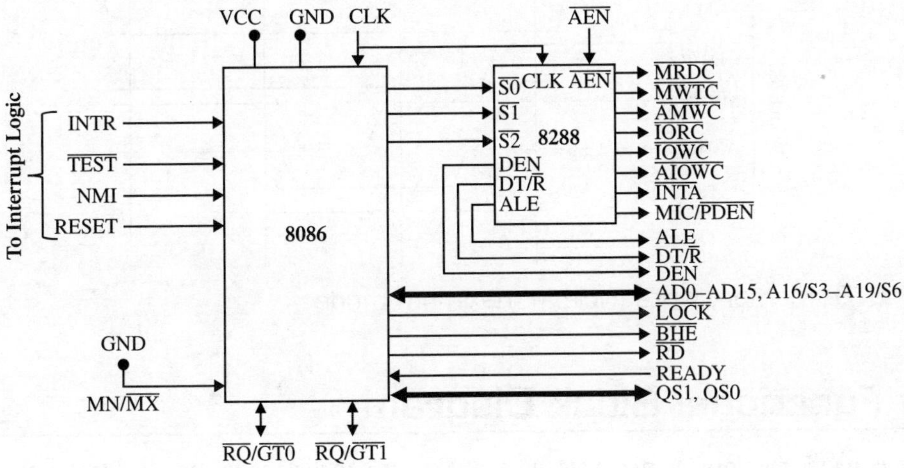

FIGURE 6.6 Maximum mode memory interfacing

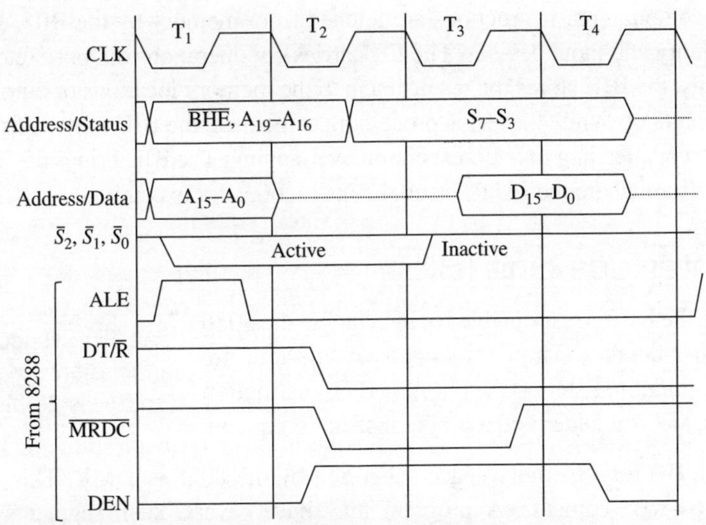

FIGURE 6.7 Memory read cycle in maximum mode

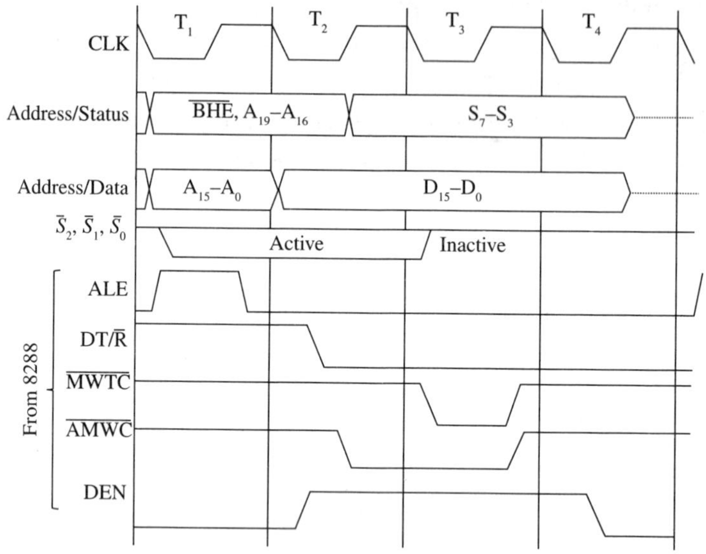

FIGURE 6.8 Memory write cycle in maximum mode

6.3 Functional Block Diagram

LO 3

After considering the memory interfacing techniques for 8086, in this section we shall look into the internal organization of 8086 processor. The internal functional block diagram of the 8086 processor has been shown in Figure 6.9. It broadly consists of two segments – the *Bus Interface Unit* (BIU) and the *Execution Unit* (EU). While the BIU is responsible for communicating with the memory chips and the I/O devices, the EU carries out the core computation. Instructions are fetched from memory by the BIU. The BIU also brings in data from the memory and input devices. The EU carries out operations on those fetched data, guided by the instruction. Finally, the BIU stores the results back to the memory locations or output devices. The BIU and the EU work in parallel. While the EU is processing some data, the BIU may be free to bring in further instructions from memory, leading to fetch-execution overlapping. The BIU brings the instruction bytes into an instruction queue from where the EU takes up the instructions for execution.

6.3.1 Bus Interface Unit (BIU)

As noted previously, the BIU is responsible for interfacing the 8086 with memory and I/O devices. The components of this unit are as follows – four segment registers (CS, DS, SS, ES), Instruction Pointer (IP) register, a 20-bit adder and a 6-byte instruction queue.

> ALU in EU performs 16-bit operations, whereas, the adder in BIU is 20-bit

Segment Registers: The total memory space supported by the 8086 is 1 MB. This whole space is logically divided into 64-KB segments. A program may have several such segments associated with it, each of them being maximum 64 KB in size. The user of the system has the liability to decide upon

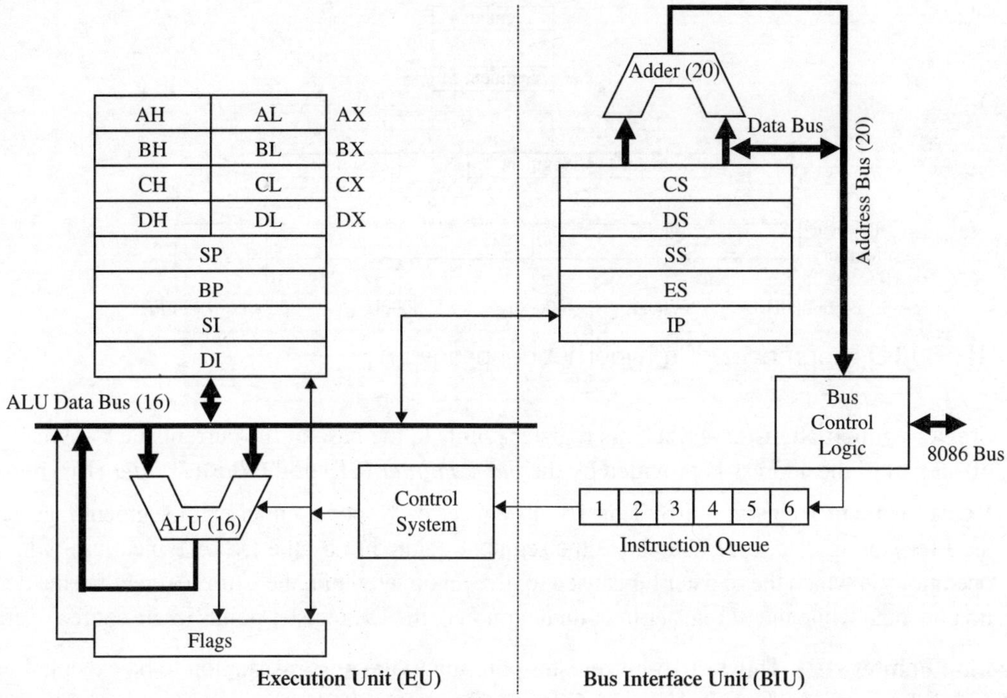

FIGURE 6.9 Internal block diagram of 8086

the program segments. A segment may correspond to a procedure, set of data structures/data elements, stack etc. However, at any point of time, the 8086 can directly address only four such segments. Figure 6.10 shows the segments of some hypothetical program. The program segments may be overlapping also, as shown in the figure. Out of all those segments, the 8086, at a time can refer to only four of them. From the organizational point of view, one of these segments may contain the code to be executed, one may be for the global data, one for the stack and another one for additional data parts. Accordingly, the designers of 8086 have named the segment registers as Code Segment (CS) register, Data Segment (DS) register, Stack Segment (SS) register and Extra Segment (ES) register. All these registers are 16-bit wide and hold the most significant 16 bits of the corresponding segment address. To generate a 20-bit address, the segment register content is left-shifted by 4 bits (equivalent to multiplication by 16) and an offset is added to it.

- **Code Segment Register – CS:** This 16-bit register holds the start address of the current code segment. The IP register provides the offset. The physical address is computed as,

$$\text{Physical address} = 16 \times CS + IP$$

- **Data Segment Register – DS:** The 16-bit DS register points to the base of the current data segment. Operands for most of the instructions are fetched from this segment. The offset part is provided by the *Source Index* (SI) register, *Destination Index* (DI) register or 16-bit displacement provided in the instruction itself. Thus, the physical address is formed as 16 × DS + SI, for example.

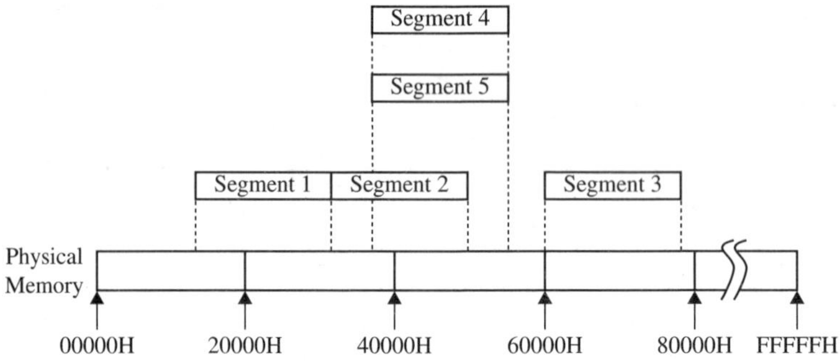

FIGURE 6.10 Segments of a hypothetical program

- **Stack Segment Register – SS:** This register points to the base of the current stack segment. The offset part of the address is provided by the *Stack Pointer* (SP) or the *Base Pointer* (BP) registers.
- **Extra Segment Register – ES:** This 16-bit register corresponds to an extra segment that may be used for storing data, over and above the segment pointed to by the DS. A typical use is in string operations in which the register DI is used to give an offset within the extra segment for the destination operand while the SI register in conjunction with the DS register points to the source operand.

Instruction Pointer (IP): This is a 16-bit register pointing to the next instruction to be executed within the current code segment. The content of the CS register is left shifted by 4 bits and added with IP, using the 20-bit adder in BIU to generate the 20-bit physical address. The IP register is automatically incremented as the successive instructions are fetched.

Instruction Queue: This is a 6-byte *First-In-First-Out* (FIFO) structure in which the successive instruction bytes, fetched from memory are stored. The EU takes the bytes one-by-one for execution. Thus, the fetch and execution of instructions get overlapped in a pipelined operation. This enhances the overall speed of operation of the processor. However, in case of control transfer instructions (like JUMP and CALL), the content of the instruction queue has to be discarded and the queue has to be reloaded with bytes from the destination address. This is called *pipeline stall.* The programmer may take special care while developing the program code to avoid such pipeline stalls by minimizing the number of program branchings.

6.3.2 Execution Unit (EU)

The EU is responsible for the execution of instructions. The components within the EU are as follows – a set of registers, a 16-bit ALU to carry out the operations and a flag register containing the status flags for the ALU operations.

Registers: There are three different types of registers in the EU – four general-purpose registers, two pointer registers and two index registers. All registers are 16-bit wide.

- **General-Purpose Register AX:** It is also known as the accumulator. It can be accessed as a 16-bit register AX or two 8-bit registers AH and AL. The AL corresponds to the lower order byte of AX while AH corresponds to the higher order byte. The I/O operations use AL or AX register, depending upon the port being 8-bit or 16-bit wide, respectively. The multiplication and division instructions also use the AX or AL register, depending upon the operand size.

- **General-Purpose Register BX:** Similar to AX, the BX register also consists of two 8-bit registers, namely, BH and BL. This is the only general-purpose register that can be used in addressing the memory. The BX register can provide the offset part of memory address with DS as segment register, by default.
- **General-Purpose Register CX:** This register consists of two 8-bit registers CH and CL. Apart from its usage as a general-purpose register, CX is also used as a counter in instructions like LOOP, ROTATE and SHIFT.
- **General-Purpose Register DX:** It consists of two 8-bit registers DH and DL. Apart from being used as a general-purpose register, DX register holds the higher order 16 bits of the results of 16×16 multiplications. For division operation, DX holds the 16-bit dividend before the operation and the 16-bit reminder after the operation.
- **Stack Pointer (SP):** The 16-bit SP is used to access data in the stack segment. It provides the offset within the current stack segment (SS being the corresponding segment register). The SP register gets updated automatically on execution of instructions like PUSH, POP and CALL
- **Base Pointer (BP):** This is another 16-bit register that can be used to access the stack segment with SS as the segment register. The BP register is used in *based addressing*, discussed later.
- **Source Index (SI):** This is a 16-bit register used for indexed addressing. For string operations, the SI is used in conjunction with the data segment register DS, to point to the source string.
- **Destination Index (DI):** This is the other 16-bit register available for indexed addressing. DI uses the ES as its default segment register. For string operations, the DI register, in conjunction with ES, points to the destination string.

Flag Register: This is a 16-bit register that holds the status of last ALU operation. The structure of the register has been shown in Figure 6.11.

The individual bits of the flag register are as follows.

- **Carry Flag (CF):** Set when there is a carry/borrow from the most significant bit position in the ALU addition/subtraction operation.
- **Parity Flag (PF):** Set to 1 if the lower byte of the result of an operation contains even number of 1's, reset to 0 otherwise.
- **Auxiliary Carry Flag (AF):** The bit is set if there is a carry/borrow from bit 3 to bit 4 during addition/subtraction operation of the ALU.
- **Zero Flag (ZF):** The flag is set to 1 if the result of computation is zero, reset otherwise.

				OF	DF	IF	TF	SF	ZF		AF		PF		CF
15	14	13	12	11	10	9	8	7	6	5	4	3	2	1	0

FIGURE 6.11 Flag register of 8086

- **Sign Flag (SF):** The flag is set to 1 if the result is negative.
- **Trap Flag (TF):** If the bit is set, the processor enters into a single-step execution mode by generating interrupts after executing each instruction.
- **Interrupt Flag (IF):** If the bit is set, external maskable interrupts can be recognized by the 8086. Clearing the flag bit disables these interrupts.
- **Direction Flag (DF):** This flag is used by the string manipulation instructions. If the bit is 0, the string is processed from its lowest to the highest address. If the bit is 1, the string is processed from the highest address towards the lowest address.
- **Overflow Flag (OF):** The OF is set if an overflow occurs during the operation. Overflow occurs if for signed 8-bit operation, the result is more than 7 bits in size. For 16-bit signed operation, the overflow occurs if the result is more than 15 bits in size.

6.4 Addressing Modes

LO 4

There are several addressing modes supported by the 8086. Table 6.5 summarizes the addressing modes. The addressing modes can be grouped into four categories. Group I consists of register and immediate operand accesses. Group II has all the memory access alternatives. Group III corresponds to I/O addressing, while in Group IV, the operands are implied by the instruction structure itself. The individual addressing modes have been discussed in the following.

Register Addressing: A register is used as the operand. Name of the register is mentioned in the instruction itself. For example, in the instruction

```
MOV CL, DH    ; CL gets content of DH
```

Both the operands are specified using register addressing.

Immediate Addressing: An 8- or 16-bit data is specified as the operand in the instruction. Typical examples of immediate addressing are as follows.

TABLE 6.5 Addressing modes of 8086

Group I	Register and immediate data	Register Addressing Immediate Addressing
Group II	Memory data addressing	Direct Addressing Register Indirect Addressing Based Addressing Indexed Addressing Based Indexed Addressing String Addressing
Group III	I/O port addressing	Direct I/O Port Addressing Indirect I/O Port Addressing
Group IV	Implied addressing	Implied Addressing

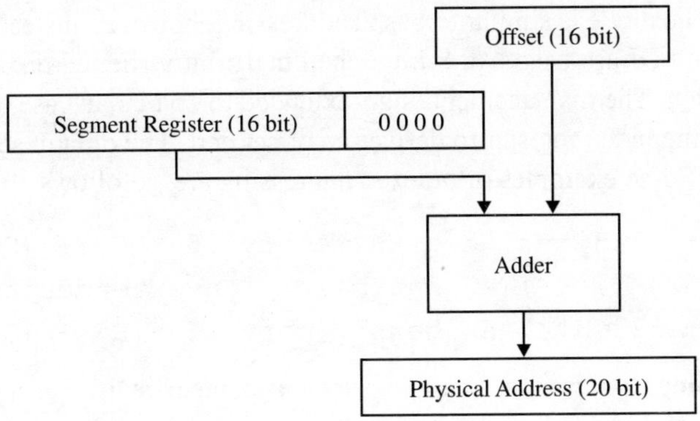

FIGURE 6.12 Memory address computation

```
MOV AH, 06H    ; AH gets 06H
MOV AX, 5497H ; AX gets 5497H
```

Direct Addressing: This mode is one type of memory access methodology. All the memory accesses, including direct addressing take place using the strategy depicted in Figure 6.12. The different addressing modes are the techniques to specify the offset part of the address. The segment part is taken care of by the segment register earmarked for the instruction. In direct addressing, the 16-bit offset is specified as part of the instruction itself. Some examples of this mode are as follows.

```
MOV AX, [1234H] ; AX gets Memory[16 × DS + 1354H]
MOV BL, [5600H] ; BL gets Memory[16 × DS + 5600H]
```

Register Indirect Addressing: In this mode, any of the registers BX, BP, SI or DI can hold the offset part of the address. For access through BP, the segment register used is SS. Similarly, for DI, the default segment register is ES. For BX and SI registers, the segment register is DS. Typical examples of instructions utilizing this addressing mode are as follows.

```
MOV AX, [BX]    ; AX gets Memory[16 × DS + (BX)]
MOV BX, [DI]    ; BX gets Memory[16 × ES + (DI)]
```

Based Addressing: In this addressing mode, a base register, along with a displacement is specified to indicate the memory address. The registers BX or BP can be used as the base register. The displacement can be a signed 8-bit number or an unsigned 16-bit number. The displacement part is added to the base register to compute the offset part of the address. The 8-bit displacement is sign-extended to 16 bits before carrying out the addition. The default segment register for BX is DS and for BP, it is SS. Some examples of based addressing are as follows.

```
MOV AX, [BX + 05H]   ; AX gets Memory[16 × DS + (BX) + 0005H]
MOV AX, [BP + 2134H]; AX gets Memory[16 × SS + (BP) + 2134H]
```

Indexed Addressing: The mode is similar to based addressing, however, instead of BX or BP registers, the index registers SI and DI are used. An 8-bit signed or 16-bit unsigned displacement value may be specified in the instruction. The displacement (sign-extended to 16 bits, in case of 8-bit displacement) is added to the content of the index register to derive the offset part. The default segment register for SI is DS and for DI, it is ES. Some examples of indexed addressing are as follows.

```
MOV AX, [SI + 0A2H]   ; AX gets Memory[16 × DS + (SI) + FFA2H]
                      ; A2 has been sign-extended to 16 bits
MOV AX, [DI + 1234H]  ; AX gets Memory[16 × ES + (DI) + 1234H]
```

Based Indexed Addressing: In this case, a base register and an index register, along with a displacement is used to specify the memory address. The displacement is computed by adding the contents of base register and index register along with the sign extended displacement value. If BP is used as the base register, the default segment register is SS, for other cases it is DS. Examples of this addressing mode are as follows.

```
MOV AX, [BX + SI + 02H]   ; AX gets Memory[16 × DS + (BX) + (SI) + 0002H]
MOV AX, [BP + DI + 1234H]; AX gets Memory[16 × ES + (BP) + (DI) + 1234H]
```

String Addressing: This mode is used for operations on string data. The source index register SI, along with segment register DS points to the source string. The destination index register DI, along with segment register ES points to the destination string. The mode will be discussed in more detail in conjunction with string manipulation instructions. An example of this addressing mode is as follows.

```
MOVSB ; Memory[16 × ES + (DI)] gets Memory[16 × DS + (SI)]
```

Direct I/O Port Addressing: This mode is used for I/O operations via the IN and OUT instructions to specify 8-bit port addresses. A typical example of the mode is as follows.

```
IN AL, [45H]  ; AL gets content of port with address 45H
IN AX, [54H]  ; AX gets content of port with address 54H
```

Indirect I/O Port Addressing: In this mode, the content of 16-bit register DX is used as the port address. A typical example is as follows.

```
OUT [DX], AX  ; Port pointed to by DX gets content of AX
```

Implied Addressing: These instructions do not contain any operand specified as part of the instruction. The operand is fixed by the semantics of the instruction itself. An example of the same is as follows.

```
CLC ; Clears the carry flag CY to zero
```

6.5 Instruction Set

LO 5

In this section, we shall discuss about the instructions supported by the 8086 processor. The instructions can be put into ten broad groups, as follows.

1. Data Transfer Instructions
2. Arithmetic Instructions

3. Logical Instructions
4. Rotate and Shift Instructions
5. Control Transfer Instructions
6. String Manipulation Instructions
7. Flag Manipulation Instructions
8. Stack Manipulation Instructions
9. Input-Output Instructions
10. Other Miscellaneous Instructions

Instructions belonging to each of the six categories have been discussed in the following.

6.5.1 Data Transfer Instructions

These instructions are used to transfer data from a source to a destination. The source operand may be specified as a register, a memory location or an immediate value. The destination must be a register or a memory location. The transfer is in terms of bytes or words, however, the source and destination sizes must match. There are various instructions belonging to this category, enumerated next. In the discussion to follow, the abbreviation *reg* stands for a processor register, *mem* stands for a memory operand and *data* stands for immediate data. The operands can be specified using the addressing modes noted in Section 6.4.

MOV: The MOV instruction is of the format "MOV *<dest>*, *<src>*". It transfers the content of source *src* to destination *dest*. The source and destination cannot both be memory locations. None of the flags get affected by the movement. Typical examples of MOV instruction are as follows.

```
MOV AX, CX      ; AX gets content of CX
MOV AX, 5697H   ; AX gets 5697H
MOV BL, [2345H]; BL gets content of Memory[16 × DS + 2345H]
MOV AL, [BX]    ; AL gets content of Memory[16 × DS + (BX)]
MOV AX, ARR[BP]; AX gets content of Memory[16 × SS + (BP) + ARR]
                ; Similar to MOV AX, [BP + ARR]
```

It may be noted that the programmer may ask the 8086 to use a different segment register, instead of the default one. For example, the instruction "MOV AL, [BX]" uses DS as the segment register. If the user wants to use the segment register ES, instead of DS, the following instruction may be used.

```
MOV AL, ES:[BX]; AL gets content of Memory[16 × ES + (BX)]
```

This is known as *segment override prefix*.

XCHG: The exchange instruction is of the format "XCHG *<op1>*, *<op2>*", one of the operands must be a register. Contents of two operands get exchanged by the execution of this instruction. Some examples of this instruction are as follows. The instruction does not affect any of the flags.

```
XCHG AX, CX      ; Exchange contents of AX and CX
XCHG AH, ARR[BX] ; Exchange AH and Memory[16 × DS + (BX) + ARR]
```

LEA: The "*Load Effective Address*" instruction is of the format "LEA <*reg*>, <*src*>". It instructs the 8086 to load the offset (also called *effective address*) of the source memory location *src* within segment into the register *reg*. This instruction also does not affect any of the flags. A few typical examples of the instruction are as follows.

```
LEA BX, XYZ    ; BX gets offset of XYZ within the data segment
                 pointed to by DS
LEA BP, XYZ    ; BP gets offset of XYZ within the stack segment
                 pointed to by SS
LEA CX, [SI]   ; CX gets offset of location pointed to by SI in DS
```

LDS: The "*Load Data Segment*" instruction, similar to LEA, loads the specified register with the content of the specified location. In addition, the data segment register DS is loaded with the contents of next 2 bytes. The instruction does not affect any of the flags. For example, the instruction "LDS BX, [2345H]" ensures that content of Memory[16 × DS + 2345H] is loaded into BL, Memory[16 × DS + 2346H] to BH, Memory[16 × DS + 2347H] to lower byte of DS, Memory[16 × DS + 2348H] to higher byte of DS.

LES: The "*Load Extra Segment*" instruction is similar to LDS, however, instead of the DS register, the ES register gets loaded, apart from the register specified in the instruction. For example, the instruction "LES BX, [2345H]" makes content of Memory[16 × DS + 2345H] loaded into BL, Memory[16 × DS + 2346H] to BH, Memory[16 × DS + 2347H] to lower byte of ES, Memory[16 × DS + 2348H] to higher byte of ES. The instruction does not affect any of the flags.

6.5.2 Arithmetic Instructions

These instructions are responsible for arithmetic computations. The operations include addition, subtraction, multiplication, division, increment, and decrement. In the following, the individual instructions have been discussed.

ADD/ADC: These instructions are of the form "ADD <*dest*>, <*src*>" and "ADC <*dest*>". The destination *dest* can be a register or a memory location. The source *src* may be a register, a memory location or an immediate value. Both the instructions add the source operand to the destination. The ADC instruction further adds the *carry flag* (CY) to the sum. It is not permitted to have memory locations as both source and destination operands. Flags affected by these instructions are AF, CF, OF, SF, ZF. Some examples of these instructions are as follows.

```
ADD AH, 56H      ; Add 56H to AH
ADD AL, BL       ; Add BL to AL
ADD BX, [SI]     ; Memory word at [16 × DS + (SI)] gets added to BX
ADC BL, XYZ[BX]  ; Memory word at [16 × DS + (BX) + offset of XYZ]
                   gets added to BL
                 ; along with the carry flag
```

SUB/SBB: These are for subtraction (SUB) and subtract with borrow (SBB) operations. The instructions are similar to ADD and ADC, respectively. Only instead of addition, the operation carried out is subtraction.

MUL: This is the multiplication instruction with format "MUL <*src*>". The source *src* can be a register or a memory location. The AL or AX register is used as the other operand for multiplication. If the *src* happens to be a byte operand, the unsigned byte is multiplied with the unsigned byte in AL. On the other hand, if *src* is a word, the unsigned word is multiplied with the unsigned word in AX. For byte multiplication, the 16-bit result is put into the AX register. For 16-bit multiplication, the most significant word of 32-bit result is put into the DX register, the least significant word in AX register. If, in the result, the most significant byte/word is zero, the flags CF and OF are reset to 0. The flags AF, PF, SF and ZF are undefined after the execution of a MUL instruction. Some examples of MUL instruction are as follows.

```
MUL BL   ; AX gets AL × BL
MUL BX   ; DX and AX get higher and lower order words of AX × BX,
           respectively
MUL XYZ  ; If XYZ is defined as a byte, AX gets AL × Memory[16 × DS
           + XYZ]
         ; If XYZ is defined as a word, DX and AX get higher and
           lower order words
         ; of AX × Memory[16 × DS + XYZ], respectively
```

IMUL: This instruction is similar to MUL, however performs signed multiplication. If in the result, the most significant byte/word is zero, excepting the sign bit, the flags CF and OF are set to 0, else those flags are set to 1.

DIV: This instruction is used to divide an unsigned word by a byte or an unsigned double word by a word. For division by byte, the word must be available in the AX register. After division, the AL register will contain the 8-bit quotient while AH will hold the 8-bit remainder. For division by word, the most significant word should be in DX, least significant word in AX. After division, AX will contain the quotient and DX will contain the remainder. The DIV instruction is of the format "DIV <*src*>" with *src* being a register or a memory location. If there is an attempt to divide-by-0 or if the quotient is too large to fit, the 8086 generates a type 0 software interrupt. All flags are undefined after a DIV operation. Typical examples of the DIV instruction are as follows.

```
DIV CH   ; Divide AL by CH; quotient in AL, remainder in AH
DIV CX   ; Divide the double word in DX:AX by CX; quotient in AX,
           remainder in DX
DIV XYZ  ; If XYZ is defined as a byte, divide AL by Memory[16 ×
           DS + XYZ]
         ; with quotient in AL, remainder in AH. If XYZ is defined
           as a word, divide
         ; DX:AX by word at Memory[16 × DS + XYZ]
         ; with quotient in AX, remainder in DX
```

IDIV: The instruction is similar to DIV, excepting that it refers to signed division. The sign of the remainder is same as the sign of the dividend.

INC: This is the increment instruction of the format "INC <*dest*>" in which the *dest* can be an 8-bit or a 16-bit register or memory location. The operation increments the destination by 1. The instruction affects flags AF, OF, PF, SF and ZF, but not the carry flag CF. Thus, if an 8-bit destination contains FFH or a 16-bit destination contains FFFFH, an increment operation will turn the content to 0, but the CF will not get affected. Some examples of the INC instruction are as follows.

```
INC AL        ; AL gets incremented by 1
INC BX        ; BX gets incremented by 1
INC XYZ       ; Byte or word at Memory[16 × DS + XYZ] gets
                incremented by 1
```

DEC: This is the reverse of the increment instruction with the format "DEC <*dest*>". The operation decrements the destination by 1. The instruction affects flags AF, OF, PF, SF and ZF, but not the carry flag CF. Thus, if an 8-bit destination contains 00H or a 16-bit destination contains 0000H, a decrement operation will turn the content to FFH or FFFFH, respectively, however, the CF will not get affected. Some examples of the INC instruction are as follows.

```
DEC AL        ; AL gets decremented by 1
DEC BX        ; BX gets decremented by 1
DEC XYZ       ; Byte or word at Memory[16 × DS + XYZ] gets
                decremented by 1
```

DAA: The instruction stands for decimal adjust after BCD addition. The instruction can be used after adding two packed BCD numbers with result in AL. Execution of DAA will rectify the content of AL, similar to the DAA instruction of 8085. The flags AF, CF, SF, PF and ZF get affected. The status bit OF is undefined.

DAS: The "*decimal adjust after subtraction*" instruction is similar to the DAA with the only difference that it can be used after subtracting two packed BCD numbers.

CBW: This instruction, "*convert signed byte to signed word*", copies the sign bit of AL register to all bits of AH. Thus, AH sign extends AL. The instruction does not affect any flag. For example, if AL = F3H, CBW will make AX = FFF3H.

CWD: The instruction, "*convert signed word to signed double word*", is similar to CBW, excepting that the instruction works on double word operands. The instruction copies the sign bit of AX to all bits of DX. The instruction does not affect any of the flags.

AAA: The instruction "*ASCII adjust for addition*" helps in adding numerical data represented in ASCII format. The ASCII codes for the decimal digits '0' to '9' are 30H to 39H. Thus, for any digit, the ASCII code has the upper nibble as 3, the lower nibble contains the digit. After addition, executing the AAA instruction corrects the output. For example, if the AL register contains 35H (ASCII of '5') and BL 39H (ASCII of '9'), after addition, the AL register content becomes 6EH. The AAA instruction converts AL to 04H (BCD 4) with carry flag CF = 1, thus representing the number 14 in decimal. The instruction works only on the AL register, updates flags AF and CF, however, the flags PF, SF and ZF are left undefined.

AAS/AAM/AAD: These are similar to AAA, however, works in conjunction with subtraction (AAS), multiplication (AAM) and division (AAD) operations. Operation of AAS is exactly same as AAA. AAM creates two unpacked BCD bytes in the AX register from the result of a multiplication of two unpacked

BCD digits. The AAD instruction has to be used before the division operation. It converts two unpacked BCD digits in AX into an equivalent binary number in AL.

6.5.3 Logical Instructions

These instructions are responsible for logical computations over the operands. The operations include AND, OR, XOR, NOT, negation, compare and test. The instructions have been detailed in the following.

AND/OR/XOR: These are all two-operand instructions of the form "AND <*dest*>, <*src*>" and so on. The source and the destination operands are similar to the ADD/SUB instructions. The flags CF and OF are both 0 after the operation, PF (only for 8-bit operands), SF and ZF depend upon the outcome, while AF is undefined. A few examples of these instructions are as follows.

```
AND AX, [SI]     ; AX gets AX AND Memory[16 × DS + (SI)]
OR BH, CL        ; BH gets BH OR CL
XOR CX, XYZ[SI] ; CX gets CX XOR Memory[16 × DS + XYZ + (SI)]
```

NOT: This instruction, of the format "NOT <*dest*>" inverts each bit specified in the destination operand, which may be of 8 or 16 bits. No flags are affected. Typical examples are as follows.

```
NOT AX           ; All bits of AX are inverted
NOT XYZ          ; All bits at byte/word Memory[16 × DS + XYZ]
                   are inverted
```

NEG: The "NEG <*dest*>" instruction replaces the content of destination operand by its 2's complement. Thus, the instruction changes the sign of the operand. The CF flag is set to 0 if the operand is 0, else 1. The OF, SF, ZF, AF and PF flags are also affected by the result. Typical examples of the instruction are as follows.

```
NEG AL           ; AL gets its 2's complement
NEG XYZ          ; Negate the byte/word Memory[16 × DS + XYZ]
```

CMP: The instruction "CMP <*dest*>, <*src*>" compares a byte/word specified in the source operand *src* with the byte/word specified in the destination operand *dest*. The source and the destination operands can be specified as in AND instruction. The flags are affected by the instruction. In particular, the flags CF, ZF and SF are indicative of the outcome of the conversion, as noted next.

	CF	ZF	SF
src = dest	0	1	0
src > dest	0	0	0
src < dest	1	0	1

Some examples of the CMP instruction are as follows.

```
CMP AH, 33H      ; Compare immediate value 33H with byte in
                   register AH
CMP BX, XYZ      ; Compare word at Memory[16 × DS + XYZ] with BX
CMP XYZ, 49H     ; Compare 49H with word at Memory[16 × DS + XYZ]
```

TEST: The instruction "TEST <*dest*>, <*src*>" computes AND of the byte/word specified at *dest* and *src*. However, neither of the operands get modified. Rather, the flags get affected as in the case of AND instruction. This is a technique to set the status flags when the source and the destination operands are ANDed. The instruction is often used in conjunction with conditional jump instructions (to be discussed later). Some examples of the TEST instruction are as follows.

```
TEST AL, BL     ; Update flags as per AND of AL and BL
TEST BX, 0234H  ; Update flags as per AND of BX and 0234H
TEST AH, XYZ    ; Update flags as per AND of AH and Memory[16 × DS
                  + XYZ]
```

6.5.4 Rotate and Shift Instructions

These instructions are responsible for rotating or shifting the destination operand by some specified number of bits. The CF becomes a part of the operand. The following are the instructions belonging to this category.

RCL/RCR: The instructions are to rotate the destination operand left (for RCL) or right (for RCR) through the CF. The format of these instructions is "RCL/RCR <*dest*>, <*count*>". The operand *dest* can be a register or a memory location. To rotate by one bit, *count* should be specified as 1. For multibit rotation, the register CL should be preloaded with the count and the operand *count* should be specified in CL. The instructions affect the status flags CF and OF only, moreover, for multibit rotation, OF is undefined. Figure 6.13 shows the execution of RCL and RCR instructions. Some typical examples of RCL and RCR instructions are as follows.

```
RCL BX, 1      ; Rotate BX left by 1 bit, MSB to CF, CF to LSB
MOV CL, 2      ; Load CL with rotate count
RCL XYZ, CL    ; Rotate Memory[16 × DS + XYZ] left by number of
                 bits specified in CL
```

ROL/ROR: These instructions are similar to RCL and RCR instructions respectively, with the difference that the carry flag CF does not become part of the operand to be rotated. However, for ROL, MSB of destination shifts to CF. For ROR, LSB shifts to the CF. Execution of ROL and ROR instructions have been shown in Figure 6.14.

SAR/SHR: These instructions are similar to rotate right, however, the LSB is not fed back to the MSB. In the SAR instruction, the old MSB is set as the new MSB also, thus maintaining the sign of the operand.

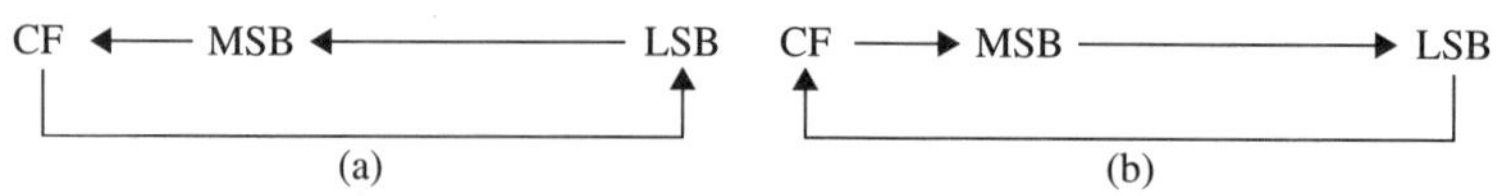

FIGURE 6.13 Execution of (a) RCL (b) RCR

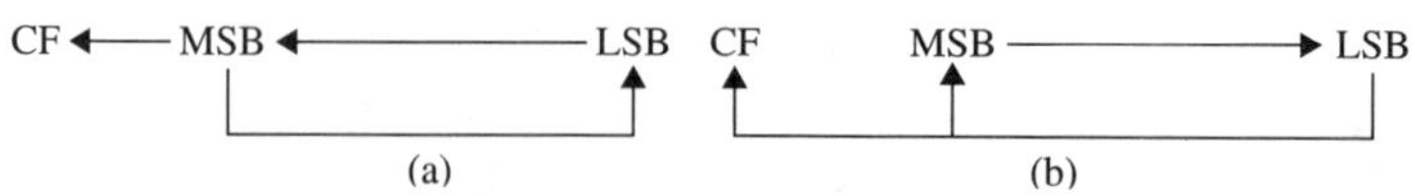

FIGURE 6.14 Execution of (a) ROL (b) ROR

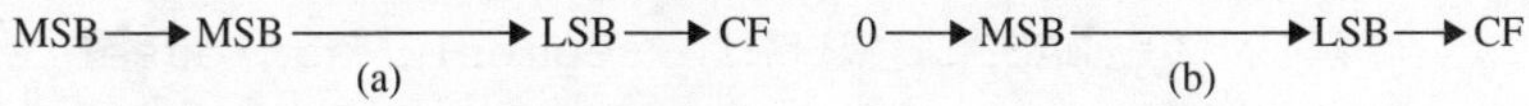

FIGURE 6.15 Execution of (a) SAR (b) SHR

FIGURE 6.16 Execution of SAL/SHL

For the SHR operation, instead of feeding the old MSB as the new MSB, the new MSB is set as zero. Thus, the shifted number becomes a positive number. The situation has been shown in Figure 6.15.

SAL/SHL: These two instructions are same. The operation is similar to the SHR instruction with the difference that shifting takes place in the left direction from LSB to the MSB side. A 0 is shifted into the LSB, MSB gets shifted to the carry flag CF. The situation has been shown in Figure 6.16.

6.5.5 Control Transfer Instructions

These instructions are responsible for changing the sequence in which the instructions get executed by 8086. The instructions include jump (both conditional and unconditional), loop, call to subroutines and return from there. In the following, the instructions have been enumerated.

JMP: This is the unconditional jump to a specified destination. The jump may be within the current code segment or to a location in a different code segment. Jumps within the current code segment are called *near jump*. Jumps to another segment are called *far jump*. In the case of a near jump, only the IP register gets modified. The difference between the current value in the IP register and the branch target is calculated and put into the instruction code. To illustrate the procedure further, consider the following branch instruction at memory location 10000H.

```
10000   JMP L1
```

If the target L1 is located at address 10200H and is within the same segment as this JMP instruction, it will be translated into a short jump. After fetching the JMP instruction during execution, the next instruction located at address 10003H will be pointed to by CS:IP. The distance from the branch target is thus 01FDH. In the machine code of the instruction, the first byte will have the code for a near JMP, while the next 2 bytes will contain 01FDH. During execution, the distance coded in the instruction is added to the IP register to ensure that the next instruction is fetched from the specified destination. This type of branching, with target specified as offset from current IP, are also called *relative branch*. In relative branching, the target address is independent of the location from where the program is loaded into memory. This type of codes are called *relocatable code*. For a far jump, both the CS and the IP registers get modified. None of the flags get affected. Another classification of JMP instruction is into *direct* and *indirect* jump classes. For direct jump, the target address is specified in the instruction itself. For indirect jump, the target address has to be retrieved from the operand specified. Some example jump instructions are as follows.

```
JMP LABEL1           ; If LABEL1 within segment: add offset with IP,
                       direct near jump
                     ; If LABEL1 in different segment: load CS and IP
                       with values coded in
                     ; instruction, direct far jump
JMP DX               ; IP gets the value of DX, indirect near jump
JMP WORD PTR [DX]    ; IP loaded with Memory[16 × DS + (DX)], indirect
                       near jump
JMP DWORD PTR [DX]   ; IP loaded with Memory[16 × DS + (DX)], CS
                       loaded with
                     ; Memory[16 × DS + (DX) +2], indirect far jump
```

J*x*: These are conditional jump instructions. The condition *x* is framed as combination of the status flags. The instructions are generally used after a comparison operation to transfer control, based upon the outcome of the last comparison. Table 6.6 shows a list of such conditional jump instructions. A restriction with the conditional jump instructions is that the target address must be within the current code segment and also in between –128 and +127 byte offsets from the current IP register content. Thus, conditional jumps are all relative to the current program location.

LOOP: The instruction of the format "LOOP <*label*>" makes the control to jump to the specified *label* if the CX register is non-zero after an auto decrement operation. The instruction is useful to repeat a series of instructions for a number of times. The number of times the instructions are to be repeated is loaded into the CX register. Every time the LOOP instruction is executed, CX is decremented by 1. Execution continues from the label if CX ≠ 0. For CX = 0, execution goes to the next instruction. The decrement operation does not affect any of the status flags. A typical code fragment utilizing LOOP has been shown next.

```
      MOV CX, 10   ; Wish to repeat for 10 times
  L1: ...

      ...
      LOOP L1      ; Loop back to L1 if CX≠0 after auto decrement
```

LOOPE/LOOPZ/LOOPNE/LOOPNZ: These instructions consider the ZF along with CX to determine the loop continuation condition. The instructions LOOPE/LOOPZ continue with the next iteration of the loop body if CX ≠ 0 and ZF = 1. On the other hand, LOOPNE/LOOPNZ continues as long as CX ≠ 0 and ZF = 0. No status flag gets affected by these instructions.

CALL: The CALL instruction is similar to JMP, excepting that the return address is saved onto the stack. For a *near* call, only the IP register is saved into the stack and the IP is loaded with the offset of the procedure address. A corresponding RET instruction at the end of the procedure will retrieve the IP from the stack. For calling a procedure on a different segment, *far* call is used. In this case, the 8086 first decrements SP by 2 and copies the CS register to the stack. The SP is then further decremented by 2 and the IP register is saved into the stack. CS and IP registers are then loaded with the appropriate segment and offset values for the procedure address. A RET instruction at the end of the procedure will restore the CS and the IP registers. Status flags are not affected by the CALL instruction. Some examples of the CALL instruction are as follows.

```
CALL LAB1             ; If LAB1 within segment: add offset with IP,
                        direct near call
                      ; If LAB1 in different segment: load CS and IP with
                        values coded in
                      ; instruction, direct far call
CALL DX               ; IP gets the value of DX, indirect near call
CALL WORD PTR [DX]    ; IP loaded with Memory[16 × DS + (DX)], indirect
                        near call
CALL DWORD PTR [DX]   ; IP loaded with Memory[16 × DS + (DX)], CS loaded
                        with
                      ; Memory[16 × DS + (DX) +2], indirect far call
```

TABLE 6.6 Conditional jump instructions in 8086

Mnemonic	Meaning	Flag Conditions
JA/JNBE	Jump if above/ not below or equal	CF = 0 and ZF = 0
JAE/JNB/JNC	Jump if above or equal/ not below/ no carry	CF = 0
JNAE/JB/JC	Jump if not above or equal/ below/ carry	CF = 1
JBE/JNA	Jump if below or equal/ not above	CF = 1 or ZF = 1
JG/JNLE	Jump if greater/ not less than or equal	ZF = 0 and CF = OF
JGE/JNL	Jump if greater than or equal/ not less than	SF = OF
JNGE/JL	Jump if not greater than or equal/ less than	SF ≠OF
JLE/JNG	Jump if less than or equal/ not greater	ZF = 1 or SF ≠OF
JE/JZ	Jump if equal/ zero	ZF = 1
JNE/JNZ	Jump if not equal/ not zero	ZF = 0
JS	Jump if signed/ negative	SF = 1
JNS	Jump if not signed/ positive	SF = 0
JP/JPE	Jump if parity/ parity even	PF = 1
JNP/JPO	Jump if no parity/ parity odd	PF = 0
JO	Jump if overflow	OV = 1
JNO	Jump if no overflow	OV = 0
JCXZ	Jump if CX register contains 0	CX = 0

RET: This instruction is used to return from a procedure. It has been already discussed in conjunction with the CALL instruction. The 8086 allows an optional number after the RET instruction. If the number is say, *n*, SP gets incremented by *n*. Thus, content of this portion of stack gets discarded by the calling program. The feature is often used to ignore the portion of stack that the calling program might have used to pass parameters to the called procedure.

6.5.6 String Manipulation Instructions

These instructions are useful for operations over string of data. The instructions in this category include block data movement between memory segments, move string data byte/word wise to/from the registers, compare between two strings stored in memory, search for a byte/word in the string with the facility to repeat the operations over strings.

MOVS/MOVSB/MOVSW: The MOVS instruction copies a byte or a word from a location in the data segment to a location in the extra segment. The source location offset should be available in the SI register, while the destination offset in the DI register. The CX register is to be used to hold the number of bytes/words to be transferred. After every transfer, CX is decremented by 1. The registers SI and DI are updated based upon the value of the direction flag, DF. If DF = 0, SI and DI are incremented by 1 for byte movement and by 2 for word movement. If DF = 1, SI and DI are decremented by 1 for byte movement and 2 for word movement. The 8086 assemblers generally distinguish byte and word movements by looking into the source and the destination operand types. However, the programmer may explicitly mention the operation to be a byte movement via MOVSB or a word movement via MOVSW. For multi byte/word transfer, the REP prefix needs to be attached before the MOVS instruction. In that case, movement of data continues until CX becomes equal to 0. Flags are not affected by the MOVS instruction. A typical example of such multibyte data movement has been noted in the following.

```
MOV SI, OFFSET SRC   ; Load offset of SRC in DS into SI
MOV DI, OFFSET DEST  ; Load offset of DEST in ES into DI
CLD                  ; Clear DF to indicate auto increment of SI
                       and DI
MOV CX, 10H          ; Load CX with number of bytes to move
REP MOVSB            ; Move string bytes until CX = 0
```

LODS/LODSB/LODSW: The instructions can be used to load string byte or word into AL (for LODSB) or AX (for LODSW), respectively. The SI register should contain the offset of the string within data segment. If DF = 0, SI is auto incremented by 1 for byte load and by 2 for word load. Similarly, if DF = 1, the SI gets auto decremented. LODSB or LODSW can be used to tell explicitly a byte or word movement. A typical usage may be as follows.

```
MOV SI, OFFSET SRC   ; Load offset of SRC in DS into SI
CLD                  ; Clear DF to indicate auto increment of SI
LODSB                ; Copy byte to AL
```

STOS/STOSB/STOSW: These instructions are just the reverse of LDOS/LDOSB/LDOSW and can be used to store a byte/word from AL/AX to a memory location in the extra segment pointed to by DI.

The register DI gets automatically incremented or decremented, depending upon the value of the direction flag. The instruction does not affect any of the flags. A typical example may be as follows.

```
MOV DI, OFFSET DEST ; Load offset of DEST in ES into DI
CLD                 ; Clear DF to indicate auto increment of DI
STOSB               ; Copy byte from AL
```

CMPS/CMPSB/CMPSW: The compare string bytes/words instruction compares a byte/word in one string (pointed to by DS:SI) with a byte/word in another string (pointed to by ES:DI). CX register holds the number of string elements. A comparison affects the flags AF, CD, OF, SF and ZF. After comparison, CX gets decremented by 1, SI and DI are auto incremented/decremented by 1 (for byte comparison) or by 2 (for word comparison) depending upon the value of the direction flag DF. The following is a code fragment demonstrating the CMPSB instruction in which comparison continues till end-of-string or reaching an unequal byte.

```
MOV SI, OFFSET STR1 ; Point SI to first string in data segment
MOV DI, OFFSET STR2 ; Point DI to second string in extra segment
CLD                 ; Clear direction flag
MOV CX, 50          ; Load CX with number of elements
REPE CMPSB          ; Repeat comparison until end or compared bytes
                      are unequal
```

SCAS/SCASB/SCASW: The SCAS instruction compares a byte (in AL) or a word (in AX) with a byte/word pointed to by ES:DI. After comparison, DI is incremented or decremented (by 1 for byte, 2 for word) depending upon the DF flag. The flags AF, CF, OF, PF, SF and ZF get affected. CX register should be loaded with the number of string elements. Putting suitable REP prefix, the scanning operation can be repeated till the desired byte/word found in the string. The following code fragment illustrates the same to search for character 'A' (ASCII 41H) in the string STR.

```
MOV DI, OFFSET STR  ; Point DI to string in extra segment
MOV AL, 41H         ; Load 'A' in AL register
MOV CX, 80          ; Use CX as counter
CLD                 ; Clear direction flag
REPNE SCASB         ; Repeat comparison until end or character 'A'
                      found
```

REP/REPE/REPZ/REPNE/REPNZ Prefix: This repeat prefix is used before a string instruction to repeat the instruction until some specified condition is met or CX becomes equal to zero. For example, as noted earlier, "REP MOVSB" will continue copying bytes from source string to destination string until CX becomes zero. The meanings of different prefixes are as follows.

Prefix	Meaning	Condition
REP	Repeat	Repeat as long as CX ≠ 0
REPE/REPZ	Repeat if equal/zero	Repeat as long as ZF = 1 and CX ≠ 0
REPNE/REPNZ	Repeat if not equal/not zero	Repeat as long as ZF = 0 and CX ≠ 0

6.5.7 Flag Manipulation Instructions

These instructions are used for manipulating the individual bits of the status flag register. In the following, the instructions have been enumerated. All instructions are of zero operand, as the flag bit is identified by the instruction itself.

STC: This instruction sets the carry flag CF to 1. No other flags are affected.

CLC: Similar to STC, but clears the carry flag.

CMC: This instruction complements the carry flag without affecting others.

STD: This instruction sets the direction flag to 1.

CLD: It is used to clear the direction flag to 0.

STI: The *set interrupt flag* instruction sets the interrupt flag to 1. This enables the INTR interrupt input for the 8086.

CLI: This instruction clears the interrupt flag. The INTR input of 8086 gets disabled.

LAHF: This instruction copies the low byte of the flag register into AH. The AX register may now be saved into stack, thus saving the flag bits into the stack.

SAHF: This instruction copies the AH register content into the low byte of the flag register. A previously stored status register content from stack may first be copied into AX and from there transferred to low byte of the flag register by the SAHF instruction. Thus, LAHF and SAHF instructions could be used to save and restore the status flags to/from stack.

6.5.8 Stack Manipulation Instructions

These instructions are used to save operands into the stack or restore operands from the stack. The programmer must ensure that the stack segment register SS and the stack pointer SP are initialized properly to some empty memory area before using these instructions, making procedure calls or enabling the interrupts. The stack manipulation instructions have been discussed in the following.

PUSH/PUSHF: The instruction "PUSH <*src*>" instruction pushes the word operand *src* into the stack. For this, the stack pointer SP is first decremented by 2 and the word is copied into Memory[16 × SS + SP]. The operand *src* can be a general-purpose register, a segment register or a memory word. Some examples of the PUSH instruction are as follows.

```
PUSH AX        ; Save AX in stack
PUSH DS        ; Save DS in stack
PUSH XYZ[BX]   ; Save Memory[16 × DS + (BX) + XYZ] in stack
```

PUSHF saves the flag register into the stack.

POP/POPF: The instruction "POP <*dest*>" is the reverse of PUSH, in which one word from the top of the stack is popped out and copied into the destination *dest*. The stack pointer is incremented by two. Some examples of the POP instruction are as follows.

```
POP AX          ; Restore AX from stack
POP DS          ; Restore DS from stack
POP XYZ[BX]     ; Restore Memory[16 × DS + (BX) + XYZ] from stack
```

POPF copies two topmost location contents from stack into the flag register.

6.5.9 Input-Output Instructions

These instructions are responsible for input/output of data from/to ports. The following are the two instructions belonging to this category.

IN: The instruction "IN <*acc*>, <*port*>" copies *port* content into *acc*. The operand *acc* is AL for 8-bit data transfer and AX for 16-bit data transfer. The *port* may be of format *fixed port* or *variable port*. For fixed port type, 8-bit port address is specified directly in the instruction. Some examples of fixed port are as follows.

```
IN AL, 56H      ; A byte read from port 56H into AL
IN AX, 45H      ; An word read from port 45H into AX
```

For variable port type, the 16-bit port address is put into the DX register, which is then used as the address in IN instruction. Some examples are as follows. No flags are affected by the IN instruction.

```
MOV DX, 1234H ; Put port address into DX register
IN AL, DX     ; A byte read from port 1234H into AL
IN AX, DX     ; An word read from port 1234H into AX
```

OUT: The instruction "OUT <*port*>, <*acc*>" copies the content of *acc* to *port*. Meaning of the operands are same as for the IN instruction.

6.5.10 Miscellaneous Instructions

The following are some other instructions available in 8086.

HLT: This instruction causes 8086 to stop instruction fetch and execution, processor enters into a halt state. An interrupt on INTR or NMI pin or a reset on the RESET input will take the processor out of halt state.

NOP: The "no operation" instruction just delays the 8086 by three clock cycles.

ESC: The "Escape" instruction is used for the instructions meant for the coprocessor 8087. The 8086 will treat this instruction as a NOP.

INT: The instruction "INT <*type*>" is the software interrupt for 8086 (similar to RST in 8085). The term *type* is a number between 0 and 255, both inclusive. The sequence of actions in INT execution are as follows.

1. Decrement SP by 2 and push flags onto stack.
2. Decrement SP by 2 and push CS onto stack.

3. Decrement SP by 2 and push offset of next instruction (IP) onto stack.
4. Load IP with Memory[4 × *type*].
5. Load CS with Memory[4 × *type* + 2].
6. Reset IF and TF, other flags not affected.

For example, for the instruction "INT 35", the IP is loaded from Memory[0008CH] and CS from Memory[0008EH].

INTO: The "*interrupt on overflow*" instruction causes interrupt if the OF is set. The instruction is equivalent to "INT 4".

IRET: The "*interrupt return*" instruction may be used at the end of an interrupt service routine. It is similar to the RET instruction, in the sense that IP and CS values are restored in both of them. The IRET instruction causes the flag register also to be restored from the stack.

LOCK: If the 8086 is connected in a multiprocessor configuration, while accessing external buses, the 8086 may like to ensure that no other processor gains access to the bus. While executing an instruction with LOCK prefix, the 8086 informs the external bus controller to prevent other processors from taking the bus. The instruction does not affect any flag. A typical situation occurs while executing the XCHG instruction that requires bus over two accesses. An example is noted next.

```
LOCK XCHG XYZ, AL    ; Exchange Memory[16 × DS + XYZ] with AL
```

WAIT: This instruction makes the 8086 to wait in idle condition until the 8086 test input pin becomes low or until an interrupt comes on pin INTR or NMI. If an interrupt comes in the idle state, after finishing the service routine, the 8086 returns to idle state again.

XLAT/XLATB: The instruction translates a byte in AL specified in some code to another code. The conversion table is pointed to by DS:BX. Before executing the XLAT/XLATB instruction, the conversion table must be preloaded into the memory. Assuming that the CODE_TABLE contains the code conversion table, the following code fragment may be used to change the code in AL.

```
MOV BX, OFFSET CODE_TABLE  ; Point BX to start of CODE_TABLE
XLATB                      ; Replace code of AL
```

6.6 Interrupts in 8086 LO 6

The interrupts in 8086 can be classified into two main categories – *Software Interrupts* and *Hardware Interrupts*. As noted earlier, software interrupts are utilized by the programmer by inserting INT instructions at desired places of the program. On the other hand, hardware interrupts are caused by external signals on the pins NMI and INTR.

Software Interrupts: The INT instruction of 8086 is dedicated for the software interrupts. It is a two byte instruction of the format "INT *n*" with the first byte containing the opcode for INT and the second byte containing the 8-bit type identifier *n*. There are 256 different software interrupts available. The execution of the software interrupt proceeds as follows.

1. Flag register value pushed into the stack.
2. CS and IP values for the return address pushed into the stack.
3. IP register loaded with the content of location Memory[$n \times 4$].
4. CS register loaded with the content of location Memory[$n \times 4 + 2$].
5. Interrupt flag (IF) and trap flag (TF) are reset to zero.
6. Control branches to the interrupt service routine.

Thus, the start address of type 0 interrupt is stored from location 00000H, for type 1 from 00004H and so on. The first five types are some dedicated interrupt pointers, as noted next.

- *Type 0*: Division by zero situation
- *Type 1*: Single-step execution during debugging
- *Type 2*: Non-maskable hardware interrupt NMI
- *Type 3*: Break-point used for debugging
- *Type 4*: Overflow situation

Software interrupt types 5–31 are reserved for upward compatibility purposes of advanced microprocessors. Interrupt types 32–255 are available to the user for other hardware and software services.

Though INT is a 2-byte instruction, there are two special software interrupts, which are one byte long – the breakpoint interrupt "INT 3" and the overflow interrupt "INT 4" (also called INTO).

The opcode for "INT 3" is CCH. Opcode for INTO is CEH. INTO is a conditional interrupt instruction. Processor executing the INTO instruction branches to the corresponding interrupt handler only if the overflow flag (OF) is set to 1.

Hardware Interrupts: There are two hardware interrupt pins in 8086 – NMI and INTR. NMI is a non-maskable interrupt while INTR is a maskable interrupt with lower priority. The pin INTA stands for interrupt acknowledge.

The NMI is a type 2 interrupt. The course of actions taking place when an interrupt on the NMI pin is received, are as follows.

1. Processor completes its current instruction.
2. Pushes flag register into stack.
3. CS and IP values for the return address are pushed into the stack.
4. IP is loaded with the content of Memory[00008H].
5. CS is loaded with the content of Memory[0000AH].
6. Interrupt flag (IF) and trap flag (TF) are reset to zero.

The maskable INTR interrupt needs to be enabled by using the set interrupt flag (STI) instruction by the processor. On getting this interrupt, the processor completes its current instruction and sends '0' on

INTA pin twice. The first '0' is to acknowledge the interrupting device while during the second '0', the 8086 receives an 8-bit value to be used as the interrupt type. This value is then used to determine the address of the interrupt service routine, as discussed in the context of software interrupts.

6.7 8086 vs. 8088

LO 7

Intel 8088 is a variant of the 8086 microprocessor. It has an 8-bit external data bus, compared to 16-bit for 8086. However, the internal registers are 16-bit wide, as in 8086. The address space is also one megabyte. The EUs of 8086 and 8088 are having the same structure, while the BIUs are different. The 8088 was introduced by Intel to ensure simple circuit boards and compatibility with many of the peripheral chips, which have mostly 8-bit data bus. Table 6.7 shows a comparison between the BIU of 8086 and 8088.

TABLE 6.7 Comparison of BIU of 8086 and 8088

BIU of 8086	BIU of 8088
16-bit data bus	8-bit data bus
Can read/write 8/16 bit data	Can read/write 8 bit data
Memory organized as two 512 KB banks	Memory organized as single 1 MB bank
Can operate at 5/8/10 MHz	Can operate at 5/8 MHz
Instruction queue 6-byte long	Instruction queue 4-byte long
Has BHE pin	BHE replaced by SSO
Pin 28 is $\overline{IO}/M$	Pin 28 is $IO/\overline{M}$, to make it compatible with 8085

SUMMARY

The 8086 processors is an advancement over the 8085. It is having 20-bit address bus and 16-bit data bus. The processor can operate in two different modes – minimum and maximum. While in the minimum mode it works as a standalone processor, in the maximum mode it can support coprocessors. The memory is divided into 64 KB segments, thus separating the code, data and stack portions of a program. To implement fetch-execute overlap, 6-byte prefetch buffer has been incorporated in the BIU. Several addressing modes are supported with the involvement of base registers and index registers. Different arithmetic and logic instructions are supported. Compared to 8085, 8086 supports multiplication and division operations. The conditional transfer instructions have been enhanced to support relative addressing, which can lead to relocatable code generation. A major improvement in 8086 over 8085 is the support of string data and string manipulation instructions. Block of data bytes could be transferred using a single instruction. The number of hardware interrupts supported has been reduced to two, compared to five in 8085. The next chapter will enumerate the programming of 8086 and interfacing of I/O devices to it.

EXERCISES

Multiple Choice Questions

1. Width of address bus of 8086 in bits is
   (a) 16 (b) 20
   (c) 24 (d) 32
2. Branch target of NMI is
   (a) 00000H (b) 10000H
   (c) FFFFFH (d) 000A8H
3. Priority of GT0 is
   (a) Higher than GT1 (b) Lower than GT1
   (c) Same as GT1 (d) None of the above
4. Instruction queue of the 8086 is of size
   (a) 6 bytes (b) 6 words
   (c) 8 bytes (d) None of the above
5. Size of a segment in 8086 is
   (a) 4 KB (b) 16 KB
   (c) 64 KB (d) None of the above
6. Higher order bits of multiplication register is stored in
   (a) AX (b) BX
   (c) CX (d) DX
7. Default segment register for BP based access is
   (a) CS (b) DS
   (c) ES (d) SS
8. Register used for indirect I/O port addressing is
   (a) DX (b) BX
   (c) CX (d) None of the above
9. The computation done in TEST is
   (a) AND (b) OR
   (c) XOR (d) None of the above
10. Register used in multibit rotation is
    (a) AL (b) BL
    (c) CL (d) DL
11. The register to hold count for LOOP instruction is
    (a) AX (b) BX
    (c) CX (d) None of the above
12. If it is required to increment SI and DI registers after every transfer via MOVS, the DF flag should be set to
    (a) 0 (b) 1
    (c) 0 followed by 1 (d) None of the above
13. For the SCAS instruction, the string is available at
    (a) DS:SI (b) ES:DI
    (c) DS:EI (d) ES:SI
14. Instruction to clear the CY flag in 8086 is
    (a) STC (b) CMC
    (c) CLC (d) None of the above
15. The instruction executed by the 8086 for instructions meant for 8087 is
    (a) TEST (b) ESC
    (c) NOP (d) None of the above
16. INTO is equivalent to
    (a) INT 0 (b) INT 1
    (c) INT 2 (d) INT 4
17. WAIT instruction will make 8086 to wait until TEST pin is
    (a) 0 (b) 1
    (c) $0 \rightarrow 1$ (d) $1 \rightarrow 0$
18. Number of hardware interrupts in 8086 is
    (a) 1 (b) 2
    (c) 3 (d) 5
19. Interrupt type of NMI Is
    (a) 0 (b) 2
    (c) 4 (d) 6
20. Number of times INTA is made 0, on getting INTR, is
    (a) 1 (b) 2
    (c) 4 (d) None of the above
21. One byte long software interrupts are
    (a) 3, 4 (b) 0, 4
    (c) 3, 0 (d) None of the above

22. After getting an NMI, the interrupt flag becomes
(a) 0 (b) 1
(c) Z (d) None of the above

23. Data bus size in bits, of 8088 is
(a) 8 (b) 16
(c) 4 (d) None of the above

24. Number of memory banks in 8088 is
(a) 1 (b) 2
(c) 4 (d) None of the above

25. Length (in bytes) of 8088 instruction queue is
(a) 16 (b) 8
(c) 6 (d) 4

Answers to Multiple Choice Questions

1. b	2. d	3. a	4. a	5. c	6. d	7. d	8. a	9. a	10. c
11. c	12. a	13. b	14. c	15. b	16. d	17. a	18. b	19. b	20. b
21. a	22. a	23. a	24. a	25. d					

Long Answer Type Questions

1. Distinguish between the minimum and the maximum mode of operation of 8086.
2. Show the physical memory organization for 8086.
3. Show the read and write memory cycles for the minimum mode operation of 8086.
4. Show the read and write memory cycles for the maximum mode operation of 8086.
5. Draw a functional block diagram of 8086 and enumerate functions of different modules in it.
6. Explain the role of segment registers in the operation of 8086.
7. Enumerate the role of different registers available in 8086.
8. List down the status flags of 8086 along with their implications.
9. Enumerate the various addressing modes of 8086 along with examples.
10. Explain different data transfer instructions of 8086.
11. Enumerate the arithmetic and logic instructions of 8086.
12. Explain different unconditional and conditional branch instructions of 8086.
13. Explain the string manipulation instructions of 8086.
14. Enumerate different software and hardware interrupts available in 8086.
15. List the structural and operational differences between 8086 and 8088.

7

Assembly Language Programming and Interfacing with 8086

LEARNING OUTCOMES

After reading this chapter, students will be able to:

LO 1 Enumerate the assembly language program structure of 8086

LO 2 Interface a hex keyboard with 8086

LO 3 Interface multiplexed seven-segment display with 8086

LO 4 Interface ADC with 8086

LO 5 Interface DAC with 8086

LO 6 Interface a stepper motor with 8086

7.1 Introduction

The assembly language of 8086 contains a large number of instructions as enumerated in the last chapter. Also, it has been noted that 8086 contains a number of segment registers. Utilizing these segment registers, for better organization, the user can divide the program into several segments, some of which are for program codes while others for data and stack. For example, the main functionalities of the program may be put into one code segment while the error routines may be kept in a different segment. Among the data used by a program, some of them may be the major ones, while the others are auxiliary in nature. Accordingly, these data items may be put into different data segments. Instructions may have operands that can also be specified through a number of addressing modes. To illustrate these concepts, in this chapter, we shall first look into the program structure as supported by most of the assemblers. A number of example programs will then be looked into. A number of peripheral I/O devices can be interfaced with the 8086, similar to 8085. The detailed hardware and software interfaces will be illustrated.

7.2 Assembly Language Program Structure (LO 1)

Due to the provision of program segmentation, the assembly language programming of 8086 is more complex than 8085. To help the programmers better organize their programs, most of the 8086 assemblers support a number of additional assembler directives, over and above those supported by the 8085 assemblers, in general. The program structure can best be understood with reference to examples. However, before looking into any assembly language program example, we need to have an understanding of the additional directives. It may be noted that the directives are specific for an assembler and thus may vary to some extent across the assemblers. In the following, some of the most commonly used directives (which are not used by 8085 assemblers) have been noted.

SEGMENT and ENDS: The SEGMENT directive is used to mark the beginning of a logical segment in the program. The directive should be preceded by a name for the segment. For example, "SEG1 SEGMENT" starts a new segment with the name SEG1. The logical end of a segment is identified by the directive ENDS. Thus, a typical segment structure may be as follows.

```
CODE SEGMENT
     ...
CODE ENDS
```

ASSUME: The directive is used to inform the assembler of the name of the logical segment to be used for a specified segment. For example, the following statement,

```
ASSUME CS: CODE, DS: DATA
```

tells the assembler that the program instructions are contained in the segment named CODE and for all references to the data segment, it should use the logical segment called DATA. It also identifies the segment register to use for access within a segment.

OFFSET: This is an operator telling the assembler to compute the offset (displacement) of a named data item from the start of the segment, containing it. For example, "MOV SI, OFFSET STR1" loads the offset of STR1 within the corresponding segment into the SI register.

PTR: The PTR operator assigns a specific type to a variable or label. For example, "INC BYTE PTR [SI]" tells the assembler that the instruction should increment the byte pointed to by SI. On the other hand, the instruction "INC WORD PTR [SI]" would increment the word pointed to by SI.

Next, we shall look into some example programs written in the assembly language of 8086 to illustrate the concepts.

EXAMPLE 7.1

The following is a program to add two 32-bit numbers A1 and A2. The 32-bit numbers can be defined using the pseudo-opcode DD (Define Double word). The result is stored in another 32-bit space RES. The final carry will be stored in the byte after RES. It may be noted that the label specified after END (label START in the program) tells the assembler to mark the corresponding instruction as the first instruction to be executed by the processor. The assembler informs the operating system to load the CS and IP registers accordingly. The programmer should initialize the rest of the segment registers (for example, DS in this program) properly. The program noted next has been organized into two segments. The first one is a data segment, named DATA. It holds the program variables A1, A2 and RES, each declared as a 32-bit double word. The next segment, named CODE, holds the code for the program. The ASSUME directive declares the segment registers to be used by the assembler in generating machine code for the assembly language statements contained in the program. The CS register is loaded automatically by the operating system, whereas the DS register has to be loaded with the absolute address of DATA segment in the memory. As segment registers cannot be loaded with immediate values, loading of DS register has been carried out via AX. To carry out the 32-bit addition operation, data of the first two bytes of A1 are loaded into AX and added to the first two bytes of A2. The result is stored at address RES. Next, the second two bytes of A1 are added to the second two bytes of A2 along with the carry generated in the previous addition. The result is stored at location RES+2. If there is a final carry (saved in DL register), it is copied to the location RES+4. The program follows next.

```
   DATA SEGMENT
        A1 DD 12345678H          ;Define A1
        A2 DD 9ABCDEF0H          ;Define A2
        RES DD ?                 ;Define space RES
   DATA ENDS

   CODE SEGMENT
        ASSUME CS:CODE, DS:DATA
 START: MOV AX, DATA
        MOV DS, AX               ;Initialize DS
        MOV DL, 00H              ;Use DL to check if result has a final
                                  carry
        MOV AX, WORD PTR A1      ;Get lower order 16-bits of A1 in AX
        ADD AX, WORD PTR A2      ;Add lower order 16-bits of A2 to AX
        MOV WORD PTR RES, AX     ;Save lower order 16-bits of result
        MOV AX, WORD PTR A1+2    ;Get higher order 16-bits of A1 in AX
        ADC AX, WORD PTR A2+2    ;Add higher order 16-bits of A2 to AX
                                  with CY
        MOV WORD PTR RES+2, AX   ;Save higher order 16-bits of result
```

```
        JNC SET_MSB              ;If no final carry, DL kept as 0
        INC DL                   ;Increment DL to indicate CY
SET_MSB: MOV BYTE PTR RES+4, DL  ;Store CY of result
    CODE ENDS
     END START
```

EXAMPLE 7.2

The following program finds the largest number in an array ARR of five positive numbers. The data segment DATA contains the array and a variable LARGE to hold the largest number. In the program code, the CX register is initialized to the number of elements in the array ARR. The BL register is used to hold the largest number temporarily. The array element currently being considered is pointed to by the SI register. For the sake of comparison, the element pointed to by SI is brought into the AL register and compared with the current maximum available in BL. The registers BL, CX and SI get updated accordingly to consider the next element. The process continues till CX becomes zero and the final result is stored in the variable LARGE. The program follows.

```
  DATA SEGMENT
       ARR DB 08H, 14H, 05H,     ; Define ARR
       0FH, 09H
       LARGE DB ?                ; Define space LARGE
  DATA ENDS

  CODE SEGMENT
       ASSUME CS:CODE, DS:DATA
START: MOV AX, DATA
       MOV DS, AX                ; Initialize DS
       MOV CX, 05H               ; Initialize CX to number of elements
       MOV BL, 00H               ; Assume 0 to be largest
       LEA SI, ARR               ; Load offset of ARR into SI
   L1: MOV AL, [SI]              ; Get next number in AL
       CMP AL, BL                ; Compare number with current largest
       JL L2                     ; If less, no change in largest
       MOV BL, AL                ; Update largest in BL
   L2: INC SI                    ; Increment SI
       DEC CX                    ; Decrement CX
       JNZ L1                    ; Repeat if all elements not seen yet
       MOV LARGE, BL             ; Store largest in RES
  CODE ENDS
   END START
```

EXAMPLE 7.3

The next program is to check whether a string variable STR1 is a palindrome or not. It may be noted that a string is called a palindrome if it is the same after reversal. For example, the string "MALAYALAM" is a palindrome as it remains the same, read in either direction. In the program, if the string is a palindrome, the location CHECK is initialized to 1, else it is initialized to 0. The program has been organized to have two data segments – one named DATA (holding the string STR1 and the byte variable CHECK to hold the result), and the other named EXTRA that holds an auxiliary string STR2. Space for STR2 is defined using the pseudo-opcode "DB 9 DUP(?)" defining an uninitialized array of size 9. The program first makes a copy of the string STR1 into the string STR2, such that the string STR2 is the reverse of the string STR1. This is followed by a byte-level comparison between the two strings STR1 and STR2. The 8086 string instructions, LODSB, STOSB and CMPSB have been used in the program. The CX register has been used as the loop counter for the repetitive loop statement "REPZ CMPSB". The program is as follows:

```
  DATA SEGMENT
       STR1 DB "MALAYALAM"         ;Define STR1
       CHECK DB 0                  ;Define CHECK, assume not palindrome
  DATA ENDS

 EXTRA SEGMENT
       STR2 DB 9 DUP(?)            ;Define STR2 of size 9 bytes
 EXTRA ENDS
  CODE SEGMENT
       ASSUME CS:CODE, DS:DATA, ES:EXTRA
START: MOV AX, DATA
       MOV DS, AX                  ;Initialize DS
       MOV AX, EXTRA
       MOV ES, AX                  ;Initialize ES
       LEA SI, STR1                ;Make SI point to STR1
       LEA DI, STR2+8              ;Make DI point to end of STR2
       MOV CX, 9                   ;Set CX to size of STR1
   L1: CLD                         ;Clear direction flag
       LODSB                       ;Get next byte of STR1 in AL,
                                    increment SI
       STD                         ;Set direction flag
       STOSB                       ;Store AL in STR2, decrement DI
       LOOP L1                     ;Continue transfers
                                   ;Now STR2 has reverse of STR1
       LEA SI, STR1                ;Make SI point to STR1
       LEA DI, STR2                ;Make DI point to STR2
```

```
        MOV CX, 9                   ;Load CX with length of string
        CLD                         ;Clear direction flag
        REPZ CMPSB                  ;Repeat as long as bytes match
        JNZ L2                      ;If not matched, not palindrome
        MOV CHECK, 1                ;Yes, palindrome
   L2:
  CODE ENDS
   END START
```

EXAMPLE 7.4

In the following, we will look into a program to search for an element NUM in an array of five numbers AR_NUM. If NUM is present in AR_NUM, the location FOUND will be set to 1, else it will be set to 0. The DATA and CODE segments have been defined accordingly. The program has been listed in the following.

```
    DATA SEGMENT
         AR_NUM DB 99H, 45H,         ;Define array AR_NUM
         0ABH, 56H, 34H
         NUM DB 45H                  ;Define number to search NUM
         FOUND DB 0                  ;FOUND reset to 0
    DATA ENDS
    CODE SEGMENT
         ASSUME CS:CODE, DS:DATA
  START: MOV AX, DATA
         MOV DS, AX                  ;Initialize DS
         MOV AL, NUM                 ;Get number to search in AL
         LEA SI, AR_NUM              ;Load address of array in AR_NUM
         MOV CX, 04H                 ;Load CX with array size less 1
  LOOP1: MOV BL, [SI]                ;Get first number in BL
         CMP AL, BL                  ;Compare
         JZ PRESENT                  ;Found? Goto set FOUND
         INC SI                      ;Point to next array index
         DEC CX                      ;Decrement CX count
         JNZ LOOP1                   ;Continue if all numbers not seen
         JMP END1                    ;Skip over setting FOUND
PRESENT: MOV FOUND, 01H              ;Set FOUND to 1
   END1:
    CODE ENDS
     END START
```

EXAMPLE 7.5

The next program sorts a set of five numbers stored in the array starting at NUMBERS, in ascending order. The program is organized into two segments – the segment DATA holds the program variables while the segment CODE holds the program code. The sorting program works by implementing two nested loops comparing the elements of the array. If two successive elements in the array are out of order, an exchange between the elements is performed to rectify it. The CH and CL registers have been used as the counters for the two loops. The program follows.

```
  DATA SEGMENT
       NUMBERS DB 99H, 45H,      ;Define NUMBERS
       0ABH, 56H, 34H
  DATA ENDS
  CODE SEGMENT
       ASSUME CS:CODE, DS:DATA
START: MOV AX, DATA
       MOV DS, AX                ;Initialize DS
       MOV CH, 04H               ;Use CH as outer loop counter
LOOP2: MOV CL, 04H               ;Use CL as inner loop counter
       LEA SI, NUMBERS           ;Make SI point to NUMBERS
LOOP1: MOV AL, [SI]              ;Get first number in AL
       MOV BL, [SI+1]            ;Get next number in BL
       CMP AL, BL                ;Compare
       JC DOWN                   ;Exchange if out of order
       MOV DL, [SI+1]
       XCHG [SI], DL
       MOV [SI+1], DL
 DOWN: INC SI                    ;Increment loop pointer
       DEC CL                    ;Decrement count for inner loop
       JNZ LOOP1                 ;Check with next number
       DEC CH                    ;Decrement count for outer loop
       JNZ LOOP2                 ;Continue with next number
  CODE ENDS
   END START
```

7.3 Interfacing

In this section, we shall look into the interfacing of commonly used I/O devices with the 8086. In particular, the devices, such as keyboard, display, ADC, DAC etc., have been taken up for connection with 8086. The principles of operation of these devices have been already discussed in Chapter 5 while enumerating the interfaces with 8085. However, the 8086 has a better structure and programming features, so it may be possible to have better design alternatives. Some interfacings, like 8255 have been skipped to avoid redundancy. However, it should be noted that instead of I/O ports directly addressed by the 8086, the 8255-based interfacings can also be used. We shall start by discussing the matrix keyboard interfacing.

7.3.1 Keyboard Interfacing

LO 2

The keyboard interfacing circuit has been shown in Figure 7.1. Compared to the interfacing discussed in Chapter 5 for 8085, in this case, both row and column values have been read by the input port. For this purpose, diodes have been introduced at the output port bits, so that a row may be read as low (0), only if the corresponding output bit is 0, turning on the diode. For example, the key ‘C’ gets detected if row-0 of the output port is 0. This is reflected by the bit-4 of the input port being 0. With row-0 as 0, pressing key ‘C’ will also make bit-3 of the input port 0. Hence, the code for key ‘C’ is 11100111 (in binary) = E7H. In a similar fashion, pressing of key ‘3’ is detected by the input pattern 7EH, ‘7’ by BEH etc. Accordingly, a data structure KEY_CODE has been used in the program to store the codes corresponding to the keys.

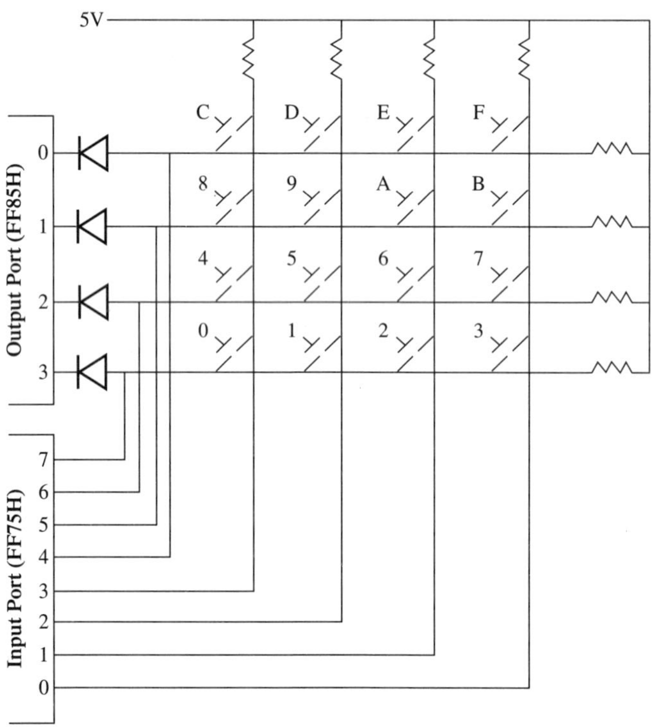

FIGURE 7.1 Interfacing a 4 × 4 hex keyboard

```
DATA SEGMENT
    KEY_CODE DB 77H, 7BH, 7DH, 7EH, ;Codes for '0', '1', '2', '3',
           0B7H, 0BBH, 0BDH,      ;      '4', '5', '6',
           0BEH, 0D7H, 0DBH,      ;      '7', '8', '9',
           0DDH, 0DEH, 0E7H,      ;      'A', 'B', 'C'
           0EBH, 0EDH, 0EEH       ;      'D', 'E', 'F'
DATA ENDS
CODE SEGMENT
    ASSUME CS:CODE, DS:DATA
START: MOV AX, DATA
    MOV DS, AX                    ;Initialize DS
    ; Send 0's to all rows
    MOV AL, 00H
    MOV DX, 0FF85H                ;Load output port address
    OUT DX, AL                    ;Send 0's
    ; Read columns to see if all keys open
    MOV DX, 0FF75H                ;Load input port address
 L1: IN AL, DX
    AND AL, 0FH                   ;Mask row bits
    CMP AL, 0FH                   ;Wait until all keys open
    JNE L1                        ;
    ; Read columns to see if a key pressed
 L2: IN AL, DX                    ;Read columns
    AND AL, 0FH                   ;Mask row bits
    JE L2
    ; Debounce key press
    MOV CX, 16FFH                 ;Delay for ~20 ms
DELAY: LOOP DELAY
    ; Read columns to check if key still pressed
    IN AL, DX
    AND AL, 0FH
    CMP AL, 0FH
    JE L2                         ;Key not pressed, go back to
                                   wait loop
```

```
          ; Locate the pressed key
          MOV AL, 0FEH              ;Initialize a row mask with
                                    ; bit 0 low
          MOV CL, AL                ;Save current mask
      L3: MOV DX, 0FF85H            ;Send a low on a row
          OUT DX, AL
          MOV DX, 0FF75H            ;Read columns and check for a
                                    ; low
          IN AL, DX
          AND AL, 0FH               ;Mask out row code
          CMP AL, 0FH               ;If low in a column then
          JNE GET_CODE              ;key column found, encode it
          ROL CL, 01                ;else rotate mask
          MOV AL, CL
          JMP L3
          ; Get code of the key pressed
GET_CODE: MOV BX, 000FH             ;Set up BX as a counter
          IN AL, DX                 ;Read row and column from
                                    ; port
      G1: CMP AL, KEY_CODE[ BX ]    ;Compare code with table
                                    ; entry
          JE DONE                   ;Hexcode available in BX
          DEC BX                    ;Point at next table entry
          JNS G1
          MOV AH, 01                ;Put an error code 1 in AH
          JMP G2
    DONE: MOV AL, BL                ;Get hex code in AL
          MOV AH, 00                ;Put a valid code 0 in AH
          ; AH, AL pair now got the correct hex code
      G2:
   CODE ENDS
    END START
```

7.3.2 Multiplexed Seven-Segment Display

LO 3

In Chapter 5 we have seen how to interface seven-segment LED displays with 8085. However, the approach often suffers from limitations in the driving currents provided by the microprocessor ports. To overcome this, the chip 7447 can be used to drive the seven-segment modules. A display driver

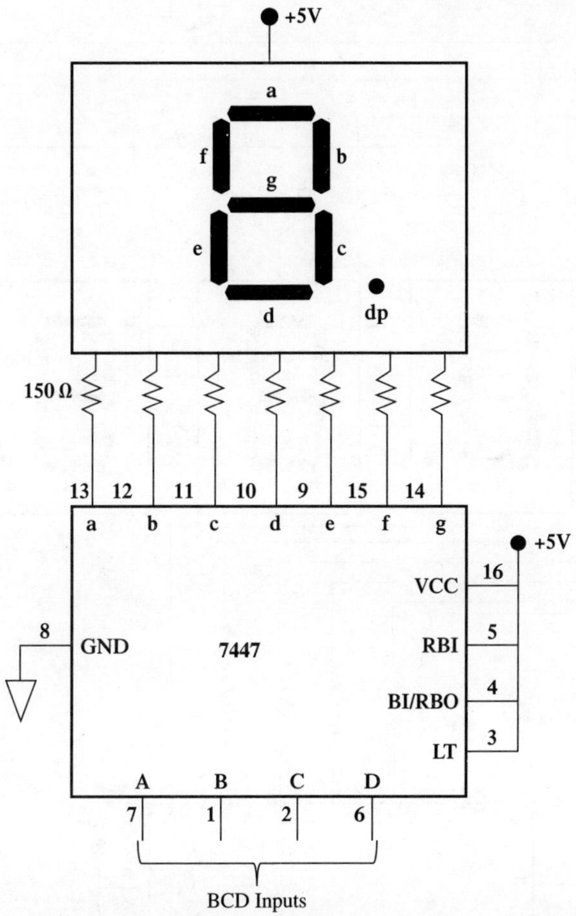

FIGURE 7.2 Interface using display driver 7447

chip IC7447 can be utilized to simplify the interface. Figure 7.2 shows the connection between a seven-segment display module and 7447. The driver IC7447 is a BCD to seven-segment decoder. It converts a 4-bit BCD input given over the input pins A, B, C and D to the seven LED segment inputs a, b, c, d, e, f and g. The other pins RBI, BI/RBO and LT have the following functions.

- BI must be high for output to function for 0 to 15.
- RBI must be high to ensure that a decimal 0 does not turn the display blank.
- BI being low turns off all LED segments.
- If BI is high and LT (Lamp Test) is low, all seven segments will be turned on. It can be used to check if all the segments are okay or not.
- RBO is typically high. If A, B, C, D and RBI are all low and LT is high, all LED segments will be off and RBO will be low.

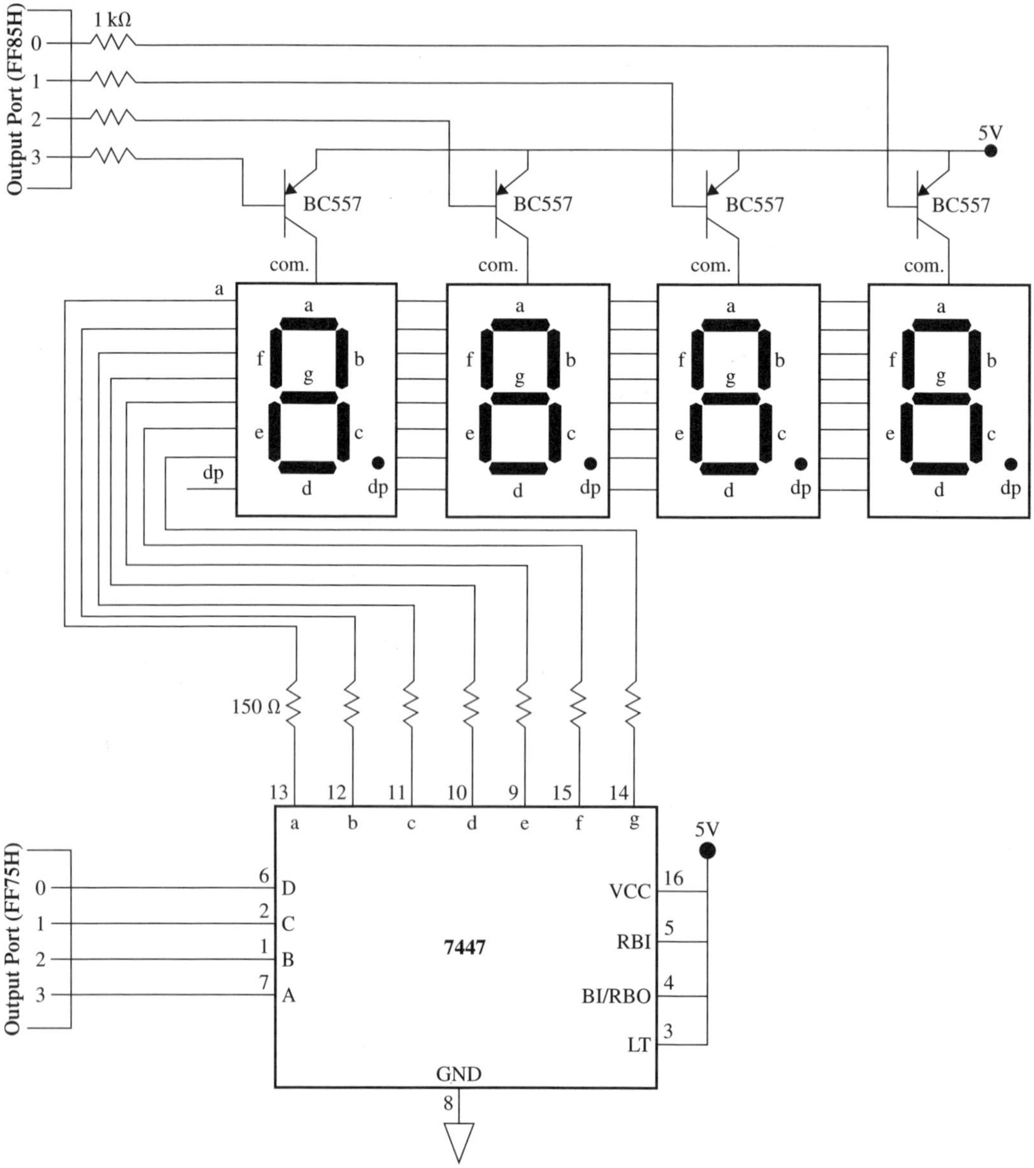

FIGURE 7.3 Four multiplexed seven-segment displays

Figure 7.3 shows the interfacing of four seven-segment display modules with 7447. The 7447 is driven by an output port (FF75H) to feed the BCD inputs. A single 7447 chip is used in a multiplexed fashion with four display modules. Individual display modules are selected via the output port FF85H. Transistors BC557 have been used to turn on individual display modules through the output port FF85H. An assembly language program to display the pattern "8086" has been detailed next.

```
   DATA SEGMENT
        PATTERN DB "8086"           ; Pattern to display
   DATA ENDS
        PORT1 EQU 0FF75H            ; Port 1 address
        PORT2 EQU 0FF85H            ; Port 2 address
   CODE SEGMENT
        ASSUME CS: CODE, DS: DATA
 START: MOV AX, DATA
        MOV DS, AX
    L1: MOV SI, PATTERN             ; Get PATTERN index into SI
        MOV AL, [SI]                ; Get first digit '8' into AL
        SUB AL, '0'                 ; Get the binary value
        MOV DX, PORT1               ; Get PORT1 address in DX
        OUT DX, AL                  ; Output pattern
        MOV AL, 0F7H                ; Select module
        MOV DX, PORT2               ; Get PORT2 address in DX
        OUT DX, AL                  ; Output module
        CALL DELAY_RTN              ; Put a small delay
                                    ; Display second digit
        INC SI
        MOV AL, [SI]                ; Get second digit '0' into AL
        SUB AL, '0'                 ; Get the binary value
        MOV DX, PORT1               ; Get PORT1 address in DX
        OUT DX, AL                  ; Output pattern
        MOV AL, 0FBH                ; Select module
        MOV DX, PORT2               ; Get PORT2 address in DX
        OUT DX, AL                  ; Output module
        CALL DELAY_RTN              ; Put a small delay
                                    ; Display third digit
        INC SI
        MOV AL, [SI]                ; Get third digit '8' into AL
        SUB AL, '0'                 ; Get the binary value
        MOV DX, PORT1               ; Get PORT1 address in DX
        OUT DX, AL                  ; Output pattern
        MOV AL, 0FDH                ; Select module
        MOV DX, PORT2               ; Get PORT2 address in DX
```

```
            OUT DX, AL                ; Output module
            CALL DELAY_RTN            ; Put a small delay
                                      ; Display fourth digit
            INC SI
            MOV AL, [SI]              ; Get fourth digit '6' into AL
            SUB AL, '0'               ; Get the binary value
            MOV DX, PORT1             ; Get PORT1 address in DX
            OUT DX, AL                ; Output pattern
            MOV AL, 0FEH              ; Select module
            MOV DX, PORT2             ; Get PORT2 address in DX
            OUT DX, AL                ; Output module
            CALL DELAY_RTN            ; Put a small delay
            JMP L1                    ; Repeat display procedure

DELAY_RTN PROC NEAR
            MOV CX, 16FFH             ; Delay for ~20 ms
     DELAY: LOOP DELAY
            RET
DELAY_RTN ENDP
      CODE ENDS
       END START
```

7.3.3 Analog-to-Digital Converter

LO 4

Figure 7.4 shows an interfacing of analog-to-digital converter ADC0808/0809 with the 8086. The output port with address FF75H is used to drive the inputs A, B, C, ALE and SC to the ADC. The input port FF85H is taken as a 16-bit port to get the digital inputs 0 through 7 and the end-of-conversion bit EOC. The assembly language program to have analog input via channel IN1 and convert to digital value has been noted next. The digital value gets stored at memory location DIG_VALUE.

```
DATA SEGMENT
     DIG_VALUE DB              ; Location to keep digital value
                                 read
DATA ENDS
     PORT1 EQU 0FF75H          ; Port 1 address
     PORT2 EQU 0FF85H          ; Port 2 address

CODE SEGMENT
     ASSUME CS: CODE, DS: DATA
```

```
START: MOV AX, DATA
       MOV DS, AX
       MOV AL, 01H                ; Select channel IN1
       MOV DX, PORT1              ; Get port address in DX
       OUT DX, AL                 ; Output channel selection
       MOV AL, 09H                ; Make ALE, SC, A high
       OUT DX, AL                 ; Output to port
       MOV AL, 01H                ; Make ALE, SC low
       MOV DX, AL                 ; Output to port
                                  ; Check if conversion over
       MOV DX, PORT2              ; Read port
   L1: IN DX, AX                  ; Get 16-bit pattern
       RCL AX, 1                  ; Get MSB (EOC) in Carry
       JNC L1                     ; If EOC≠1, read EOC again
       IN DX, AX                  ; Read digitally converted value
       MOV DIG_VALUE, AL          ; Save digital value
  CODE ENDS
   END START
```

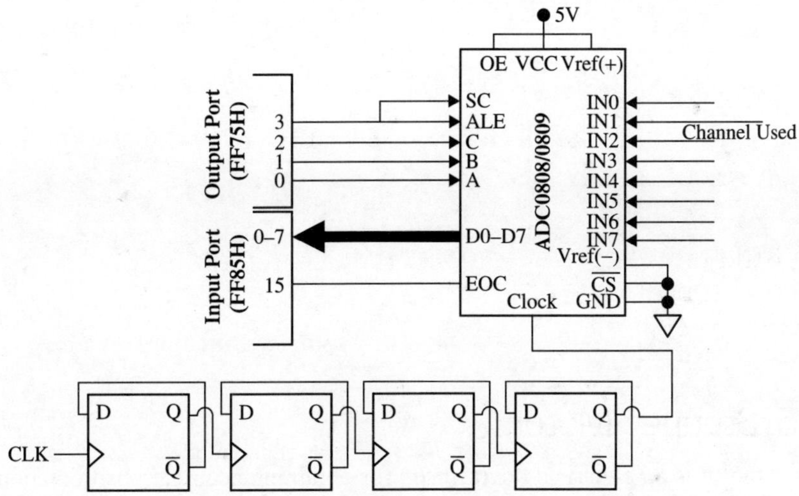

FIGURE 7.4 Interfacing ADC0808/0809 with the 8086

7.3.4 Digital-to-Analog Converter

LO 5

To interface a Digital-to-Analog Converter (DAC) with the 8086, the DAC has to be connected to an 8-bit output port, say with address FF75H. A program code to generate a triangular wave has been noted next. For this, the values 0–255 are output to port FF75H followed by decrementing up to 0. The digital values, when converted by the DAC will provide an analog output with a triangular waveform.

```
          PORT1 EQU 0FF75H        ; Port 1 address
     CODE SEGMENT
          ASSUME CS: CODE
   START: MOV AL, 0
          MOV DX, PORT1
      L1: OUT DX, AL              ; Output first pattern
          CALL DELAY_RTN          ; Call delay routine
          CMP AL, 0FFH            ; Check if maximum value reached
          JZ L2                   ; Maximum reached, start down slope
          INC AL                  ; Increment AL
          JMP L1                  ; Output next pattern
      L2: DEC AL                  ; Decrement AL
          OUT DX, AL              ; Output pattern
          CALL DELAY_RTN          ; Put a small delay
          CMP AL, 01H             ; Check if AL has reached 1
          JNZ L2                  ; If not repeat
          MOV AL, 0               ; Else clear AL
          JMP L1                  ; Repeat from beginning

DELAY_RTN PROC NEAR
          MOV CX, 16FFH           ; Delay for ~20 ms
   DELAY: LOOP DELAY
          RET
DELAY_RTN ENDP
     CODE ENDS
      END START
```

7.3.5 Stepper Motor

LO 6

A stepper motor is an electrical device used for obtaining accurate position control in applications like disk drives, dot matrix printers, robotic movements etc. to name a few. It translates electrical pulses into mechanical movements. The stepper motor consists of a permanent magnet *rotor*, surrounded by a *stator*. Most common stepper motors have four stator windings and are known as *four-phase unipolar* stepper motors. Also, the rotor has a number of teeth. The structure has been shown in Figure 7.5(a). To rotate the shaft, a sequence of pulses are applied to the stator windings. The teeth of the stator and rotor lock with each other to fix the position of the shaft. When a pulse is applied to the winding input, the rotor rotates by one teeth position, thus rotating by an angle, say $x°$. If there are N number of rotor

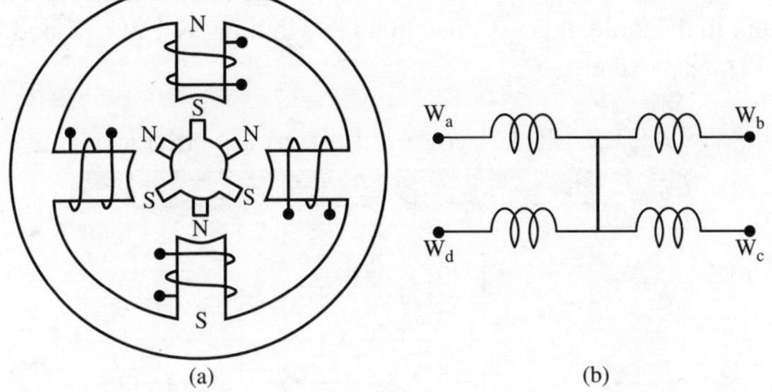

FIGURE 7.5 Stepper motor: (a) internal schematic and (b) winding arrangement

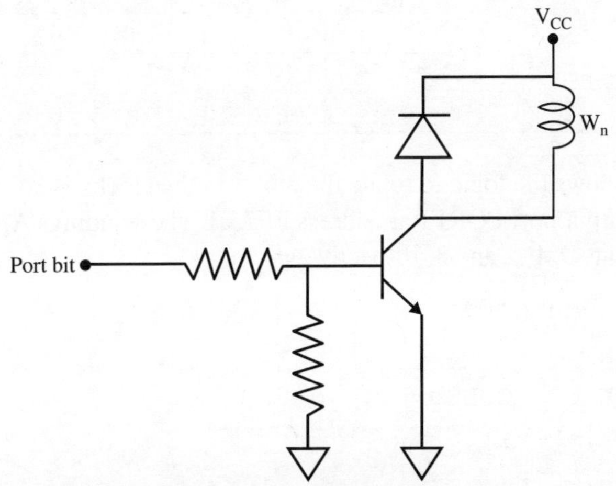

FIGURE 7.6 Interfacing winding W_n with a port bit

teeth, the angle x is given by, $x = 360°/N$. The value of N also determines the number of steps per revolution. Typical parameters of a stepper motor are as follows.

```
Operating Voltage      : +12 V
Current rating         : 0.2 A
Step angle             : 1.8°
Steps per revolution   : 200
Torque                 : 3 kg-cm
```

Figure 7.5(b) shows the winding arrangements for the stator. Binary pulses are applied to the winding inputs to make the shaft rotate. The circuit to apply the pulses through the port bits of 8086 and feeding the windings has been shown in Figure 7.6. For rotating the shaft in a clockwise/anticlockwise direction, the windings are to be excited in a particular sequence. Such a sequence/scheme is also known as *wave scheme*. For example, Table 7.1 shows two such schemes – one for clockwise rotation and the other for anticlockwise rotation. In the table, it is assumed that the windings W_a, W_b, W_c and W_d are driven by the port bits A, B, C and D, respectively.

TABLE 7.1 Wave schemes for rotation

Direction	Step	A	B	C	D
Clockwise	1	1	0	0	0
	2	0	1	0	0
	3	0	0	1	0
	4	0	0	0	1
	5	1	0	0	0
Anticlockwise	1	1	0	0	0
	2	0	0	0	1
	3	0	0	1	0
	4	0	1	0	0
	5	1	0	0	0

The following program shows the logic to rotate the motor in the clockwise direction. It is assumed that the motor is driven by output port PORT1 at address FF75H. The windings A, B, C and D are assumed to be driven by the port bits 0, 1, 2 and 3, respectively.

```
         PORT1 EQU 0FF75H         ; Port 1 address
    CODE SEGMENT
         ASSUME CS: CODE
  START: MOV DX, PORT1
     L1: MOV AL, 01H
         OUT DX, AL               ; Energize winding A
         CALL DELAY_RTN           ; Call delay routine
         MOV AL, 02H
         OUT DX, AL               ; Energize winding B
         CALL DELAY_RTN           ; Call delay routine
         MOV AL, 04H
         OUT DX, AL               ; Energize winding C
         CALL DELAY_RTN           ; Call delay routine
         MOV AL, 08H
         CALL DELAY_RTN           ; Call delay routine
         JMP L1                   ; Repeat from beginning
```

```
DELAY_RTN PROC NEAR
          MOV CX, 16FFH          ; Delay for ~20 ms
   DELAY: LOOP DELAY
          RET
DELAY_RTN ENDP
     CODE ENDS
      END START
```

SUMMARY

This chapter has looked into the programming and peripheral device interfacing issues of 8086. Compared to the assembly language programs of 8085, in the case of 8086, the programmers have the flexibility to organize their programs in modules. A module may correspond to a code fragment or data. Each module may be contained into separate segments. The assembler directives for 8086 programs include few additional ones, compared to 8085 assemblers. In the interfacing part, several device interfaces have been revisited. It has started with the interfacing of the matrix keyboard. For the display part, the BCD segment driver chip 7447 has been utilized to connect to multiplexed seven-segment LEDs. The 8086 program provides the BCD code of the digit to be displayed and 7447 generates the corresponding LED segment drive bits. The interfacing of ADC and DAC has been introduced. Finally, the interfacing with the stepper motor has been discussed. Thus, 8086 can be utilized to develop complex programs and interface devices with varying functionalities.

EXERCISES

Multiple Choice Questions

1. SEGMENT directive marks beginning of a
   (a) Logical segment
   (b) Physical segment
   (c) Physical memory
   (d) None of the above
2. End of a segment is marked by
   (a) END
   (b) END SEGMENT
   (c) ENDS
   (d) None of the above
3. The CS register is loaded by
   (a) Program
   (b) Operating system
   (c) User
   (d) None of the above
4. DD directive stands for
   (a) Define Double Word
   (b) Define Double Data
   (c) Define Double Domain
   (d) None of the above

5. Input to 7447 is
   (a) BCD
   (b) Binary
   (c) Hexadecimal
   (d) Octal
6. Output from 7447 is
   (a) Segment inputs
   (b) Segment outputs
   (c) Binary numbers
   (d) None of the above
7. For a decimal 0 to be displayed on seven-segment, the RBI of 7447 should be
   (a) High
   (b) Low
   (c) Tristated
   (d) None of the above
8. For lamp testing of LED display through 7447
   (a) BI = 0, LT = 0
   (b) BI = 0, LT = 1
   (c) BI = 1, LT = 0
   (d) BI = 1, LT = 1
9. In stepper motor, the permanent magnet part is
   (a) Stator
   (b) Rotor
   (c) Mixer
   (d) None of the above
10. In stepper motor, teeth are made on
    (a) Stator
    (b) Rotor
    (c) Both stator and rotor
    (d) None of the above

Answers to Multiple Choice Questions

1. a	2. c	3. b	4. a	5. a	6. a	7. c	8. b	9. b	10. b

Long Answer Type Questions

1. Write an assembly language program to square each number stored in an array ARR of size 10. Assume each result to be confined to one byte.
2. Write an assembly language program to identify the decimal digits of a number stored at NUMBER. The result be stored in an array starting at ARR.
3. Write an assembly language program to reverse a string STR1. The result be stored at STR2.
4. Write an assembly language program to search for character CH in a string STR.
5. Show the hardware and software to interface a keyboard with 8086.
6. Show the interface of four seven-segment display modules through 7447 for 8086.
7. Show the hardware and software to interface an ADC with 8086.
8. Develop a program code to generate a sinusoidal wave through DAC interfaced with 8086.
9. Show the interface of a stepper motor with 8086.
10. Develop an 8086 program code to rotate a stepper motor through 8086 in anticlockwise direction.

8

The 8051 Microcontroller Architecture

LEARNING OUTCOMES

After reading this chapter, students will be able to:

LO 1 Enumerate the pin distribution of 8051

LO 2 Draw and state the internal structure of 8051

LO 3 State the strategies to apply clock and reset signals to 8051

LO 4 Enumerate the I/O port structure of 8051

LO 5 Interface external program and data memory chips with 8051

LO 6 Draw the timing diagrams for accessing external program and data memory modules

8.1 Introduction

Microcontrollers are single-chip microcomputers housing the CPU, RAM, ROM and I/O ports on the same chip. Apart from that, most of the microcontrollers possess timers/counters, ADCs and DACs for interacting with the outside world. One of the earliest popular microcontroller is Intel's 8051, belonging to the family MCS-51. It was introduced by the Intel Corporation, in the year 1981. The resources available with the 8051 chip have been enumerated in Table 8.1.

As it can be noted, it has 4 kB on-chip ROM which may be sufficient to hold the machine code for small-to-medium-sized applications of 8051. For larger programs, externally interfaced program memory of size up to 64 kB can be utilized. The on-chip data memory part is restricted to 128 bytes

TABLE 8.1 Features of 8051 processor

CPU	8 bit
On-chip program memory (ROM)	4 kB
On-chip data memory (RAM)	128 bytes
Timer/Counter	2, 16-bit
Digital I/O	32 pins
Serial Port	1
Interrupts	6
Program memory address space	64 kB
Data memory address space	64 kB

that also contains the CPU registers in it. This memory, also called *scratch pad*, contains the CPU registers as well, making time to access a memory location the same as that for a register. However, this memory being limited in size, can be augmented by connecting external data memory of size up to 64 kB.

The MCS-51 family has two more members in it, introduced by Intel. The first one, named 8031, possesses features similar to the 8051 except that it does not contain any on-chip data memory. Thus it will always need external memory to be connected to hold additional data items. The second one, named 8052 is an advanced version of 8051 with several augmented features. In particular, 8052 has 8 kB on-chip ROM, 256-bytes RAM, 3 timers and 8 interrupt sources. It may be noted that several variants of the 8051 microcontrollers are available in the market, fabricated by different manufacturers. Each of these products comes with different part numbers. The variation comes primarily due to variations in the memory technology used for the ROM part. Some such variants have been enumerated in the following.

- *8751*: The 8751 chip realizes the on-chip ROM using UV-EPROM technology. A PROM burner can be used to program the memory. However, for reprogramming, the previous content needs to be erased using a UV-EPROM eraser. The erasing process may take time up to 20 minutes.
- *AT89C51*: This is a flash-memory-based version. To program the on-chip ROM, a ROM burner is needed. However, eraser is not required, saving the reprogramming time.
- *DS89C4x0*: This is a family of microcontrollers using flash ROM. The value of *x* in the name varies from 2 to 5. The advantage of the device is the availability of on-chip loader that can be interfaced with a PC for uploading the program memory through serial communication ports.
- *OTP versions*: To come up with cheaper variants of 8051, *one-time-programmable* (OTP) versions of 8051 are available. Due to lower cost, these devices are often used in the mass production of systems. More costly flash versions of 8051 may be utilized to develop and debug the program logic. However, once the program logic has been debugged and finalized, mass production may utilize the OTP variants.

On-chip program memory can be programmed only once.

8.2 The 8051-Pin Distribution

LO 1

The 8051 chip follows a 40-pin dual in-line package (DIP) organization. The pin layout of the chip has been shown in Figure 8.1. The pins can be grouped into several clusters based on their functionalities. Figure 8.2 shows the categorization of pins into groups. Next, we shall look into the functionalities of each pin, in the following.

- **V_{CC}** (Pin 40) and **V_{SS}** (Pin 20): These two pins constitute the power supply for the 8051 chip. As per the datasheet, the pin VCC should have a supply of +5 V with a current rating of 125 mA and a maximum power dissipation of 1 W. VSS is the ground pin.
- **XTAL2** (Pin 18) and **XTAL1/CLKIN** (Pin 19): It is recommended to connect a crystal of 12 MHz frequency between the pins 18 and 19 along with two 30 pF disc capacitors, as shown in Figure 8.2. In the absence of a crystal, an external clock source can be directly connected to XTAL1 pin. The XTAL2 pin is left unconnected, in that case.
- **RST** (Pin 9): This is the pin useful to reset the operation of 8051. The pin is active high. The chip needs to be reset on power-on. For this, a power-on reset circuitry, as shown in Figure 8.2, is utilized. The reset circuitry consists of a 10-μF capacitor and an 8.2-kΩ resistor. Further, the push-button switch connected in parallel to the capacitor acts as the asynchronous reset input. To reset the 8051 at any time after powering on, the switch must be closed for at least two machine cycles. This generates a proper reset pulse for the processor.
- **$\overline{EA}$** (Pin 31): The 8051 processor can access program memory – internal or external. The setting of the EA pin decides the *external access* by the 8051. If the pin is set to high, the internal ROM is used to get the instructions. The pin is set to low to force the 8051 to access the external memory.

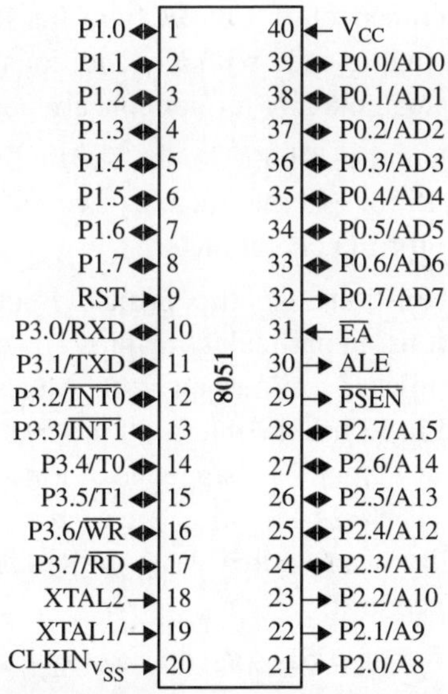

FIGURE 8.1 8051 pin diagram

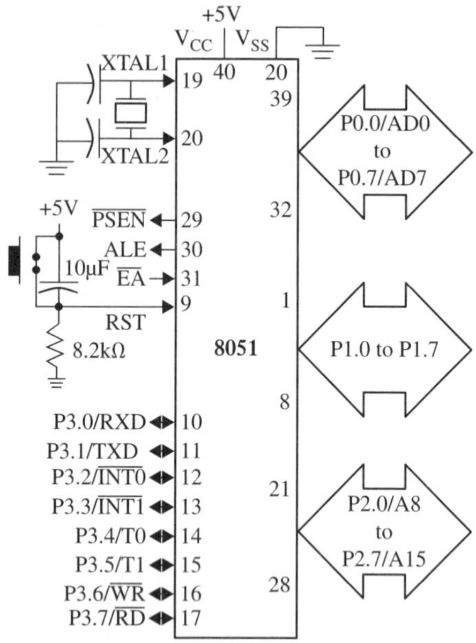

FIGURE 8.2 Grouping of 8051 pins

- **Port 0** (Pins 32–39): Port 0 is an 8-bit I/O port. As shown in the figure, the pins also provide the address-data bus lines AD7–AD0 in a multiplexed fashion. It may be noted that the port bit lines are open-drain. This necessitates external pull-up resistors to be connected for their proper operation, controlling the current flow.
- **Port 1** (Pins 1–8): Port 1 is another set of 8-bit I/O port lines. It may be noted that Port 1 is the only port whose lines are not multiplexed with any additional functions. Thus, the lines can be used to connect to I/O devices without any further concern about system functionality.
- **Port 2** (Pins 21–28): Port 2 is another 8-bit I/O port. The bit lines are also used to provide higher order address bus lines A8–A15 for external memory access. Thus, if external memory is interfaced, the port is not available for I/O operations.
- **Port 3** (Pins 10–17): Port 3 constitutes another 8-bit I/O port. However, the port bits are also assigned some alternate functions. In particular, the pins 10 and 11 are used for serial I/O operations. Pins 12 and 13 are multiplexed as interrupt lines. An active low signal on pin 12 provides an interruption on line $\overline{\text{INT0}}$. Pin 13 has the interrupt line $\overline{\text{INT1}}$ multiplexed onto it. On the other hand, pin 14 has the timer control input T0 assigned to it. Pin 15 has the additional role of providing control input for timer T1. The pins 16 and 17 are used to provide read/write control. Pins 16 has the additional responsibility of providing write ($\overline{\text{WR}}$) control for external memory access. Pin 17 provides the read ($\overline{\text{RD}}$) control.
- **$\overline{\text{PSEN}}$** (Pin 29): This output line of 8051 is used as a strobe for accessing the external program memory.

- **ALE** (Pin 30): Similar to 8085 and 8086 processors, the ALE pin in 8051 is used to demultiplex the lines AD7–AD0 into the lower order address bus A0–A7 and data bus D0–D7.

8.3 Internal Block Diagram

LO 2

Figure 8.3 shows the internal organization of the 8051 processor at the block diagram level. A detailed discussion of the individual blocks has been noted in the following subsections. Since all these modules (such as on-chip memories, timers and I/O ports) are on the same chip as the processor, communication to/from the processor is very fast.

8.3.1 CPU

Similar to other processors, the 8051 CPU consists of an *Arithmetic Logic Unit* (*ALU*), *Instruction Decoder and Control* and a number of *registers*. The ALU can handle 8-bit data and has the capability to perform several arithmetic and logic operations. One dedicated register, named *Program Counter* (*PC*) points to the memory address from where the next instruction is going to be fetched. From memory, an instruction is first brought into a register called the *Instruction Register*. On receiving the instruction, it is decoded by the *Instruction Decoder* to generate a relevant sequence of control signals needed to execute the instruction properly. The programmer does not have much control over these two registers. To facilitate programming, a set of general- and special-purpose registers are included in 8051, similar to other processors. However, all these registers are mapped onto RAM locations. Hence, discussion on them has been taken up in conjunction with RAM.

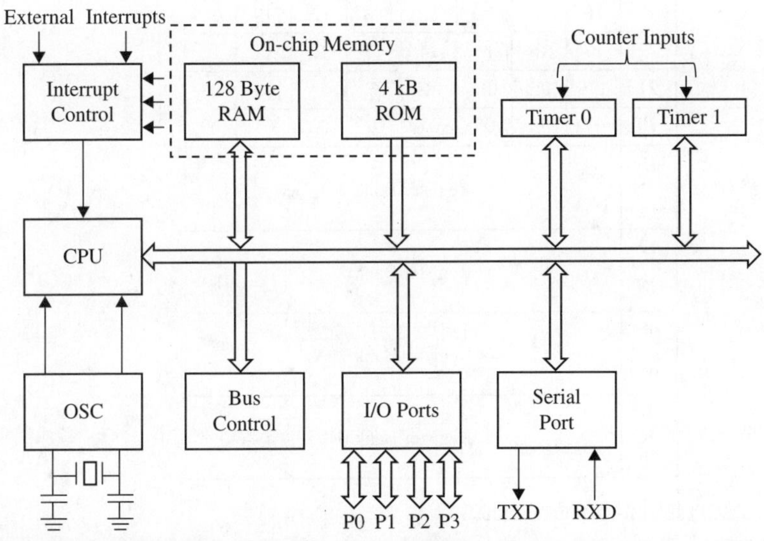

FIGURE 8.3 Internal block diagram of the 8051

8.3.2 Memory

The on-chip program memory, available in 8051, is of size 4 kB. The memory is implemented using EPROM/flash technology. Apart from that, it has 128-byte data memory (RAM) spanning over the address range 00H to 7FH. The internal RAM structure has been shown in Figure 8.4. There are RAM locations beyond 7FH (up to FFH), however, these addresses correspond to some special function

Address									
FF–80	Special Function Registers								
7F–30	General-Purpose RAM								
2F	7F	7E	7D	7C	7B	7A	79	78	Bit Addressable RAM
2E	77	76	75	74	73	72	71	70	
2D	6F	6E	6D	6C	6B	6A	69	68	
2C	67	66	65	64	63	62	61	60	
2B	5F	5E	5D	5C	5B	5A	59	58	
2A	57	56	55	54	53	52	51	50	
29	4F	4E	4D	4C	4B	4A	49	48	
28	47	46	45	44	43	42	41	40	
27	3F	3E	3D	3C	3B	3A	39	38	
26	37	36	35	34	33	32	31	30	
25	2F	2E	2D	2C	2B	2A	29	28	
24	27	26	25	24	23	22	21	20	
23	1F	1E	1D	1C	1B	1A	19	18	
22	17	16	15	14	13	12	11	10	
21	0F	0E	0D	0C	0B	0A	09	08	
20	07	06	05	04	03	02	01	00	
1F–18	Register Bank 3								
17–10	Register Bank 2								
0F–08	Register Bank 1								
07–00	Register Bank 0								

FIGURE 8.4 Internal RAM locations

registers. Only a few of the address locations between 80H and 0FFH are used for this purpose, while the rest of the locations are not usable. The components in the internal RAM are as follows.

1. **Register Banks 0–3:** This part spans over the memory addresses 00H to 1FH. These 32 bytes of RAM are grouped into 4 banks of 8 locations each, namely Bank-0, Bank-1, Bank-2 and Bank-3. The registers in a bank are named as R0, R1, R2, R3, R4, R5, R6 and R7. The instruction under execution can use only one bank of registers at a time. Thus, among the four banks, only one is active at a time. The Processor Status Word (PSW) register has got two dedicated bits to control the bank activation. Since these registers are memory locations only, individual instructions can refer to a byte by the name of a register or its corresponding memory address. For example, the RAM address 05H and register R5 in Bank-0 refer to the same memory location 05H.
2. **Bit-Addressable RAM:** The 16-byte memory locations 20H to 2FH are marked as bit–addressable RAM. The individual bits located within these 16 bytes can be accessed by the bit-manipulating instructions. These instructions can set/clear a bit, complement it or check the value. Apart from these RAM locations, some of the *Special Function Registers* (*SFR*s) are also bit-addressable. Details of these registers have been presented later.
3. **General-Purpose RAM:** The RAM locations 30H to 7FH constitute the 80 bytes of general byte-addressable RAM. The locations can be used by the individual programs to hold the program variables. However, a precaution is necessary for programs utilizing procedure calls and interrupts, as quite often, the stack is also maintained in this region. After accommodating the stack, the space available for general-purpose storage may get constrained further.
4. **Special Function Registers (SFRs):** These are registers with some special purposes. The registers occupy some of the addresses in the range of 80H to 0FFH. Though there is a total of 128 locations available in this range, the SFRs correspond to only 21 of them, the remaining addresses being undefined. Results will be unpredictable if a user program attempts to access those addresses. The special function registers are noted next.
   - **Accumulator:** This register with RAM address 0E0H is used in all the arithmetic and logic instructions to hold one of the source operands and also the destination operand. The register is also referred to as *A* register in the instructions. The accumulator register is 8-bit wide and is bit-addressable in nature.
   - **B Register:** This is another general-purpose 8-bit register with address 0F0H. It is also bit-addressable in nature. The register is primarily used as an operand in arithmetic and logic instructions and also as the destination of multiplication and division operations.
   - **Processor Status Word:** Similar to other processors, this register (nicknamed as PSW) holds various status information about the CPU. The memory address is 0D0H. This is an 8-bit register that contains 4 bits corresponding to the outcome of the last ALU operation. Two bits control the register bank to be used by the instructions. One bit is available to the user for programming, while one more bit is kept undefined. Table 8.2 shows the bit usage of the PSW register.

TABLE 8.2 Bits of PSW register

<table>
<tr><th>Bit</th><th>Functionality</th></tr>
<tr><td>0 (PSW.0)</td><td>Parity. Set if there are odd number of 1s in the accumulator, reset otherwise.</td></tr>
<tr><td>1 (PSW.1)</td><td>Kept undefined.</td></tr>
<tr><td>2 (PSW.2)</td><td>Overflow. Set if there is an overflow in signed arithmetic operations, that is, the result is beyond the range -128 to +127.</td></tr>
<tr><td>4 (PSW.4), 3 (PSW.3)</td><td>Also called RS1 (PSW.4) and RS0 (PSW.3), used to select the register bank.
<table>
<tr><th>RS1</th><th>RS0</th><th>Bank Selected</th><th>Address Range</th></tr>
<tr><td>1</td><td>1</td><td>3</td><td>18H-1FH</td></tr>
<tr><td>1</td><td>0</td><td>2</td><td>10H-17H</td></tr>
<tr><td>0</td><td>1</td><td>1</td><td>08H-0FH</td></tr>
<tr><td>0</td><td>0</td><td>0</td><td>00H-07H</td></tr>
</table></td></tr>
<tr><td>5 (PSW.5)</td><td>Named as Flag 0 (F0), the bit is available to the programmer.</td></tr>
<tr><td>6 (PSW.6)</td><td>Auxiliary Carry. Set if there is a carry from bit-3 to bit-4 in addition or a borrow from bit-4 to bit-3 in subtraction, as in 8085.</td></tr>
<tr><td>7 (PSW.7)</td><td>Carry. The bit is set if there is a carry/borrow in 8-bit addition/subtraction operation, clear otherwise.</td></tr>
</table>

- **Stack Pointer (81H):** The register with address 81H points to the top of the stack. When the 8051 is reset, the stack pointer contains 07H. Hence, the stack starts from this memory address. It may be noted that the stack is implemented in the RAM, overlapping with the register bank area. The stack pointer is not bit-addressable.
- **Data Pointer:** The data pointer is a 16-bit register at an address spanning over 82H and 83H. It consists of two 8-bit parts – *Data Pointer High* (*DPH*) at RAM address 83H and *Data Pointer Low* (*DPL*) at RAM address 82H. The DPH and the DPL registers can be accessed separately as well. DPTR is not bit-addressable.
- **Interrupt Related:** There are two interrupt-related registers in 8051 – *Interrupt Enable Control* (*IE*) and *Interrupt Priority Control* (*IP*). The two registers are located at RAM addresses 0A8H and 0B8H, respectively. Also, both the registers are bit-addressable. The interrupt enable register, IE is used for enabling/disabling certain interrupts. The IP register settings are used to define the relative priorities between the enabled interrupts.
- **Timer/Counter Related:** The following are the various registers related to timer/counter operations.
  - Timer Value Registers *TL0*, *TH0*, *TL1*, *TH1*: These 8-bit registers hold the current values for the timers/counters. For *timer-0*, the lower and the higher byte are stored in registers *TL0* (memory address 8AH) and *TH0* (8CH), respectively. For *timer-1*, the corresponding registers are *TL1* (8BH) and *TH1* (8DH).
  - Timer Control Register *TCON*: The *TCON* register, located at memory address 88H controls the operation of all timers/counters in 8051. The register is bit-addressable, in nature.
  - Mode Set Register *TMOD*: Located at memory address 89H, the register configures the modes of the timers/counters.

- **Serial I/O Related:** In 8051, the serial input-output is controlled by the following two registers.
    - Serial Control Register *SCON*: This is a bit-addressable register located at RAM address 98H controlling the serial I/O operations.
    - Serial Buffer Register SBUF: The *SBUF* register at memory address 99H holds the data received/transmitted serially through the RXD and TXD pins.
- **Power Control Register PCON:** The *PCON* register at memory location 87H is used to set the power control to be applied. The register is not bit-addressable.

8.3.3 Digital I/O Ports, Timers/Counters

The 8051 contains four 8-bit parallel ports – Port 0, Port 1, Port 2 and Port 3. Each bit of the ports can be programmed independently to act as input or output. Excepting Port 1, other ports have additional functions multiplexed onto them. Ports 0 and 2 provide the address and data lines for external memory access. Some of the bits of Port 3 are to be used as the read/write control signals. There are two pins TXD and RXD, available for serial mode digital communication. This facilitates the serial transmission of data.

Two on-chip timers are available in the 8051 microcontroller – *Timer-0* and *Timer-1*. The timers can be driven by external events occurring on pins *T0* and *T1* also, to act as counters. The timer/counter modules can be configured to function as 16-bit, 13-bit and 8-bit ones.

8.4 Clock Circuit

LO 3

There are two alternatives to feed clock signal to the 8051 chip – (i) using crystal and (ii) direct clock input.

(i) **Crystal to generate clock:** The 8051 allows a crystal of suitable frequency to be connected between the pins XTAL1 and XTAL2. It may be noted that different manufacturers support up to certain maximum clock frequencies for the crystal, hence the datasheet of the part actually used in a design should be consulted to know the frequency. Figure 8.5(a) shows the connection diagram for the circuit. Two 30 pF capacitors are also needed in it.

(ii) **Direct clock input:** In the absence of crystal, an alternate frequency source (such as a TTL controller) can be connected to XTAL1 with XTAL2 left unconnected, to provide a clock signal to the 8051. It has been shown in Figure 8.5(b).

Similar to other processors, the 8051 also needs a certain number of machine cycles to execute an instruction. The datasheet of the 8051 chip lists the number of machine cycles needed for each instruction. There are many different manufacturers of 8051 and these variants support different maximum clock frequencies. Further, the number of clock cycles per machine cycle and the number of machine cycles for an instruction, also vary across these implementations. A MOV instruction in the original 8051 chip needs 1 machine cycle with 12 clock cycles in it. On the other hand, for the chip implemented as DS89C4x0, the same MOV instruction needs 2 machine cycles with each machine cycle requiring just one clock cycle.

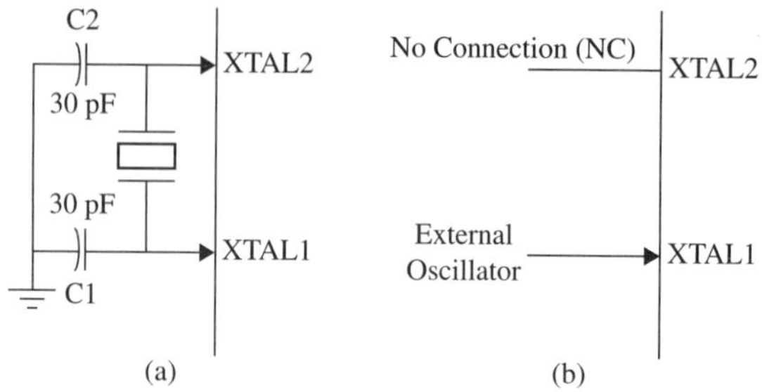

FIGURE 8.5 Clock circuit – (a) crystal and (b) direct

EXAMPLE 8.1

Consider that the crystal has a frequency of 11.0592 MHz. To illustrate the concept discussed, let us compute the time needed to execute (i) MOV (ii) MUL instructions, for 8051 and DS89C4x0 chips. With the stated crystal frequency, the duration of a clock cycle is $1/11.0592 = 0.0904$ μS. As in the original 8051 chip one machine cycle contains 12 clock cycles, the duration of a machine cycle for it is $12 \times 0.0904 = 1.0848$ μS. On the other hand, for DS89C4x0, the machine cycle duration is $1 \times 0.0904 = 0.0904$ μS, as each machine cycle contains only one clock cycle. From the datasheet, it can be found that the number of machine cycles for MOV in 8051 is 1 and for DS89C4x0, it is 2. Hence, the time needed to execute a MOV instruction in 8051 is 1.0848 μS. For DS89C4x0, the time needed is $2 \times 0.0904 = 0.1808$ μS. For the multiplication instruction MUL, four machine cycles are needed for 8051 and nine for DS89C4x0. Thus, the corresponding time requirements for a MUL instruction are $4 \times 1.0848 = 4.3392$ μS for 8051 and $9 \times 0.0904 = 0.8136$ μS for DS89C4x0.

8.5 Reset Circuit

LO 3

Resetting the 8051 chip can be achieved via the active high RESET input (marked as RST, Pin 9). For normal operation, the input pin is held low. A high pulse of duration equal to at least two machine cycles has to be applied to RST to reset the chip. Since the 8051 variants have different durations for a machine cycle, the duration of the high pulse to reset the chip also varies. The following are the two situations in which the 8051 needs to be reset.

1. *Power-on Reset*: Figure 8.6 shows the circuit used for the purpose of resetting the 8051. As the power is turned on, the RST pin is held high for a duration determined by the value of the capacitor and the charging rate. For most of the 8051 variants, a capacitor of 10 μF and a resistor of 8.2 kΩ suffice to keep the RST pin high for at least two machine cycles.
2. *Reset Switch*: The *Reset Switch S* can be used for asynchronous reset. The switch is connected in parallel to the capacitor. Closing the switch takes the logic value of RST to high. *To reset the 8051, the switch should be kept pressed for at least two machine cycles.* On reset, the important registers in the 8051 chip contain the values noted in Table 8.3.

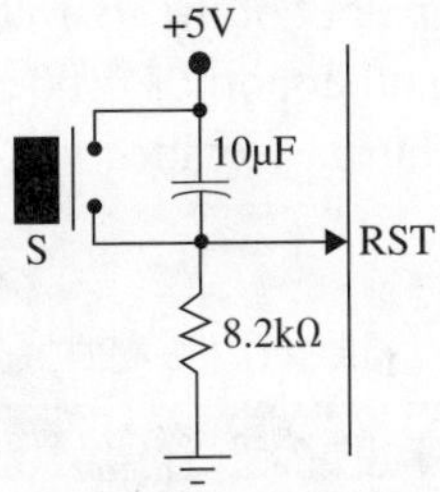

FIGURE 8.6 Reset circuit

TABLE 8.3 Register values on reset

Register	Value (Hex)
Accumulator	00
B	00
PC	0000
DPTR	0000
PSW	00
SP	07
Ports P0-P3	0FF

8.6 Stack and Stack Pointer

Stack is a part of RAM used for implementing subprograms. The return addresses and parameters are passed through the stack. Similar to 8085, the 8051 CPU has a stack pointer, SP – an 8-bit register. On resetting the 8051, the SP register is initialized to 07H. Thus, the stack in the 8051 commences from the RAM location 08H. It may be noted that this stack can grow in RAM locations from 08H to 1FH. Since the RAM locations 20H-2FH are reserved for bit-addressable memory, the stack can be of size 24 bytes only. If a program requires a larger sized stack, the RAM locations 30H to 7FH could be used, this requires the SP to be initialized accordingly.

Also, the default stack locations 08H onwards clash with the portion of RAM used by the registers R0–R7 in Bank-1. Hence, a program using the default stack space, should not use the registers in Bank 1. For using Bank 1 registers, the SP register should be reassigned to create a stack at some other area, such as 30H onwards.

8.7 Input-Output Ports

LO 4

The 8051 possesses four ports, named as Port 0, Port 1, Port 2 and Port 3, that can be used for input-output operations. Each port is bidirectional and 8-bit wide. All port pins can be programmed individually to act either as input or as output. There is a D-latch associated with each pin. The latch holds the value of its corresponding port pin. It may be noted that out of the four ports, only Port 1 can be considered to have pure I/O responsibilities assigned. Among other ports, Port 0 pins act as multiplexed lower order address and data buses. The higher order address bus is provided via Port 2. The multiplexed responsibilities to Port 3 have been noted in Table 8.4.

TABLE 8.4 Alternate functions of Port 1

Bit No.	Alternate Function
0	RxD
1	TxD
2	$\overline{\text{INT0}}$
3	$\overline{\text{INT1}}$
4	T0
5	T1
6	$\overline{\text{WR}}$
7	$\overline{\text{RD}}$

8.7.1 Hardware Structure of I/O Pin

Figure 8.7 shows the generic I/O pin hardware structure. The hardware is more specific to the pins of Port 1, as these pins are not having any additional functionality. The overall structure of an I/O pin is as follows.

- **Data bus connection:** Each port bit is connected to one of the data bus lines. As a port is 8-bit wide, 8-bit data can be communicated via the bus to the port bits and vice versa.
- **D-latch:** One D-latch is used to hold the value that is output from the CPU to the pin. To accomplish the same, the *Latch_Enable* signal should be made high.
- **Tristate buffers:** There are two tristate buffers TB1 and TB2 controlling the flow of information from the port to the data bus line. Tristate buffer TB1 is entrusted with controlling the availability

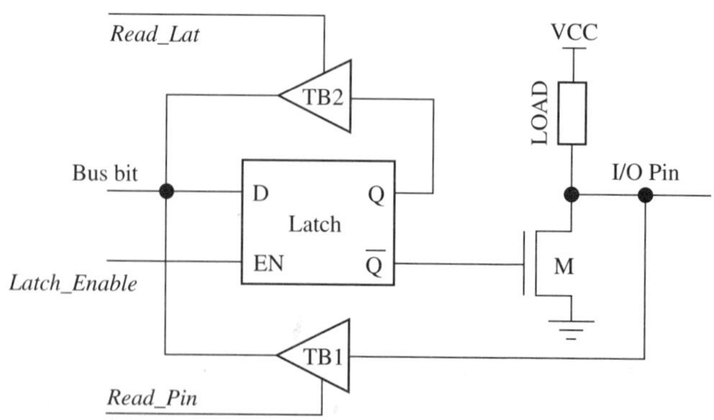

FIGURE 8.7 Hardware structure of I/O pin

of I/O pin input to the CPU bus line. On the other hand, TB2 controls the access of the latch content into the data bus line. TB1 is controlled by the input *Read_Pin*, while TB2 is controlled by *Read_Latch*. These two control signals should not be activated simultaneously. Accordingly, 8051 has different instructions for the two categories of operations.

- **Gate transistor:** The gate transistor controls the output data at the pin. Enabling the gate control input to the nMOS transistor, turns it on, making the output low at the pin. Disabling the transistor pulls the logic level of the port pin to high.

Write Operation: The write operation is relatively simple and is affected by the gate transistor. To write a high (1) to the I/O pin, the CPU turns the bus bit high and enables the *Latch_Enable* signal. This sets the Q output of the latch to 1 and $\overline{Q}$ to 0. As a result, the transistor M gets turned off and the port pin gets charged to 1. To output a 0, the bus bit is set to 0. This makes $\overline{Q}$ to become 1, turning on the transistor M. The port pin line discharges through M, output logic becomes 0.

Read Operation: Two types of I/O read (input) operations are available in 8051. The first one is to read the content of the D-latch. For this, it is necessary to activate the *Read_Latch* and deactivate *Read_Pin*. The other input category reads the value available at the port pin as input. This requires the *Read_Pin* to be activated and *Read_Latch* deactivated. The input operation requires some additional precautions to be taken. To understand the issue, consider that the D-latch has been set previously with the value $Q = 0$, making $\overline{Q} = 1$. With $\overline{Q}$ set to 1, the transistor M is turned on. This forces the port pin to get discharged, even if the user has set it to 1. The problem does not occur if the input value at the pin is also 0. Thus, though reading a 0 input at the port pin is error-free, to read a 1 properly, the D-latch must be set to $Q = 1, \overline{Q} = 0$, at the beginning. Through software, this can be ensured by outputting a 1 to the port bit. As the pin input value is dependent on the value fed by the environment/user, before every input operation, to ensure the correct result, it is required that an "output 1" operation is performed at the port bit. Such an operation will be set $\overline{Q}$ to 0, disabling the transistor M. Thus the following two operations are to be performed in sequence.

1. Output 1 to the port bit
2. Input the bit from the port

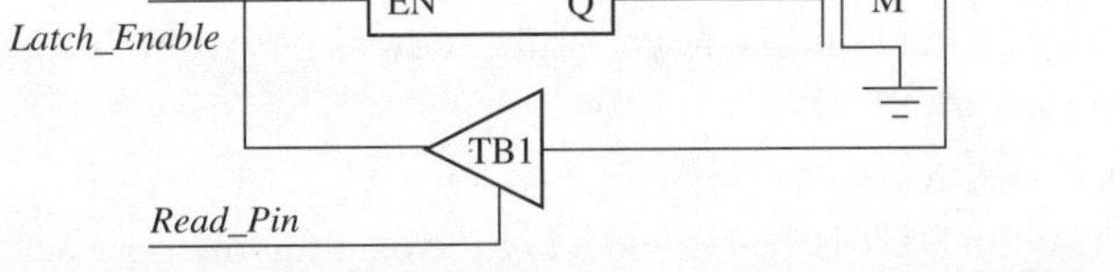

FIGURE 8.8 Hardware structure Port 0 pin

To accomplish the above two tasks, any input operation will require two instructions. The first instruction writes a 1 to the port pin, disabling the transistor M, while the second instruction performs the actual input read operation.

8.7.2 Port 0 Structure

Compared to other port bits, the structure of Port 0 pins is a bit different. A typical Port 0 pin structure has been shown in Figure 8.8. The point to be noted is that there is no internal pull-up resistance available with Port 0 pins. These port bits are *open-drain*. When a 1 is written to the D-latch, the transistor M gets turned off. This causes the pin to float. The port can continue functioning as a multiplexed address/data bus. However, to use it for I/O purposes, an external pull-up resistance must be connected to source current to the pin. Since for address/data multiplexing, the port bit is connected to some latch (for example, 74LS373), external pull-up is not needed in that case. With this open-drain configuration, Port 0 can drive more number of fanouts, compared to other ports. As noted in the 8051 datasheet, the ports Port 1, Port 2 and Port 3 can drive up to 4 TTL inputs, while Port 0 can drive up to 8 TTL inputs.

8.8 Memory Organization (LO 5)

The basic 8051 chip contains internal 4 kB ROM and 128-byte RAM. The on-chip storage has the fastest access from the CPU. However, this space may be insufficient in certain applications. For example, out of the total 4 kB on-chip ROM space, certain parts may be used by the system designer to accommodate the core routines to interface the devices that are needed for the system. This may leave only a little or no space for further user routines to be integrated with it. This is particularly true for systems with medium to high complexity. The situation is more acute with the RAM space. Out of the 128-byte RAM space, a major part is occupied by the general- and special-purpose registers. Only the locations 30H-7Fh are available for general-purpose usage. This memory range is commonly known as *scratch-pad*. Among these 80 bytes also, sometimes the stack has to be accommodated for programs implementing several nested subprogram calls with a good number of parameters. This scarcity of RAM space often leads to the requirement of interfacing external RAM for keeping data values. In 8051, the program and data memory size can be extended up to 64 kB each, with the support of external memory chips. The 8051 uses an altogether different set of instructions for accessing external program memory than those used for the data memory access. Hence, the program and the data memory can be individually of size 64 kB each. This is a marked difference from many earlier processors, such as 8085, that support 16-bit address bus connecting up to 64-kB memory, however, without any distinction between program and data. The important signal lines participating in the interface are as follows.

1. **Port 0:** The 8-bit Port 0 lines provide the multiplexed signal lines AD0–AD7. The 8-bit data bus is multiplexed with the lower order address bus. Similar to 8085, a latch along with the ALE signal from the 8051 can be used to separate out the lower order address bus from the data bus. The multiplexing saves the number of pins for the 8051.
2. **Port 2:** Port 2 provides the 8-bit higher order address lines A8-A15 for memory access. As noted earlier, the *program counter* (*PC*) in 8051 is 16-bit wide. For the external memory access, the higher order byte of the PC is sent to Port 2 and the lower order byte to Port 0. For data memory access, generally another

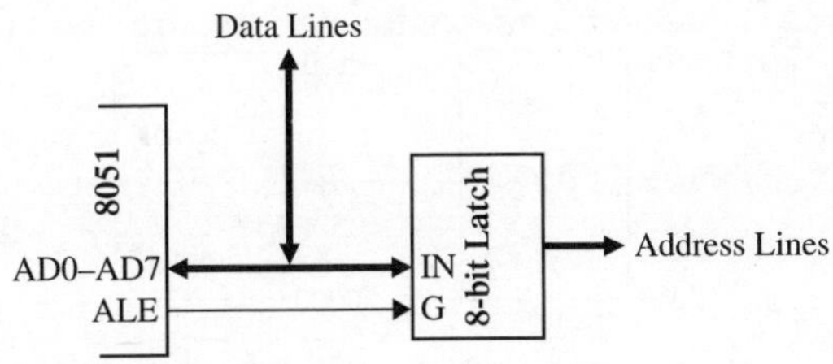

FIGURE 8.9 Demultiplexing of address and data lines

16-bit register DPTR is used. The DPTR consists of two 8-bit registers DPH and DPL. The higher order 8 bits from DPH are copied to Port 2, while DPL is copied to Port 0 for data memory access.

3. **EA:** The pin *External Access* (*EA*) decides whether the on-chip ROM or an externally connected memory chip be used for getting program instructions. It is an active low-input signal. If the signal line is kept high, the on-chip ROM is used as the memory to source program instructions. On the other hand, to use external program memory for fetching instructions, the signal has to be set to low.
4. **ALE:** Similar to many other processors, the *Address Latch Enable* (*ALE*) is an active high output signal of the 8051 chip. The signal is made high whenever the lower order address byte is put on AD0–AD7 (that is, Port 0). The signal should be used to latch the lower order address lines, even after the ALE signal has been withdrawn. Generally, 8051 activates the ALE signal for two clock cycles. The demultiplexing of address bus and data bus has been shown in Figure 8.9.
5. **PSEN:** The signal, *Program Strobe Enable* (*PSEN*) is used for accessing the external program memory. This is an active low-output signal. The ROM chip, generally used to implement external program memory contains an *Output Enable* (*OE*) pin. The PSEN line of 8051 is connected to the OE pin of the ROM.
6. **RD:** The active low *Read* (*RD*) signal is utilized for reading data from the external data memory. As noted earlier, the data memory can be up to 64 kB in size. The DPTR register is used to provide the data memory address.
7. **WR:** To write onto the external data memory, the *Write* (*WR*) signal is used. It is also an active low output signal from 8051 to be connected to the write control of the memory.

8.8.1 External Program Memory Interface

Figure 8.10 shows the alternatives by which the program memory may be interfaced. The access mechanism is based upon the EA input setting. As shown in Figure 8.10(a), when the EA pin is grounded, for accessing program instructions, the 8051 uses only the external program memory. Thus, the internal ROM space is not used as program memory. Setting EA to 1, enables the 8051 to use both the on-chip and the external program memories to get the program instructions. In the basic 8051, the on-chip program memory is size-restricted to 4 kB. Hence, as the CPU generates program addresses in the range 0000H to 0FFFH, the on-chip memory is used. For addresses beyond 0FFFH, that is in the range 1000H to FFFFH, the external ROM is accessed. This situation has been shown in Figure 8.10(b).

Figure 8.11 shows the signals of 8051 side to interface with external program memory chip. The EA pin should be set to high or low by the user, as discussed. The higher order address bus lines are provided

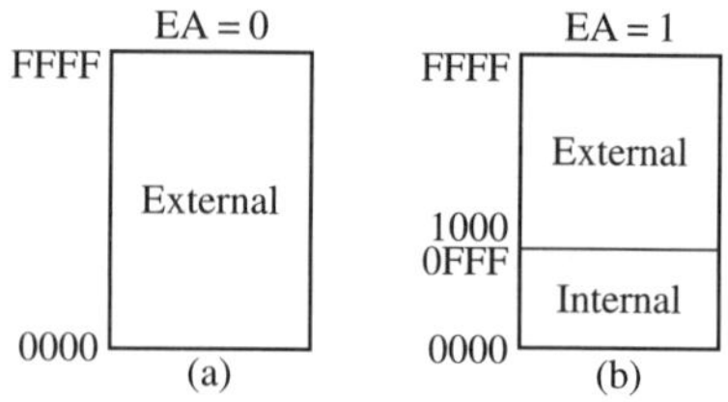

FIGURE 8.10 External program memory with (a) EA = 0, (b) EA = 1

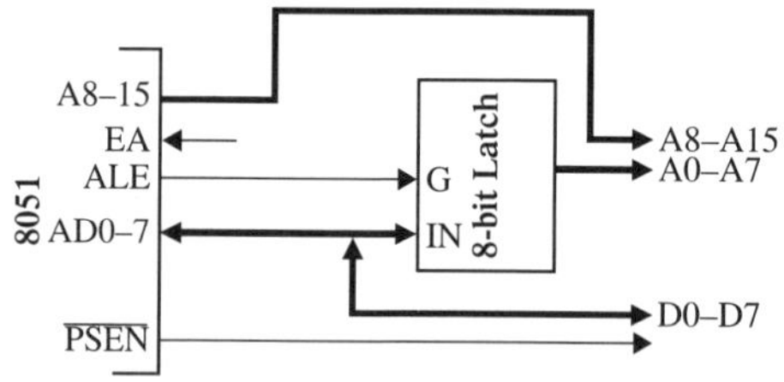

FIGURE 8.11 Signal lines for program memory interface

by the 8-bit Port 2. Port 0 lines AD0–AD7 are demultiplexed using the ALE signal to get the lower order address lines. The PSEN line should be connected to the OE pin of the external ROM to control memory read operations from it.

EXAMPLE 8.2

Figure 8.12 shows the interfacing of 64 kB external program memory to the 8051. The EA line has been grounded leading to the setting that the on-chip ROM space will not be used by the system. Individual ROM chips are assumed to be of size 16 kB each. Four such memory chips are needed to cover the 64 kB space.

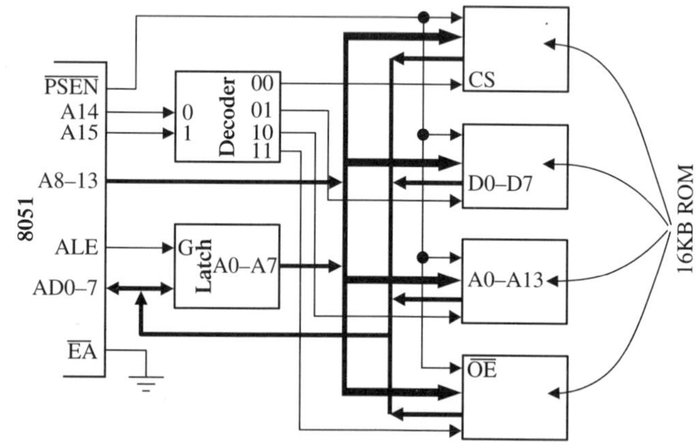

FIGURE 8.12 Example program memory interface

8.8.2 External Data Memory Interface

Apart from program memory, the 8051 provides an option to interface up to 64 kB of external data memory. Figure 8.13 shows the 8051 signal lines that can be utilized to interface external data memory. Compared to the program memory access lines, the differences are as follows.

- No dependency on EA line. The 8051 instruction determines the case when the external data memory be used instead of internal RAM.
- PSEN connection is not required, as it is not accessing program memory.
- Port 0 and Port 2 lines are used to provide 16-bit address and 8-bit data-bus lines, similar to program memory interface.

Read (RD) and Write (WR) signals from Port 3 are used for read and write operations.

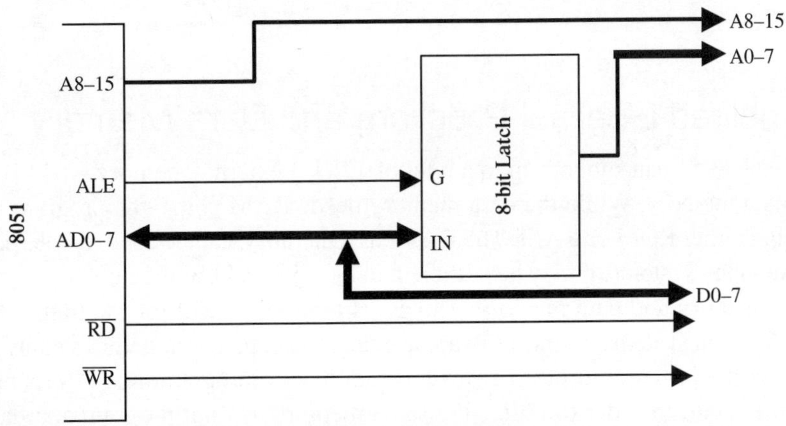

FIGURE 8.13 Signal lines for data memory interface

EXAMPLE 8.3

An external data memory interface with the 8051 chip has been shown in Figure 8.14. Here, 64 kB external data memory space has been interfaced. Four memory chips, each of size 16 kB are used to cover the 64 kB address space. The address- and data-bus connections are made in a manner, similar to that for Example 8.2. However, the PSEN and EA lines are not required. Instead of them, RD and WR controls have been used to interface with individual RAM chips.

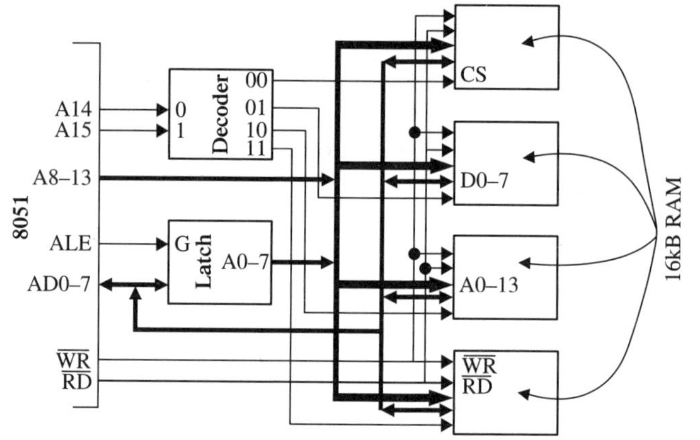

FIGURE 8.14 Example data memory interface

8.8.3 Combined External Program and Data Memory

As noted earlier, the 8051 can support up to a total of 128 kB external memory – 64 kB program memory and 64 kB data memory. While the data memory needs RAM chips, the program memory can be implemented using either ROM or RAM. The RAM, as a memory, can be used to hold program code, as well. However, in many systems, it may be sufficient to have just 64 kB of external memory, functioning both as program memory and data memory. This is particularly useful for the trainer kits and system design facilities. In such systems, as the software is being developed, it undergoes many changes during debugging. Storing the program in RAM saves the additional reprogramming efforts needed in ROM, for such an environment. In order to utilize the same memory as joint program and data memory, the $\overline{\text{PSEN}}$ and $\overline{\text{RD}}$ signals from the 8051 are AND-ed and connected to the $\overline{\text{RD}}$ line of RAM. On the other hand, the $\overline{\text{WR}}$ line of the 8051 should be connected to the $\overline{\text{WR}}$ input of RAM. The rest of the lines from 8051 are interfaced as per the usual requirements. Figure 8.15 shows the combined program and data memory interface.

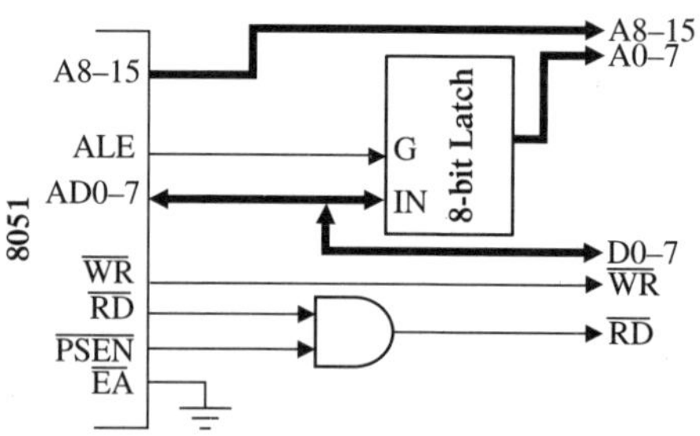

FIGURE 8.15 Combined program and data memory

8.9 Timing Diagrams and Execution Cycles

LO 6

Similar to other processors, the 8051 also needs several machine cycles to execute an instruction. The number of machine cycles needed to execute a specific instruction may differ across the variants of 8051. In the original 8051 chip, designed by Intel, four machine cycles are required to execute a MUL instruction, one machine cycle for a MOV instruction etc. In the original 8051, one machine cycle consists of 12 clock pulses. This has been shown in Figure 8.16. Two pulses of the clock form a *state* of the processor. Hence, a total of six states constitute a machine cycle. It may be noted that there are two ALE pulses per machine cycle. The first one is between *State 1* and *State 2*, while the second one is generated between the states 4 and 5. Each ALE pulse spans over two clock pulses.

Figure 8.17 shows the timing diagram of a read operation in 8051 from the external program memory. Here, 2 bytes are being read from the memory. The whole operation takes one machine cycle. The first byte is available at pulse 2 of State 4. The other byte is read at pulse 2 of State 1 in the next machine cycle.

Figure 8.18 considers a special situation in which a 1-byte instruction is accessed from the external program memory. The instruction, in turn, needs one more byte to be read from the external data memory. The external data memory access is performed by the MOVX instruction. The operation takes two machine cycles. The first machine cycle accesses the external program memory with values of PCH and PCL made available on the lines of Port 2 and Port 0 for address. The content is read from the program memory.

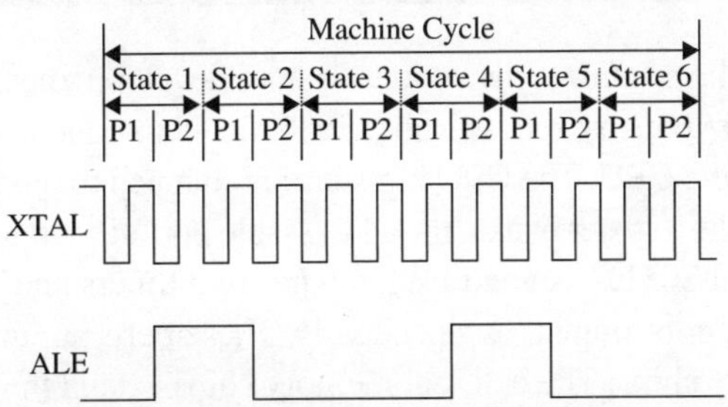

FIGURE 8.16 Machine cycle of the Intel 8051 CPU

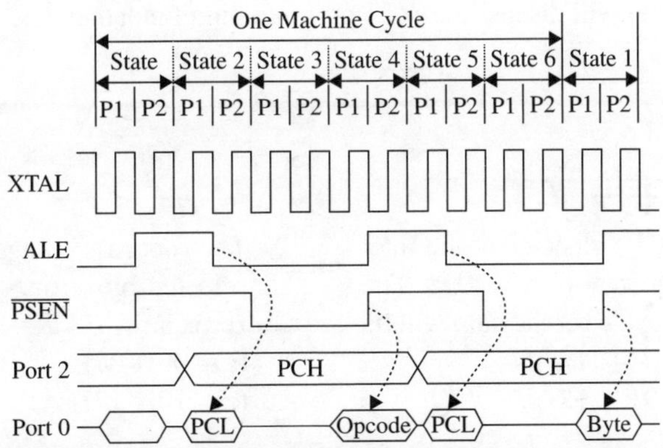

FIGURE 8.17 Timing diagram for external program memory access

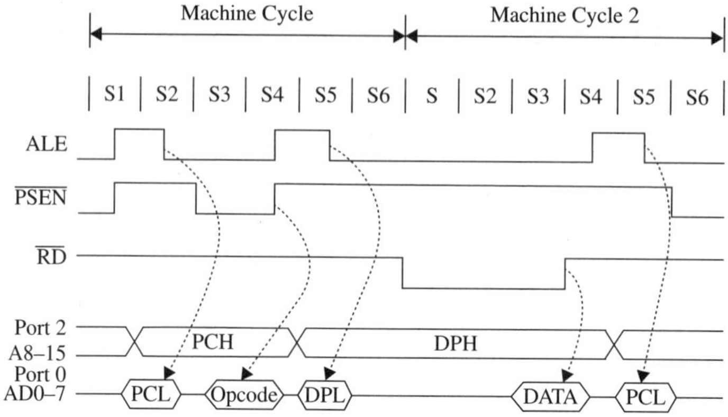

FIGURE 8.18 Timing diagram with external program and data memory access

At the rising edge of $\overline{PSEN}$, the content of program memory location is read. The next machine cycle puts the content of DPH on Port 2 and DPL on Port 0, to provide the address. To provide the read control, a low-to-high transition of the $\overline{RD}$ line is used as a strobe.

SUMMARY

Over the years, the 8051 has been utilized in various system developments. It is one of the most popular microcontrollers. Compared to a microprocessor like 8085, the chip contains on-chip ROM, RAM, timers and I/O ports, over and above the CPU. The CPU registers are mapped to on-chip RAM. The RAM space is divided into several regions – the register banks, bit-addressable portion, scratch pad, stack space and special function registers. The original 8051, designed by Intel, has two timers and four, 8-bit I/O ports. The port bits can be programmed as input or output, independently. These ports are to be used for interfacing external program- and data-memory chips. The external memory chips extend the total memory available to the programmers. Instruction execution in 8051 consists of several machine cycles with an individual machine cycle divided into a fixed number of clock cycles. This chapter has given a detailed view of all these aspects of the 8051. The next chapter will discuss the 8051 programming fundamentals.

EXERCISES

Multiple Choice Questions

1. Assume that the crystal connected to the Intel 8051 chip has a frequency of 10 MHz. The length of its machine cycle for the chip will be
   (a) 1.4 μs **(b)** 1.2 μs
   (c) 1.0 μs **(d)** 1.6 μs

2. The address range of the register bank selected by setting the bits RS0 and RS1 to zeros is
   (a) 00H-07H **(b)** 08H-0FH
   (c) 10H-17H **(d)** 18H-1FH

3. Identify the register bank address range when RS0 and RS1 are set to 1 and 0, respectively.
(a) 00H-07H (b) 08H-0FH
(c) 10H-17H (d) 18H-1FH

4. In order to select the address range 10H-17H as the present register bank, the bits RS0 and RS1 be set to
(a) RS0 = 1, RS1 = 1
(b) RS0 = 1, RS1 = 0
(c) RS0 = 0, RS1 = 1
(d) RS0 = 0, RS1 = 0

5. Identify the byte address in 8051 in which the bit address 42H is located.
(a) 26H (b) 28H
(c) 2AH (d) None of the above

6. Identify the one which is NOT a valid status word bit in 8051.
(a) Z (b) OV
(c) CY (d) AC

7. Identify the one which is NOT a valid status word bit in 8051.
(a) P (b) OV
(c) CY (d) None of the above

8. What is the value of SP in 8051, on reset?
(a) 07H (b) 00H
(c) 0FH (d) 1FH

9. To be used for I/O operations, identify the port of 8051 that needs external pull-up resistances to be connected.
(a) P0 (b) P1
(c) P2 (d) P3

10. Suppose that the programmer forgets to write '1' to a port bit before an input operation. Then, reading of which value may create a problem?
(a) 0 (b) 1
(c) Both 1 and 0 (d) Neither of 0 nor 1

11. Identify the byte address in 8051 to which the bit address 67H belongs.
(a) 2CH (b) 2DH
(c) 2FH (d) None of the above

12. Identify the byte address to which the bit address 42H belongs to n 8051 RAM.
(a) 26H (b) 28H
(c) 2AH (d) None of the above

13. The IP register in 8051 is used to control
(a) Instruction parity
(b) Interrupt parity
(c) Interrupt priority
(d) None of the above

14. An 8051 program uses the MSB and the LSB of port P0 as output. The statement for this can be
(a) MOV P0, #81H (b) MOV P0, #FFH
(c) MOV P0, #8FH (d) None of the above

15. In 8051, the number of interrupt pins is
(a) 1 (b) 2
(c) 3 (d) 4

16. The errors in signed arithmetic are indicated by the 8051 status flag bit
(a) P (b) OV
(c) CY (d) None of the above

17. Mark the meaning of the 8051 instruction "MOV A, 12H".
(a) 12H is stored in the accumulator
(b) Fifth bit of accumulator set to 1
(c) Content of memory location 12H copied to the accumulator
(d) None of the above

18. The facility *External Access* in 8051 is meant for
(a) Peripherals (b) Power supply
(c) ALE (d) Memory interfacing

19. The range of addresses for SFRs in 8051 is
(a) 80H to 0FEH (b) 00H to 0FFH
(c) 80H to 0FFH (d) 70H to 80H

20. For external memory interfacing, the I/O ports in 8051, used as address and data buses are
(a) P1 and P2 (b) P1 and P3
(c) P0 and P2 (d) P0 and P3

21. The bit-level operation possible in 8051 is
(a) Complement **(b)** Set
(c) Reset **(d)** All of the above

22. The control signal(s) in 8051, used to access off-chip devices, are
(a) ALE **(b)** PSEN
(c) RD, WR **(d)** All of the above

23. Identify the 8051 register that does not have an on-chip RAM address.
(a) Stack pointer **(b)** Program counter
(c) Data pointer **(d)** Timer

24. The byte address range of the bit-addressable memory range in 8051 is
(a) 00H to 0FFH **(b)** 01H to 7FH
(c) 00H to 7FH **(d)** 80H to 0FFH

25. The 8051 port bits that are multiplexed as external counter inputs are
(a) P3.4, P3.5 **(b)** P2.4, P2.5
(c) P3.0, P3.1 **(d)** None of the above

Answers to Multiple Choice Questions

1. b	2. a	3. b	4. c	5. b	6. a	7. d	8. a	9. a	10. b
11. a	12. b	13. c	14. d	15. b	16. b	17. c	18. d	19. c	20. c
21. d	22. d	23. b	24. c	25. a					

Long Answer Type Questions

1. Draw the block diagram of 8051 and explain its architecture.
2. Identify the components in the internal RAM of 8051.
3. Draw the circuit diagrams corresponding to (a) clock and (b) reset operations in 8051 and explain their operations.
4. Explain the necessity of writing a '1' to a port bit before reading it.
5. How does the EA pin control the external memory interfacing in 8051?
6. Identify the 8051 signals related to external memory interfacing.
7. Explain the external data memory access with an appropriate timing diagram.
8. Explain with a timing diagram the external program memory access procedure in 8051.
9. Describe the I/O pin structure of 8051 ports.
10. Why do Port 0 pins need external resistors to be connected?
11. Explain the register bank switching procedure in 8051.
12. Interface an external memory system with a 16 kB program and 8 kB data-memory to the 8051.
13. Assume that a crystal of frequency 12 MHz has been connected to the Intel 8051 chip. Determine the minimum duration for which the reset pin needs to be activated to reset the 8051.
14. What is the maximum size of the stack of 8051 such that the stack does not interfere with the register banks?

9

The 8051 Instruction Set and Programming

LEARNING OUTCOMES

After reading this chapter, students will be able to:

LO 1 Enumerate the addressing modes of 8051

LO 2 Enumerate the instructions of 8051

LO 3 Write programs in the assembly language of 8051

LO 4 Write programs in embedded C language targeted to 8051

LO 5 Design application systems with 8051 using its timers and counters

LO 6 Design systems utilizing 8051 interrupts

LO 7 Transmit and receive characters serially from/to 8051

9.1 Introduction

A large number of instructions of different sizes are supported by the 8051 processor. There are 49 single-byte instructions, 45 two-byte instructions and 17 three-byte instructions, making a total of 111 instructions available for programming the 8051 processor. The instruction set is uniform across all variants of 8051. As 8051 is an 8-bit processor, the instruction set is most suitable for the 8-bit operations, though operations at bit-level are also very popular in applications. It may be noted that the CPU registers are part of the on-chip RAM. Hence, accessing a register is similar to accessing the corresponding memory location. As a result, it may be possible to write the same instruction in different formats using different versions to specify the register operand. Different addressing modes and operations are supported in 8051. Compared to the processors like 8085 and 8086, the 8051 processor supports bit-level operations. This is particularly important in hardware interfacing applications in which individual port bits need to be controlled separately. In the following, we shall discuss about the addressing modes and instruction set of 8051, in detail.

9.2 Addressing Modes

LO 1

The following are the addressing modes supported by the 8051.

- Immediate addressing
- Register addressing
- Direct addressing
- Register indirect addressing
- Bit direct addressing

The addressing modes have been discussed in detail, along with examples, in the following.

9.2.1 Immediate Addressing Mode

In immediate addressing, the operand is an immediate value, available as part of the instruction itself. Thus, as soon as the instruction has been fetched into the CPU, the operand is available to work upon. No further memory access is needed to get the operand. The constant value is mostly identified by a preceding '#' symbol, in most of the assemblers. Some examples of the usage of immediate addressing mode are as follows.

```
MOV A, #46H        ; load 46H into the A register
MOV R7, #5AH       ; load 5AH into the R7 register
MOV B, #3BH        ; load 3BH into the B register
MOV P1, #55H       ; Output 55H to Port 1
MOV DPTR, #9A43H   ; load 9A43H into the register DPTR
```

As DPTR is a 16-bit register, the instruction "MOV DPTR, #9A43H" is equivalent to loading the two constituent 8-bit registers DPH and DPL using the following two instructions, taken together.

```
MOV DPH, #9AH
MOV DPL, #43H
```

9.2.2 Register Addressing Mode

In this mode, a register is named as the operand. For this, a register from the set R0–R7 belonging to the current register bank, registers A, B or other special function registers can be mentioned. A few examples are as follows.

```
MOV A, R3  ; A gets a copy of R3
MOV R6, A  ; R6 gets a copy of A
ADD A, R2  ; A gets A + R2
```

As a restriction, an instruction cannot use two registers in it where both belong to the current register bank. To explain it further, the instruction "MOV R1, R2" is invalid. It may be noted that it does not cause a serious constraint to the programmer, as all registers are mapped to their corresponding memory locations and the 8051 does not put any restriction to move data between memory locations.

9.2.3 Direct Addressing Mode

The on-chip memory locations can be specified as operands in this mode. Some examples of the same are as follows.

```
MOV R6, 45H  ; R6 gets the content of the RAM location 45H
MOV 56H, R2  ; RAM location 56H gets the content of R2
MOV 30H, 20H ; RAM location 30H gets the content of RAM location 20H
```

The immediate and direct operands appear to be similar in the instructions. Only the '#' character introduced by some assemblers makes them easily distinguishable. The programmers need to consult the assembler manual to learn the distinction used by it. In the last example, noted above, both the operands have been specified using direct addressing mode. This can be utilized to circumvent the limitation of not allowing multiple register operands in an instruction. For example, to move the content of register R3 to register R5 of Bank-0, the instruction "MOV 5, 3" could be used. However, the instruction "MOV R5, R3" with a similar meaning, is not valid in 8051 (as both the operands have been specified as registers).

9.2.4 Register Indirect Addressing Mode

This register mode is similar to the usage of pointers in high-level languages. The operand is a memory location. However, the address of the memory location is not made available directly in the instruction itself. Instead of that, registers R0 or R1 can be used to act as the address provider. The other registers R2–R7 cannot act as pointers. In order to use the register R0 (or R1) as a pointer in the instruction, the register should be preceded by a '@' character. Some examples of register indirect addressing mode have been noted next. Here it has been assumed that the registers R0 and R1 contain values 45H and 55H, respectively.

```
MOV A, @R0 ; A gets the content of memory location 45H pointed to by R0
MOV B, @R1 ; B gets the content of memory location 55H pointed to by R1
```

The registers R0 and R1 can be used to index into the on-chip RAM only. The off-chip RAM can be accessed indirectly via the DPTR register. It may be noted that the MOV instruction cannot use the DPTR register for external access. For this, the MOVX instruction has to be used, as illustrated next.

```
MOVX A, @DPTR   ; A gets the content of the external RAM
                  location pointed to by DPTR
```

9.2.5 Bit Direct Addressing Mode

This mode is applicable for the bit operands. It is similar to the direct addressing mode in which the bit address is specified as part of the instruction. The on-chip RAM locations 20H to 2FH and most of the special function registers are bit-addressable, as noted in Chapter 8. Each bit location possesses a unique address. A few examples of bit-direct addressing have been noted next.

```
SETB 35H  ; Sets bit location 35H to 1
CPL 46H   ; Complements bit location 46H
```

9.3 Instruction Set

LO 2, 3

The 8051 processor supports a rich instruction set. The instructions can be grouped into four major categories, as shown next.

- Data transfer instructions
- Branching instructions
- Arithmetic and logic operation instructions
- Other miscellaneous instructions

Next, we shall have a detailed look at instructions belonging to each of these categories.

9.3.1 Data Transfer Instructions

There are a number of instructions supported by the 8051 that can be used for data transfer between source and destination. The transferred data can be of size bit, byte or even two bytes. Data may also be exchanged between two places mentioned in the source and destination operands. The following are the instruction opcodes that come under this category – MOV, MOVC, MOVX, PUSH, POP, XCH and XCHD. Each of the opcodes has been elaborated in the discussion to follow.

1. **MOV:** This is the most widely used opcode for data transfer. The instruction framed with it can be used for the movement of 1-bit, 8-bit or 16-bit data. Since the majority of data transfers are between 8-bit registers and memory locations, the commonly used variant of MOV instruction looks like this.

```
MOV <dest-byte>, <source-byte>
```

The operands *dest-byte* and *source-byte* can be specified using various addressing modes, discussed earlier. Based on the addressing mode used for the operands, several subvariants of the MOV instruction are possible.

a. *Accumulator A as destination*: In this case, the *A* register acts as the destination of data movement. The source operand can be specified using any of the available addressing modes. Some examples of the subcategory have been shown in Table 9.1.

TABLE 9.1 MOV with A as destination

Addressing Mode	Format	Example	Meaning
Immediate	MOV A, #data	MOV A, #55H	A = 55H
Register	MOV A, Rn	MOV A, R3	A = Content of R3
Direct	MOV A, direct	MOV A, 43H	A = RAM[43H]
Indirect	MOV A, @Ri	MOV A, @R1	A = RAM[(R1)]

Note: *"MOV A, A" not allowed. Only the bank registers can be used for* Rn

b. *Accumulator A as source*: This is otherwise similar to the previous case, except that the A register is the source of data movement. The destination operand can be specified using any of the following modes – register, direct or register indirect. Some such examples have been shown in Table 9.2.

TABLE 9.2 MOV with A as source

Addressing Mode	Format	Example	Meaning
Register	MOV Rn, A	MOV R3, A	R3 = Content of A
Direct	MOV direct, A	MOV 43H, A	RAM[43H] = A
Indirect	MOV @Ri, A	MOV @R1, A	RAM[(R1)] = A

c. *Bank register as destination*: In this case, any of the registers R0–R7 in the current bank can act as the destination. The source operand can be an immediate value, the Accumulator register *A* or an on-chip memory address. Some such examples have been noted in Table 9.3.

TABLE 9.3 MOV with bank register as destination

Addressing Mode	Format	Example	Meaning
Register	MOV Rn, A	MOV R3, A	R3 = Content of A
Immediate	MOV Rn, #data	MOV R6, #47H	R6 = 47H
Direct	MOV Rn, direct	MOV R1, 34H	R1 = RAM[34H]

d. *Direct memory address as destination*: In this case, the destination operand is specified using the direct addressing mode, while the source may be specified in any of the addressing modes, including direct addressing. Table 9.4 shows the cases individually.

TABLE 9.4 MOV with direct memory address as destination

Addressing Mode	Format	Example	Meaning
Register	MOV direct, Rn	MOV 42H, R1	RAM[42H] = R1
Immediate	MOV direct, #data	MOV 42H, #22H	RAM[42H] = 22H
Direct	MOV direct, direct	MOV 1, 6	RAM[1] = RAM[6]
Indirect	MOV direct, @Ri	MOV 44H, @R1	RAM[44H] = RAM[(R1)]

e. *Indirect memory address as destination*: In this variant, the registers R0 and R1 can be used to provide the destination operand address in RAM. The source operand can be specified using any of the addressing modes. Some examples of the variant have been shown in Table 9.5.

TABLE 9.5 MOV with indirect memory address as destination

Addressing Mode	Format	Example	Meaning
Register	MOV @Ri, A	MOV @R1, A	RAM[(R1)] = Content of A
Immediate	MOV @Ri, #data	MOV @R1, #22H	RAM[(R1)] = 22H
Direct	MOV @Ri, direct	MOV @R1, 6	RAM[(R1)] = RAM[6]

f. *Bit operands*: This variant of MOV is useful to implement data transfers between bit-level operands. The source bit is copied into the destination bit. The addressing mode for the operands used in the instruction is bit direct addressing.

```
MOV C, PSW.0 ; Bit-0 (Parity bit) copied to Carry bit (PSW.7)
MOV P1.6, C  ; Carry copied to Port 1, bit 6
```

g. *16-bit data movement*: The MOV instruction can also be used for 16-bit data movement. In this connection, the 16-bit DPTR register is one of the widely used registers to access the external memory. To initialize the DPTR register, 16-bit data transfer via MOV can be done. An example of such an initialization has been noted next.

```
MOV DPTR, #143AH  ; DPTR gets 143AH
```

2. **MOVC:** The MOV instruction can move data to/from on-chip RAM. In contrast, the MOVC instruction can be utilized to move a code byte from the on-chip ROM to the A register. Programmers often store the lookup tables needed in their applications in the on-chip ROM, as part of the program itself (as the table content is not going to change). The MOVC instruction is typically utilized to access such on-chip lookup tables. Two variants of this instruction are available, depending upon the 16-bit register utilized to hold the base address of the table. The registers DPTR or PC can be used for this purpose. The offset of the table entry to be accessed should be available in the A register. Thus, the A register acts as the index of the data within the table to be accessed. The data from the table is copied into A.

a. *DPTR as base register*: The instruction in this case is,

```
MOVC A, @A+DPTR
```

It is clear that the programmer must have initialized the DPTR register with the base value of the table, before issuing this instruction. Example 9.1 shows the usage of DPTR for external memory access.

EXAMPLE 9.1

The first ten odd numbers (indexed as 0, 1, ... 9) are 1, 3, 5, 7, 9, 11, 13, 15, 17, 19. Assuming that these numbers are stored in the ROM starting at location 300H and the register R2 contains the index of the odd number we are looking for, the following code fragment using the instruction "MOVC A, @A +DPTR" could be used to get the odd number in A register.

```
        MOV DPTR, #300H      ; DPTR holds the base
        MOV A, R2            ; A contains the index
        MOVC A, @A+DPTR      ; A contains the odd
        ORG 300H
ODD     DB 1, 3, 5, 7, 9,    ; The lookup table
        11, 13, 15, 17, 19
```

b. *PC as base register*: In Example 9.1, the Table ODD can be located anywhere in the memory, not necessarily with the rest of the program code. In the specific case, the table has been kept from address 300H. However, in many cases, the programmer may find it more logical to keep it immediately following the code, rather than at a distant location. Since the program code is accessed via the Program Counter (PC), in such cases, it may be advantageous to use the PC as the base register. The A register continues to contain the offset part for the entry. The format for the instruction is "MOVC A, @A+PC". Example 9.2 shows the usage of MOVC instruction.

EXAMPLE 9.2

Using the above principle, Example 9.1 can be rewritten as follows. It may be noted that when the CPU is executing the instruction "MOVC A, @A+PC", the PC is actually pointing to the next instruction RET. The RET being a one-byte instruction, to reach the proper offset in the table, the register A (containing the index of table entry being accessed) needs to be incremented by 1, beforehand.

```
    MOV A, R3                    ; A contains the index
    INC A                        ; Adjust offset
    MOVC A, @A+PC                ; A contains the odd
    RET
ODD DB 1, 3, 5, 7, 9, 11,        ; The lookup table
    13, 15, 17, 19
```

3. MOVX: Unlike MOV, the MOVX instruction is used to access the external memory. The instruction can be used to transfer data to/from the A register. The external memory be accessed using both

16- and 8-bit addresses. To specify a 16-bit address, the DPTR register is used. For 8-bit addresses, indirect access via the registers R0 and R1 can be used. It has been illustrated in the following.

a. *Address by DPTR*: The MOVX instruction in conjunction with DPTR could be used for this purpose.

```
MOVX A, @DPTR   ; A gets the external memory byte pointed to by DPTR
MOVX @DPTR, A   ; External memory location pointed to by DPTR gets A
```

b. *Address by R0, R1*: The MOVX instruction can be used in conjunction with R0 and R1 to specify an 8-bit address in the external memory.

```
MOVX A, @Ri    ; i = 0, 1. A gets the external memory byte pointed
                 to by Ri
MOVX @Ri, A    ; i = 0, 1. External memory location pointed to by
                 Ri gets A
```

EXAMPLE 9.3

In the following, the program written in Example 9.1 has been revisited in which the odd numbers are assumed to have been stored in the external memory from address 2500H. The code fragment for the same is noted next. The index of the odd number to be accessed is specified in R2.

```
       MOV DPTR, #2500H       ; DPTR points to base
NEXT:  MOVX A, @DPTR          ; Get next odd in A
       CJNE R2, 0, L1         ; Goto L1 if not this index
       RET                    ; Got the odd, return
  L1:  INC DPTR               ; DPTR points to next odd
       DJNZ R2, NEXT          ; Decrement counter, get next
       ORG 2500H
  ODD  DB 1, 3, 5, 7, 9,      ; The lookup table
       11, 13, 15, 17, 19
```

4. PUSH and POP: These instructions work in conjunction with a stack for data transfer. The PUSH instruction copies a byte operand to the on-chip RAM location pointed to by the *Stack Pointer* (*SP*). The SP register gets incremented by 1. Only the direct addressing mode is supported to specify the operand. The following is an example of PUSH instruction. It saves the content of A register (memory address 0E0H) to stack memory location 43H, assuming that the current value of SP is 42H. The SP register is incremented to 43H.

```
PUSH 0E0H
```

Instructions specifying operands in other modes, such as, "PUSH A", "PUSH @R0" and "PUSH R2" are invalid.

The POP instruction is the reverse of PUSH. It copies the content from the top of the stack (pointed to by SP) to the specified operand. In this case also the operand can be specified using the direct addressing mode only. For example, assuming that the content of SP is 43H, the instruction "POP 0E0H" will copy the content of RAM location 43H to A register. The SP register will get decremented to 42H.

5. **XCH and XCHD:** These instructions are used to exchange the A register with the specified operand. The entire byte or the lower 4-bit nibble takes part in the exchange operation. The XCH instruction can be used to swap the 8-bit content of A register with a source byte, which can be specified using any addressing mode. The generic instruction is of the format "XCH <*byte*>". The XCHD (exchange digit), similar to the XCH, swaps only the lower nibbles (lower order 4 bits) between the source byte and A register. The higher order nibbles remain unaltered. For example, if A = 54H and R2 = A7H, "XCHD A, R2" will convert A to 57H and R2 to A4H. Table 9.6 shows a few examples of the utilization of XCH and XCHD instructions.

TABLE 9.6 XCH and XCHD instructions

Addressing Mode	Format	Example	Meaning
Register	XCH A, Rn XCHD A, Rn	XCH A, R2 XCHD A, R2	Contents of R2 and A exchanged Lower nibbles of R2 and A exchanged
Indirect	XCH A, @Ri	XCH A, @R1,	Contents of RAM[(R1)] and A exchanged
Direct	XCH A, direct	XCH A, 50H	Contents of RAM[50H] and A exchanged

9.3.2 Branching Instructions

There are three categories of branching instructions in 8051, namely, unconditional branching, conditional branching and subprogram call/return.

1. *Unconditional branching*: There are three instructions belonging to this category – LJMP, AJMP and SJMP. The instruction LJMP is of size 3 bytes, while the other two (AJMP and SJMP) are 2 bytes wide. The instructions have been detailed in the following.

   - **LJMP:** This is the *long jump* instruction with the following format.

```
LJMP <16-bit addresss>
```

   The availability of a 16-bit branching address enables the target to be located anywhere within the 64 kB address space.

   - **AJMP:** This is a 2-byte jump instruction, named "absolute jump". The branching target is specified using an 11-bit field. The instruction is of the syntax,

```
AJMP <11-bit addresss>
```

   While executing the branch, the 11-bit address specified in the instruction replaces the lower order 11 bits of the 16-bit Program Counter (PC). The higher order five bits remain unaltered. Hence, the target address has to be located within a range of 2 kB address range from the current PC location.

- **SJMP:** This is the short jump instruction and is of size two bytes. The first byte holds the code for SJMP while the second byte contains an offset (the distance of the target address from the current value of PC). This is a relative branching with the instruction format as follows.

```
SJMP <8-bit offset>
```

2. *Conditional Branching*: These instructions can be used to transfer the program control to the designated address, subject to the satisfaction of the condition specified in the instruction. Similar to SJMP, the conditional branches are all relative to the PC. The variations available are as follows.

- **J<*condition*>**: There are several instructions in this category, the identification of the *condition* makes the full mnemonic. Table 9.7 shows the instructions with their meanings.

TABLE 9.7 J<condition> instructions

Instruction	Meaning	Example
JZ <target>	Jump if A = 0	JZ LABEL1
JNZ <target>	Jump if A ≠ 0	JNZ LABEL1
JC <target>	Jump if CY = 1	JC TARGET
JNC <target>	Jump if CY = 0	JNC TARGET
JB <bit>, <target>	Jump if <bit> = 1	JB P1.4, TARGET
JNB <bit>, <target>	Jump if <bit> = 0,	JNB ACC.1, LABEL1
JBC <bit>, <target>	Jump if <bit> = 1, clear bit	JBC ACC.5, LABEL1

It may be noted that out of all the conditional branching instructions, JBC is a special one, as it also clears the bit to zero.

- **CJNE:** The instruction, *compare and jump if not equal*, is a complex instruction. It compares a source byte with a destination byte. The carry flag (CY) becomes 1 if the destination is less than the source and 0 otherwise. The format of the instruction is as follows.

```
CJNE <dest-byte>, <source-byte>, <target>
```

Table 9.8 shows some examples of the CJNE instruction using different addressing modes to specify the source and the destination bytes.

TABLE 9.8 CJNE instructions

Addressing Mode	Format	Example	Meaning
Immediate	CJNE A, #data, target	CJNE A, #55, L1	Jump to L1 if A ≠45
Direct	CJNE A, direct, target	CJNE A, 52H, L1	Jump to L1 if A ≠ RAM[52H]
Register	CJNE Rn, #data, target	CJNE R1, #55, L1	Jump to L1 if R1 ≠55
Indirect	CJNE @Ri, #data, target	CJNE @R1, #55, L1	Jump to L1 if RAM[(R1)] ≠55

EXAMPLE 9.4

A typical use of the CJNE instruction is to implement repetitive program logic for some operations, till some condition is met. As an example, the following program repeatedly reads port P1. If the value read is less than 100, it is output to port P2. If the value is more than 100, it is output to port P0. However, if the value is equal to 100, the program comes out of this repetition and proceeds to the instruction labelled NEXT. Here, port P1 is assumed to have been configured as input, beforehand.

```
L0: MOV A, P1         ; Read P1 into A
    CJNE A, 100, L1   ; If A = 100, jump to NEXT
    SJMP NEXT         ;
L1: JNC L2            ; If A > 100
    MOV P0, A         ; Output to P0
    SJMP L0           ; Read next input
L2: MOV P2, A         ; A < 100, Output to P2
    SJMP L0           ; Read next input
NEXT:
```

- **DJNZ:** The mnemonic stands for decrement and jump if not zero. As evident from the mnemonic, the instruction first decrements a specified byte. If the result is nonzero, the program control branches to the mentioned target address. An important deviation in the behaviour of the instruction is that none of the status flags get affected. The format of the instruction is as follows.

  DJNZ *<byte>*, *<target>*

  Table 9.9 shows some examples of the DJNZ instruction using different addressing modes for the byte operand.

TABLE 9.9 DJNZ instructions

Addressing Mode	Format	Example	Meaning
Direct	DJNZ direct, target	DJNZ 52H, L1	Decrement RAM[52H], if result is nonzero, jump to L1
Register	DJNZ Rn, target	DJNZ R5, L1	Decrement R5, if result is nonzero, jump to L1

EXAMPLE 9.5

Typical usage of the DJNZ instruction is in implementing loops. A general-purpose register or a memory location can be used as the loop counter. With every iteration, the counter is decremented by 1, till it becomes zero. An example has been noted next. Here the register R2 has been assumed to act as the counter. In the body of the loop, the input value read from port P1 has been copied into port P2 as output. It is assumed that port P1 has been already configured as an input.

```
L0: MOV A, P1     ; Read P1 into A
    MOV P2, A     ; Output to P2
    DJNZ R2, L0   ; Decrement counter and
                  ; repeat if the counter R2 is
                  ; nonzero
```

3. *Subprogram Call/Return*: Two instructions are available in the 8051 instruction set for subprogram calls – LCALL and ACALL.
   - **LCALL:** This instruction is used to call a subroutine located anywhere in the 64 kB address space of 8051. It is a three-byte instruction with the first byte containing the code for LCALL while the remaining two bytes contain the 16-bit address. The instruction is in the following format.

     LCALL *<16-bit address>*

     While executing the LCALL instruction, the return address is saved into the stack. For example, if the LCALL instruction is at memory location 1000H, the next instruction following it will be at 1003H (as LCALL occupies three bytes). Assuming that the SP contains 32H, the return address 1003H is stored in RAM locations 33H and 34H. The SP gets updated to 34H. After saving the return address, the 16-bit address specified in the instruction is copied into the PC. As a result, the control branches to the subprogram.
   - **ACALL:** The ACALL instruction is similar to AJMP as far as the target address specification is concerned. Here also, the least significant 11 bits of the target address is specified in the instruction. ACALL is a two-byte instruction in the following format.

     ACALL *<11-bit address>*

     The 11-bit address replaces the least significant 11 bits of PC while the most significant 5 bits of PC remains unaltered. As a result, the target subroutine must be located within the 2 kB address range from the current PC value (that is, current instruction). Similar to LCALL, the 16-bit return address is saved onto the stack and SP is incremented by two.
   - **RET and RETI:** These are single-byte instructions, to be used at the end of a subprogram. The instructions cause the execution to be restarted from the instruction following the last LCALL/ACALL. To return from the subprogram, the top two bytes are popped from the stack to load into

the PC. The SP is decremented by two, accordingly. The RETI instruction is used specifically in the subprograms corresponding to interrupt service routines. Similar to RET, RETI also restores the content of the PC from the stack. In addition, RETI restores the interrupt logic to enable it to accept any additional interrupts. In case of pending interrupts, RETI causes the corresponding service routine to be taken up for execution after executing one instruction at the return address.

EXAMPLE 9.6

To illustrate the subprogram structure of 8051, consider a system that has got some LEDs connected to Port P2. The program makes the LEDs blink continually. To accomplish this, the program first outputs 00H to the port and then FFH with a certain delay. The sequence is repeated continually to achieve the blinking effect. In the program, the delay routine has been designed as a subprogram.

```
    L1: MOV A, #00H   ; Set 00H in A
        MOV P2, A     ; Output to P2
        LCALL DELAY   ; Call Delay Routine
        MOV A, #0FFH  ; Set 0FFH in A
        MOV P2, A     ; Output to P2
        LCALL DELAY   ; Call Delay Routine
        SJMP L1       ; Repeat for ever
 DELAY: MOV R1, #0FFH ; R3 used as counter
    L2: DJNZ R1, L2   ; Loop if R1 ≠ 0
        RET           ; Return from subprogram
```

9.3.3 Arithmetic and Logic Instructions

Similar to other processors, the 8051 CPU supports several instructions to carry out arithmetic and logic operations. The instructions can be classified into three broad categories – arithmetic instructions, logical instructions and rotate instructions. The instructions have been discussed in detail, in the following.

1. *Arithmetic instructions*: The arithmetic instructions supported by the 8051 include operations like addition (ADD, ADDC, DA), subtraction (SUBB), multiplication (MUL) and division (DIV).
    - **ADD, ADDC, DA:** These are 8-bit addition instructions. Both unsigned and signed arithmetic could be performed using these instructions. While ADD and ADDC instructions require two operands, DA requires only one operand. The accumulator register A must be used as the first operand for ADD/ADDC instructions, while register A happens to be the sole operand of DA instruction. The second operand of ADD/ADDC is a byte, as noted next.

```
ADD A, <source byte>
```

Different addressing modes can be used to specify the operand source byte, as shown in Table 9.10.

TABLE 9.10 ADD with different addressing modes

Addressing Mode	Format	Example	Meaning
Register	ADD A, Rn	ADD A, R1	A = A + R1
Immediate	ADD A, #data	ADD A, #23H	A = A + 23H
Direct	ADD A, direct	ADD A, 4	A = A + RAM[4]
Indirect	ADD A, @Ri	ADD A, @R1	A = A + RAM[(R1)]

Due to the addition operation, the status flags carry (CY), auxiliary carry (AC) and overflow (OV) are affected. The addition operation may be signed or unsigned. In unsigned operation, the operands and results are in the range 00H to FFH. If the result of addition is beyond FFH (255 in decimal), the CY flag becomes 1. In the case of signed operation, the valid operands and results are in the range of –128 to +127. For results beyond this range, the overflow flag (OV) gets set.

To add the carry flag (CY) as well, the instruction ADDC could be used that stands for "Add with carry". The format of the instruction is as follows.

```
ADDC A, <source byte>
```

By executing this instruction, the A register content gets modified to A + <*source byte*> + CY. The rest of the features of ADDC instruction are the same as ADD.

Similar to 8085, BCD number additions could be facilitated by the DA (Decimal Adjust) instruction. The format of the instruction is as follows.

```
DA   A
```

EXAMPLE 9.7

To illustrate the usage of ADD and ADDC instructions, let us consider the problem of adding a set of 10 numbers stored from RAM location 55H onwards. The result of the addition will be stored at location 60H. The following program performs the task.

```
   L1: MOV R0, #55H    ; Set memory pointer in R0
       MOV R2, #10     ; Set counter in R2
       CLR A           ; Clear A
       CLR C           ; Clear CY
 NEXT: ADDC A, @R0     ; Add next number
       DJNZ R2, NEXT   ; All numbers added?
       MOV 60H, A      ; Store result at 60H
```

EXAMPLE 9.8

For the example program noted in Example 9.7, if the numbers are in BCD format, DA instruction has to be used to rectify the results of addition operations. It has been accomplished in the following program that stores the result at memory location 60H. If the final result is more than 99 (in decimal), the third most significant digit gets stored at location 61H. However, provision has been made for results limited to three decimal digits only, that is, the sum is less than 1000.

```
   L1: MOV R0, #55H      ; Set memory pointer in R0
       MOV R2, #10       ; Set counter in R2
       CLR A             ; Clear A
       CLR R3            ; R3 holds the 3rd digit
 NEXT: ADD A, @R0        ; Add next number
       DA  A             ; Adjust for BCD
       JNC L1            ; If CY = 0, do not change R3
       INC R3            ; Increase 3rd digit
       DJNZ R2, NEXT     ; All numbers added?
       MOV 60H, A        ; Store result at 60H
       MOV 61H, R3       ; Store result at 61H
```

- **SUBB:** This is the 8051 instruction to be used to perform subtraction operation. The instruction format is as follows.

  SUBB A, *<source byte>*

  Similar to addition operation, this instruction subtracts the source byte and the carry flag (CY) from the A register. Unlike ADD, there is no simple SUB instruction available in the 8051 instruction set. Hence, for simple subtraction (without borrow), the carry flag must be cleared beforehand. For example, the following code fragment subtracts 49H from the accumulator.

```
CLR C
SUBB A, #49H
```

  It may be noted that the SUBB instruction carries out subtraction in 2's complement format. Thus, the exact sequence of operations carried out is as follows.

  1. Take 2's complement of the source byte

  2. Add it to A register

  3. Invert carry flag CY

EXAMPLE 9.9

In the following, the operation of subtracting 14H from 17H has been illustrated. The carry bit has been assumed to be reset already.

$$
\begin{array}{lcll}
17H & = & 00010001 & \\
14H & = & \underline{+\,11110010} & \text{(2's complement)} \\
 & & 1\ 00000011 & \\
03H\ (03_{10}) & = & 0\ 00000011 & \text{(Invert carry)}
\end{array}
$$

On the other hand, subtraction of 17H from 14H will proceed as follows.

$$
\begin{array}{lcll}
14H & = & 00001110 & \\
17H & = & \underline{+\,11101111} & \text{(2's complement)} \\
 & & 0\ 11111101 & \\
FDH\ (-3_{10}) & = & 1\ 11111101 & \text{(Invert carry)}
\end{array}
$$

- **MUL:** The 8051 supports multiplication operation, however, it multiplies only its register A with register B. As the registers A and B are 8-bit wide, the result of multiplication will be 16-bit wide. The lower-order byte of the 16-bit result is stored in A register. The B register holds the higher-order byte of the result. The instruction is as follows.

```
MUL AB
```

  Assuming that the register A contains 100_{10} and B 200_{10}, after executing the multiplication, the result becomes 20000_{10} = 4E20H. Now, the A register will contain 20H of the result while register B will hold 4EH. Among the status flags, the MUL instruction affects the carry (CY) and the overflow (OV) flags only. As documented in the manual, the CY flag always gets reset. If the result is more than 255, the OV flag gets set to 1, else reset to 0.

- **DIV:** The instruction divides the byte operand available in A register by the byte operand of B. The format of the instruction is as follows.

```
DIV AB
```

  To avoid the divide-by-zero error, the programmer must ensure that the content of the B register is nonzero. If the B register contains zero, the result of division will be undefined. After division, the quotient is made available in register A and the remainder in B. For example, if A = 42_{10} and B = 5_{10}, after executing "DIV AB", the contents of A and B registers will be 8_{10} and 2_{10}, respectively. Compared to the MUL instruction, the execution of the DIV instruction clears the CY flag. In case of an attempted divide-by-zero, the OV flag gets set to 1, else the OV flag is reset to zero.

- ***Handling Overflows*****:** Most of the operations in 8051 use 8-bit signed arithmetic. With 8-bit, the range of valid numbers is –128 to +127. Whenever the result of an operation is beyond this range, the overflow flag OV gets set to 1, marking the occurrence of an error. Thus, the programmer must check the OV flag after an operation, wherever necessary. The OV flag gets set in either of the two situations, as noted next.

  (i) *There is a carry from D6 to D7 but no carry from D7*: As there is no carry from D7, the carry flag is zero. However, as there is a carry from D6 to D7, the result could not be contained in 7 bits. The D7 bit of both the operands must have been zero, indicating positive numbers. However, in the result, the bit is 1. It indicates the number to be negative. Hence, there is an overflow.

  (ii) *There is a carry from D7 but no carry from D6 to D7*: Here, there is no carry from D6 to D7, but there is a carry from D7. It implies that both the operands have D7 = 1, indicating both to be negative numbers. After the operation, D7 becomes zero, indicating a positive number. Hence it is also a case of overflow.

2. *Logic instructions*: The instruction set of 8051 contains a number of instructions to carry out logical operations on its operands. The instructions can perform the Boolean operations AND, OR, XOR and complement.

- **AND:** The instruction corresponding to the AND operation is ANL (AND logical). The following is the format of the instruction.

```
ANL <dest-byte>, <source-byte>
```

The instruction performs bit-by-bit ANDing of the two bytes. The result is put into the destination byte. Table 9.11 shows modes in which the two bytes could be specified in the ANL instruction.

A bit level AND operation can be performed in conjunction with the carry flag (CY) There is a special version of ANL instruction that works with bit operands. The CY flag acts as the destination. The instruction is in the format noted next.

```
ANL C, <source-bit>
```

A variant of the ANL bit-level instruction allows the usage of the invert of source-bit as the operand to be ANDed with the carry flag. The instruction is as follows.

```
ANL C, /<source-bit>
```

TABLE 9.11 ANL with different addressing modes

Addressing Mode	Format	Example	Meaning
Register	ANL A, Rn	ANL A, R1	A = A AND R1
Immediate	ANL A, #data	ANL A, #30H	A = A AND 30H
Direct	ANL A, direct ANL direct, A ANL direct, #data	ANL A, 6 ANL 6, A ANL 45, #42H	A = A AND RAM[6] RAM[6] = RAM[6] AND A RAM[45] = RAM[45] AND 42H
Indirect	ANL A, @Ri	ANL A, @R1	A = A AND RAM[(R1)]

EXAMPLE 9.10

To illustrate the usage of ANL instruction, consider a code fragment noted next. It sets the memory location 60H to 0FFH if the input port bits P2.3 is 1 and P2.6 is zero. Else, the location is set to 77H.

```
          MOV 60H, #0FFH     ; Set memory location 60H
          MOV C, P2.3        ; Get bit P2.3 into CY
          ANL C, /P2.6       ; AND with an invert of P2.6
          JNC L1             ; No change of 60H
          MOV 60H, #77H      ; Set memory location 60H
      L1:
```

- **OR, XOR:** The OR and XOR operations can be carried out in a similar fashion as AND. The mnemonics for these operations are ORL (OR logical) and XRL (XOR logical), respectively. The supported addressing modes for the source and destination operands are the same as in ANL instruction. In the following, some examples have been noted.

```
ORL A, R3     ; A = A OR R3
XRL A, #30H   ; A = A XOR 30H
```

- **CPL:** The instruction can be used to invert the content of the accumulator A. It computes the 1's complement of the content of A. The instruction is as follows.

```
CPL A
```

  A variant of the CPL instruction works with the bit-level operands. The specified bit gets complemented. The instruction is in the following format.

```
CPL <bit>
```

  For example, the instruction "CPL P2.5" will complement port 2's bit number 5.

EXAMPLE 9.11

The following program fragment uses the CPL instruction to toggle the bits of port P1 continually. Initially, a pattern consisting of alternate 0's and 1's is put into port P1 that sets alternate bits to 1's and 0's. After that, the bits flip continually.

```
    MOV A, #55H    ; Set A to 01010101
L1: MOV P1, A      ; Output to P1
    CPL A          ; Complement A
    SJMP L1        ; Repeat forever
```

3. *Rotate instructions*: The rotate instructions are useful to rotate the content of the accumulator register A towards left or right. There are four variants of rotations – RR, RL, RRC and RLC. In the following, each of these has been discussed in detail.
   - **RR:** This instruction performs right rotation of the accumulator by one-bit position. On execution of the instruction, the bits, starting with MSB, get shifted right by one position. The LSB is fed back to the MSB. The situation has been shown in Figure 9.1. The format of the instruction is as follows.

```
RR A
```

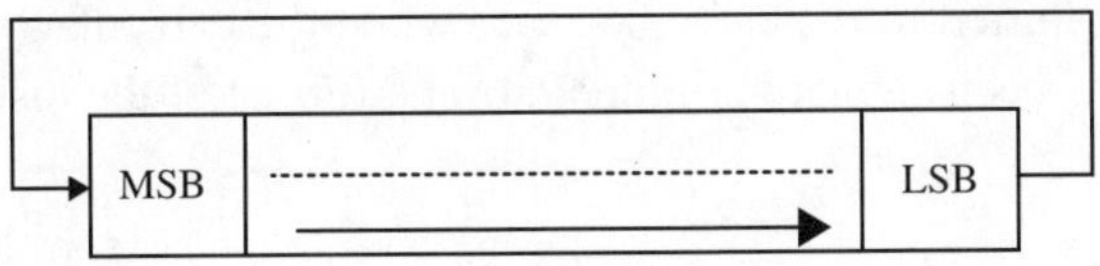

FIGURE 9.1 RR instruction

   - **RL:** This is the rotate left instruction. By executing this, the content of the accumulator gets rotated left by one-bit position. The bits are shifted left from the LSB towards the MSB, while the MSB is fed back to the LSB. The operation has been shown in Figure 9.2. The instruction has the following format.

```
RL A
```

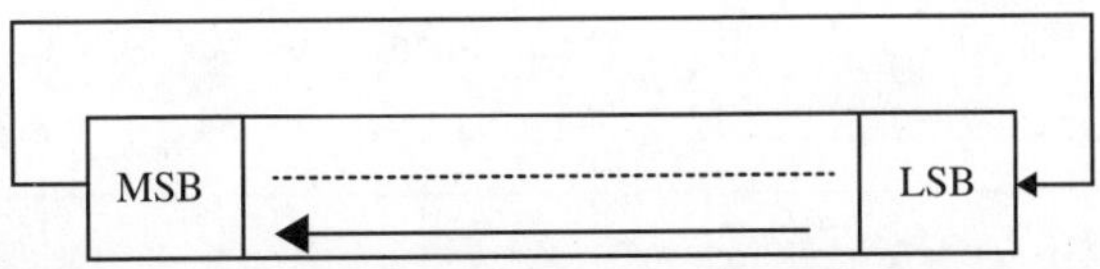

FIGURE 9.2 RL instruction

   - **RRC:** This instruction carries out rotation through the carry bit. The operation can be considered to be similar to RR with the carry flag (CY) behaving as an extension to the least significant side of the accumulator. The bits are shifted by one position from MSB towards the LSB. Finally, the LSB gets transferred to CY and the CY gets rotated to the MSB of A. The operation has been illustrated in Figure 9.3. The format of the instruction is as follows.

```
RRC A
```

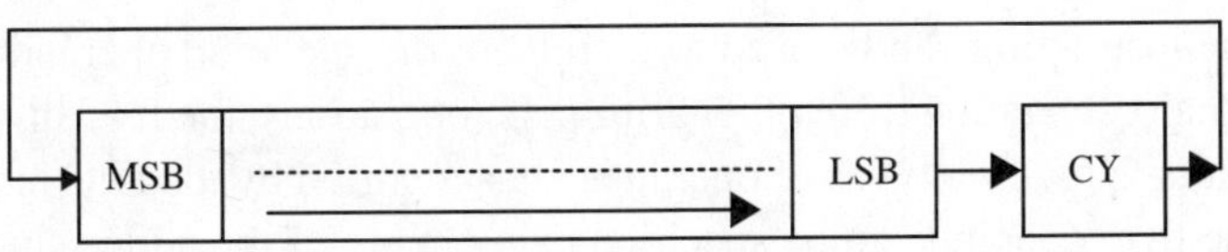

FIGURE 9.3 RRC instruction

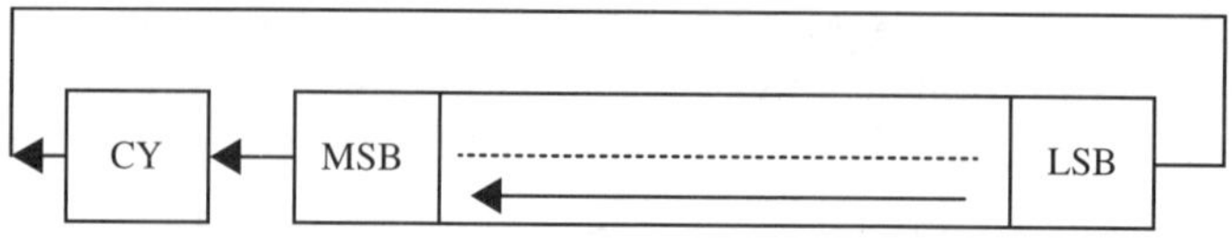

FIGURE 9.4 RLC instruction

- **RLC:** The instruction rotates the accumulator left through the carry flag. The RLC instruction is just the reverse of RRC. The A register gets extended by one bit on the MSB side via the carry flag (CY). While executing this instruction, the bits of A get shifted left from LSB to the MSB side. The MSB gets shifted to the carry flag CY, while the carry flag reaches the LSB of A. The execution of RLC has been shown in Figure 9.4. The format of the instruction is as follows.

  RLC A

EXAMPLE 9.12

To illustrate the usage of RRC instruction, consider the following code fragment that can transfer the bits of the accumulator serially through bit 0 of port P2.

```
      CLR C          ; Clear carry
      MOV R4, #8     ; R4 holds the count 8
NEXT: RRC A          ; Get next bit in carry
      MOV P2.0, CY   ; Output carry to P2.0
      DJNZ R4, NEXT  ; Check if all bits transferred
```

9.3.4 Other Miscellaneous Instructions

Apart from the different classes of instructions, discussed in previous sections, there are some special instructions in 8051 with specific functionalities. The instructions have been discussed in the following.

- **CLR:** The accumulator register A or a specific bit can be cleared via this instruction. Accordingly, there are two variants of this instruction, as noted next.

```
CLR A     ; Clears the accumulator, no status flags affected
CLR <bit> ; Clears the specified bit
```

- **INC:** The instruction can be used to increment the specified operand. It increments the operand by 1. In major deviation from other arithmetic operations, the INC instruction does not affect any of the status flags. Two types of operands are supported by the INC instruction – 8-bit and 16-bit. While the byte operand can be specified using many of the addressing modes, for a 16-bit operand only the DPTR register can be utilized. Table 9.12 illustrates the addressing modes supported by the instruction.

TABLE 9.12 INC with different addressing modes and operands

Addressing Mode	Format	Example	Meaning
Accumulator	INC A	INC A	A = A + 1
Register	INC Rn	INC R5	R5 = R5 + 1
Direct	INC direct	INC 6	RAM[6] = RAM[6] + 1
Indirect	INC @Ri	INC @R1	RAM[(R1)] = RAM[(R1)] + 1
DPTR (16 bit)	INC DPTR	INC DPTR	DPTR = DPTR + 1

- **DEC:** As the inverse of the INC instruction, the DEC instruction decrements the specified operand by 1. The DEC instruction cannot be applied to the 16-bit DPTR register. Only 8-bit operands are allowed. It may be noted that "DEC DPL" instruction is allowed. The instruction does not affect the status flags. Table 9.13 illustrates the various addressing modes supported by the DEC instruction.

TABLE 9.13 DEC with different addressing modes

Addressing Mode	Format	Example	Meaning
Accumulator	DEC A	DEC A	A = A – 1
Register	DEC Rn	DEC R5	R5 = R5 – 1
Direct	DEC direct	DEC 6	RAM[6] = RAM[6] – 1
Indirect	DEC @Ri	DEC @R1	RAM[(R1)] = RAM[(R1)] – 1

- **NOP:** Similar to 8085, the *No Operation* instruction is supported by the 8051, as well. The execution of this instruction does not affect any of the CPU registers or memory locations. Thus, the instruction can be used by the programmers to introduce small delays in the execution. The instruction has the format noted next.

```
NOP
```

- **SWAP:** The SWAP instruction swaps the nibbles of the A register. That is, the bits D3–D0 of the A register get exchanged with the bits D7–D4 of the A register itself. The format of the instruction is as follows.

```
SWAP A
```

For example, the instruction "SWAP A" with A = 65H changes A to 56H.

EXAMPLE 9.13

To illustrate programming in the assembly language of 8051, next we shall consider the problem of sorting a set of numbers in the ascending order. The array of N numbers to be sorted starting is assumed to be present at external memory address 3000H. The RAM location 30H has been used as a temporary space for interchange operations.

```
        ORG 1000H          ; Clear carry
        TEMP EQU 30H       ;
        N EQU 10           ; Array size
        MOV R2, #N - 1     ; Load N - 1 into R2
L1:     MOV DPTR, #3000H   ; Set DPTR to start of array
        MOV R3, R2
L2:     MOVX A, @DPTR      ; Get next number
        MOV R1, A          ; Copy A into R1
        INC DPTR           ; Point to next number
        MOVX A, @DPTR      ; Get next number
        MOV TEMP, A        ; Copy A into TEMP
        MOV A, R1          ; Get previous number in A
        CJNE A, TEMP, L3   ; A ≠ (TEMP), goto L3
        SJMP L4            ; A = (TEMP), goto L4
L3:     JC L4              ; A < (TEMP), goto L4
        MOVX @DPTR, A      ; A > (TEMP), exchange
        MOV A, TEMP        ; two locations
        DEC DPL            ;
        MOVX @DPTR, A      ;
        INC DPL            ;
L4:     DJNZ R3, L2        ; Repeat until  R3 is zero
        DJNZ R2, L1        ; Repeat until R2 is zero
        END
```

9.4 Embedded C Language Programming for 8051

LO 4

The assembly language programming enables the programmer to write programs in the machine language of a processor. However, the individual instructions in the assembly language are very simple. As a result, writing a program with complex logic in assembly language may be very cumbersome. Microcontrollers like 8051 are used in developing various electronic systems. The majority of such electronic systems have dedicated functionality, for example, digital camera, mobile phone, car and refrigerator. Such electronic systems with dedicated functionality are commonly known as *embedded systems*. The programming requirements for such system development need to utilize the constructs like loop, if-then-else and switch-case. The constructs are not available in the assembly language.

It necessitates writing programs in high-level languages, supporting such constructs. This software, commonly called *embedded software* is written in a modified version of high-level languages supporting direct hardware access. Accordingly, the programming language ANSI C has been modified and augmented with hardware-oriented features, often called the *embedded C*.

The choice of C for embedded hardware design provides the following benefits.

- Compared to their assembly language counterparts, C programs are generally small and compact. C programs are also easier for human understanding. Efficient programming tools including compilers are easily available for the C language. Finally, there are a good number of experienced C programmers available who can develop such applications efficiently.
- C programs can be ported easily across platforms. To retarget an application to a new microcontroller may just need a recompilation of the source code. On the other hand, the assembly language programs will need a complete redesign and rewriting of the code.
- From its inception itself, C language supports constructs and data types which are close to the hardware.

Embedded C, primarily used for embedded system design, is a language covering a subset of the features of full C language. Due to the limitation in the resources (such as memory and I/O ports), the programs are to be careful in using them. Programs written in embedded C are generally compiled using *cross-compilers*. A cross-compiler executing on a processor generates machine code for a different target processor. With the target processor being 8051, the created machine code files are downloaded onto the 8051 for execution. In the following, the features of embedded C have been discussed along with some code fragments as examples.

9.4.1 Data Types

The embedded C language for 8051 supports the following data types, over and above those supported by the standard C. These additional data types are widely used in embedded system programming.

Unsigned char: This is an 8-bit data type and can support values in the range of 0–255. Unsigned character happens to be the most fundamental data type in processors like 8051.

EXAMPLE 9.14

Consider the following C language program using the unsigned char data type. It outputs the values 00-7FH sequentially to port P1 of 8051.

```
#include <reg51.h>
void main (void)
{
     unsigned char x;
     for (x = 0; x < 0x80; m++)
          P1 = x;
}
```

Signed char: The signed character is also an 8-bit data type that can store values in the range –128 to +127. By default, all variables in an embedded C program are of this type. Hence, to enable a variable to hold values in the range 0 to 255, it must be explicitly declared as "unsigned char".

EXAMPLE 9.15

The following C language program operates on an array a of 10 signed characters. The program adds those numbers and puts the result in another variable b.

```
#include <reg51.h>
void main (void)
{
        signed char a[] = { -6, -5, 10, 40, -3, 1, -1, 45, 55, -16 };
        signed char b = 0;
        unsigned  char j;
        for (j = 0; j < 10; j++)
            b += a[ j ];
}
```

Unsigned int: The unsigned integer is 16-bit wide and thus, can represent numbers in the range 0 to 65535 (FFFFH). The 16-bit numbers have specific applications, such as memory address and counter values. As 8051 is an 8-bit processor with internal operations and memory locations being 8-bit wide, storing 16-bit numbers is a bit complex. It takes two memory bytes to store the number. As a result, access to a 16-bit number in memory is more complex than a byte access. This is often reflected in the size of the compiled code. For programs using a 16-bit unsigned int data type, the compiled machine code may be significantly larger than the 8-bit version of the same program.

EXAMPLE 9.16

The following is a very simple program that uses an unsigned integer variable to introduce some delay in the execution. Apart from that, the program toggles the bits of port P2 continually.

```
#include <reg51.h>
void main (void)
{
        unsigned int n;
        for (; ;) {
            P2 = 0x55;
            for (n = 0; n < 65535; n++);
            P2 = 0xAA;
            for (n = 0; n < 65535; n++);
        }
}
```

Signed int: This is a 16-bit data type. The variables belonging to this type can hold values in the range –32768 to +32767. It may be noted that signed int happens to be the default integer data type in embedded C.

Single bit: Bit happens to be a very important data type for embedded C language. Two bit data types are supported by the 8051 compilers. The first one, *bit* refers to the individual bit locations in the bit-addressable RAM address range 20H–2FH. The second one, called *sbit* can be used to refer to the individual bits of *special function registers*.

EXAMPLE 9.17

The following program utilizes the bit data type of embedded C. The program flips the bit-2 of port P2 continually.

```
#include <reg51.h>
sbit BIT2 = P2^2;          // Bit-2 of Port P2
void main (void)
{
     unsigned int m;
     for (; ;) {
         BIT2 = 0;
         for (m = 0; m < 65535; m++);
         BIT2 = 1;
         for (m = 0; m < 65535; m++);
     }
}
```

Special function register: The 8051 processor contains a good number of *special function registers* (*sfr*s). A byte data type *sfr* has been created corresponding to the special function registers of 8051. It may be noted that *sfr* data type corresponds to byte-size special function registers, whereas *sbit* defines variables at the bit level.

EXAMPLE 9.18

The following program uses the sfr data type to flip the bits of port P2 continually.

```
sfr P2 = 0xA0;
void main (void)
{
     unsigned int m;
     for (; ;) {
         P2 = 0x55;
```

```
        for (m = 0; m < 65535; m++);
        P2 = 0xAA;
        for (m = 0; m < 65535; m++);
    }
}
```

EXAMPLE 9.19

In the following we shall look into an assembly language program for 8051 that reads ports P0 and P1 continually, computes AND of the values read and outputs to port P2. The program will then be written in embedded C to understand the advantages of high-level programming.

```
    ORG 0000H      ;
    MOV P0, #0FFH  ; Set P0 as input
    MOV P1, #0FFH  ; Set P1 as input
L1: MOV A, P1      ; Read next value from P1
    MOV R2, A      ; Save value read in R2
    MOV A, P0      ; Read next value from P0
    ANL A, R2      ; Perform AND operation
    MOV P2, A      ; Output result to P2
    SJMP L1        ; Continue
```

In the following, the same program has been written in embedded C.

```
#include <reg51.h>
unsigned char d1, d2, result;
void main (void)
{
    while (1) {
        d1 = P1;
        d2 = P0;
        result = d1 & d2;
        P2 = result;
    }
}
```

EXAMPLE 9.20

The assembly language program noted next reads port bits P0.0 and P1.4 continually. It then computes AND of the values read and outputs to port P2.1.

```
    ORG 0000H       ;
    MOV P0, #0FFH   ; Set P0 as input
    MOV P1, #0FFH   ; Set P1 as input
L1: MOV A, P1       ; Read next value from P1
    RR A            ; Rotate right A 4 times
    RR A
    RR A
    RR A
    MOV R1, A       ; Copy content of A in R1
    MOV A, P0       ; Read next value from P0
    ANL A, R1       ; Perform AND operation
    RL A            ; Rotate A left
    MOV P2, A       ; Output result to P2
    SJMP L1         ; Continue
```

The same program when written in C will be as follows.

```
#include <reg51.h>
  bit data1 = P0.0;
  bit data2 = P1.4;
  bit result = P2.1;
  void main (void)
  {
          while (1) {
                  result = data1 & data2;
          }
  }
```

9.5 Timers and Counters

LO 5

As noted in Chapter 8, the 8051 contains within it modules that can act as timers and/or counters. The same hardware is utilized for both timer and counter. For timer operation, the clock signal is provided internally, while for counter operation, the event pulses are fed from the outside. In this section, we shall look into the techniques to program the timer/counter modules in different modes.

9.5.1 Timer Programming

The 8051 contains two timers, namely, *Timer 0* and *Timer 1*. These are 16-bit timers, however they can be configured to work as 8-bit, 13-bit or 16-bit timers.

THO								TLO							
D15	D14	D13	D12	D11	D10	D9	D8	D7	D6	D5	D4	D3	D2	D1	D0

(a)

TH1								TL1							
D15	D14	D13	D12	D11	D10	D9	D8	D7	D6	D5	D4	D3	D2	D1	D0

(b)

FIGURE 9.5 Timer registers (a) Timer 0 (b) Timer 1

The registers associated with the operation of timers are as follows.

1. *Timer registers*: There are two 8-bit registers, TH*x* and TL*x*, associated with each timer. Together, the registers hold 16-bit time/count. The registers TH0 and TL0 correspond to Timer 0, while Timer 1 has the registers TH1 and TL1. The structure of these registers has been shown in Figure 9.5. The registers may be loaded with values via the MOV instruction, as shown next.

```
MOV TL0, #55H   ; Moves 55H to TL0
MOV R2, TH1     ; Moves TH1 to R2
```

2. *Timer mode register TMOD*: The 8-bit TMOD register is used to control both the timers. Bits 0 to 3 of the TMOD register control the operation mode of Timer 0, while the bits 4 to 7 control Timer 1. Figure 9.6 shows the structure of the TMOD register. Each part has the following components.
   - **M1, M0:** The mode of operation of a timer is controlled by these two bits. Table 9.14 enumerates the mode settings. Most of the applications use either of the modes 1 and 2. In the following, we shall explore these two modes in detail.
   - **C/T:** The C/T bit is responsible for determining whether the module will be used as a timer or as a counter. When C/T bit is set to 0, it implies timer operation. On the other hand, setting C/T to 1 enforce counter operation. For timer operation, the clock ticks are derived from a crystal. The timer frequency is 1/12th the crystal frequency, as shown in Figure 9.7. Assuming that the crystal is of frequency 12 MHz, the frequency of operation for the timer is 1 MHz. Most of the

Timer 0				Timer 1			
GATE	C/T	M1	M0	GATE	C/T	M1	M0

FIGURE 9.6 TMOD register structure

TABLE 9.14 M1, M0 bit settings

M1	M0	Mode	Operation
0	0	0	• 13-bit timer/counter • TH*x* holds 8-bit count • TL*x* is a 5-bit prescalar
0	1	1	16-bit timer/counter
1	0	2	• 8-bit auto reload • TH*x* holds reload value • TL*x* holds count
1	1	3	Split timer

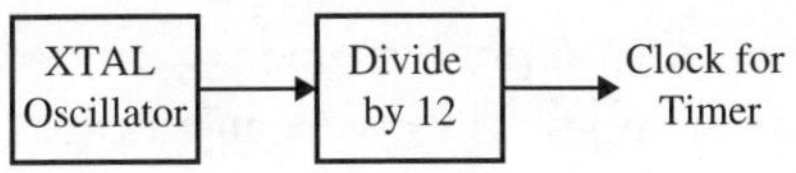

FIGURE 9.7 Derivation of timer clock

8051 based systems use crystals of frequency 11.0592 MHz. A primary reason for this is to make the serial data transfer rate match with the then existing IBM PC. Using this crystal causes the timer frequency to become 921.6 kHz, resulting into the time period of 1.085 μS. With C/T=1 (counter operation), count pulses are sensed from outside the chip. The count pulses are fed from *T*0 (P3.4) and *T*1 (P3.5) for Timer 0 and Timer 1, respectively.

- **GATE:** The GATE bit is used to specify the mechanism by which the timers will be controlled. In 8051, the starting and stopping of timers can be controlled via both software and hardware means. If the GATE bit is reset to 0, only software based control can be used to start and stop the corresponding timer. On the other hand, setting the GATE bit (making GATE = 1) provides combined software and hardware control to start or stop the timer.

The software based control of timer operations are done by setting/resetting of TCON register bits. Figure 9.8 shows the structure of the TCON register. A part of the register (the most significant four bits) controls the timers while the other part (with remaining four bits) controls the external interrupts (to be discussed in conjunction with interrupts). Each timer has a flag bit that gets set when the timer rolls back from the maximum value to zero. The bits are *TF*0 for Timer 0 and *TF*1 for Timer 1. Bits 5 and 7 are dedicated to *TF*0 and *TF*1, respectively. It may be noted that apart from roll back, the TF bits can also be set/reset using instruction as follows.

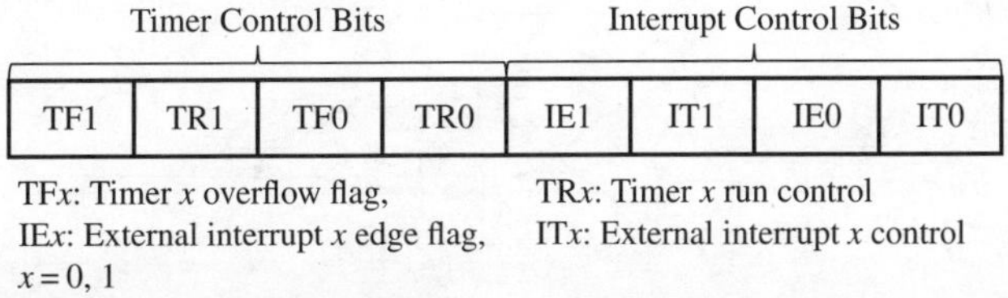

FIGURE 9.8 TCON register structure

```
SETB TF0 ; Set Timer 0 flag
CLR TF0  ; Clear Timer 0 flag
SETB TF1 ; Set Timer 1 flag
CLR TF1  ; Clear Timer 1 flag
```

The TCON register also contains the bits *TR*0 and *TR*1 to control the timers, as noted next.

```
SETB TR0  ; Start Timer 0
CLR TR0   ; Stop Timer 0
SETB TR1  ; Start Timer 1
CLR TR1   ; Stop Timer 1
```

If the GATE bit of a timer is set to 1, to run the timer, apart from setting the TR bit, the external interrupt pin is also required to be set high. Thus, we can enforce both hardware and software controls over a timer. Figure 9.9 shows the combined control for a timer. With GATE = 0, the setting of bit TRx starts the Timer x, providing pure software control of the timer operation. On the other hand, when GATE input is set to 1, to run the Timer x, apart from setting the TRx bit of TCON, the INTx pin should be set to 1. It provides both hardware and software control on timer operation. Thus, with GATE = 1, the timer can be stopped by resetting the TRx bit via software or by pulling the INTx pin low, in hardware.

Next, we shall be looking into the modes of operation of the two timers in 8051. It may be noted that the programmers mostly utilize the modes 1 and 2 for their applications. Hence, our discussions will be restricted to only these two modes.

1. *Mode 1 timer operation*: In Mode 1, a timer operates as a 16-bit one. Before starting the timer, it can be initialized to a value between 0000H and FFFFH. The 16-bit initial value can be loaded into the register pair TH-TL via MOV instructions. To start the timer, the instruction "SETB TR0" or "SETB TR1" has to be used for Timer-0 or Timer-1, respectively. Once the TR bit has been set, the timer starts incrementing its value with each clock tick. After reaching the count value FFFFH, on arrival of the next clock tick, the count value rolls over to zero and the corresponding TF bit of TCON register gets set to high. Figure 9.10 illustrates the timer operation in mode. With a crystal frequency is 11.0592 MHz, the timer operates at 921.6 kHz having clock time period of 1.085 μS. Hence, with the initial value equal to XXYY (in hexadecimal), the overflow will occur after (FFFF – *XXYY* + 1) × 1.085 μS, On the other hand, in order to determine the initial value to be loaded into a timer for generating a time

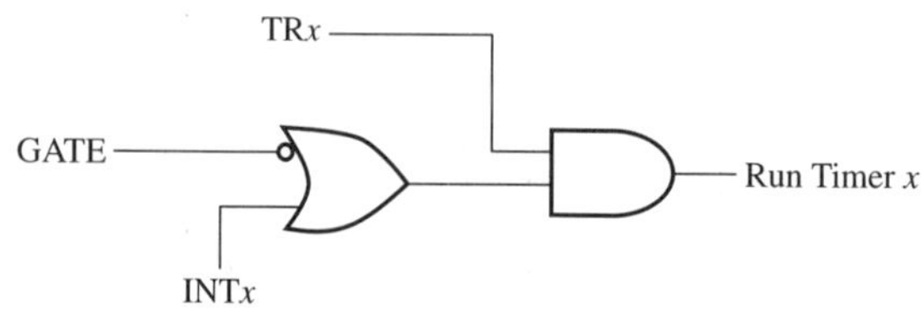

FIGURE 9.9 Run control for timer

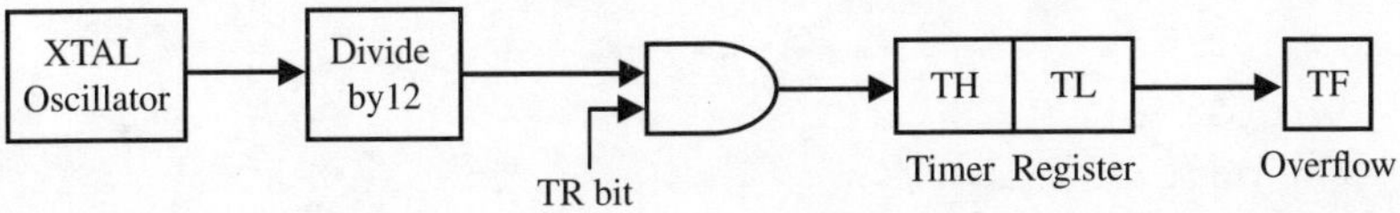

FIGURE 9.10 Timer in mode 1

delay of D μS, first, the value D is divided by 1.085 μS. Let the result of the division be n. From this, $(65536 - n)$ is computed. Let the result of subtraction be hexadecimal *xxyy*. To initialize the timer, the value *xx* should be loaded in its TH register while *yy* be loaded into TL.

EXAMPLE 9.21

To illustrate the usage of timers, next we shall consider the problem of generating a square wave with 50% duty cycle. The ON/OFF periods are taken to be 20 mS each. The program needs to have a time delay of 20 mS. To ensure the accuracy of the waveform, the delay has been generated using the Timer 0 in mode 1.

As per the preceding discussion, for initializing the timer, the value of n should be 20 ms ÷ 1.085 μS = 18434. Thus, the initial value of the timer will be 47102 (B7FEH). The TH and the TL registers should be loaded with B7H and FEH, respectively. Port bit P2.5 has been used to generate the square wave. The assembly language program has been noted next.

```
    ORG 0000H          ;
    CLR P2.5           ; Clear port bit P2.5
    MOV TMOD, #01H     ; Set Timer 0 in Mode 1
L1: MOV TL0, #0FEH     ; Load TL
    MOV TH0, #0B7H     ; Load TH
    CPL P2.5           ; Complement P2.5
    SETB TR0           ; Start timer
L2: JNB TF0, L2        ; Check roll over of TF0 flag
    CLR TR0            ; Stop timer
    CLR TF0            ; Clear timer flag
    SJMP L1            ; Repeat
    END
```

The C language program for the task will be as follows.

```
#include <reg51.h>
 void Delay_Rtn (void);
 sbit p2_5 = P2^5;
 void main (void)
```

```
{
    while (1) {
       p2_5 = ~p2_5;               // Toggle P2.5
       Delay_Rtn();                // Call delay
    }
}
void Delay_Rtn (void)
{
     TMOD = 0x01;                  // Timer 0, Mode 1
     TL0 = 0xFE;   TH0 = 0xB7;     // Load TL0 and TH0
     TR0 = 1;                      // Start timer
     while (TF0 == 0);             // Wait for TF0 rollover
     TR0 = 0;   TF0 = 0;           // Turn off timer and clear TR0
}
```

In 8051, the operation of timers in Mode 0 is similar to their operation in Mode 1. The only deviation is that in Mode 0, the timers are 13-bit wide. Thus, the value held by the TH-TL pair is in the range of 0000H to 1FFFH. The timer roles over on the next clock tick, once it has reached the value 1FFFH. The TF flag also gets set accordingly.

2. *Mode 2 timer operation*: In Mode 2, the timers operate as 8-bit ones with *auto-reload* facility. The 8-bit count value is held in the TL register and is incremented at every clock tick. Upon reaching the value 255 (FFH), the timer rolls back, raising the TF flag. However, the TL register, instead of becoming 0, gets reloaded with the value from the TH register. The loading of the TL register with the content of TH is automatic and does not require any intervention from the programmer. However, to restart the timer after a roll over, the TF flag needs to be cleared by the programmer. Figure 9.11 explains the situation. It is similar to Figure 9.10, excepting the reload part. The auto-reload feature of Mode 2 makes it attractive in applications requiring repetitive usage of timer with equal periodicity. However, due to 8-bit count value limitation, the maximum count can be only up to 255. This precludes the usage of Mode 2 from generating large delays. In the following, Example 9.21 has been revisited in Example 9.22 with the ON/OFF periods of 20 μS.

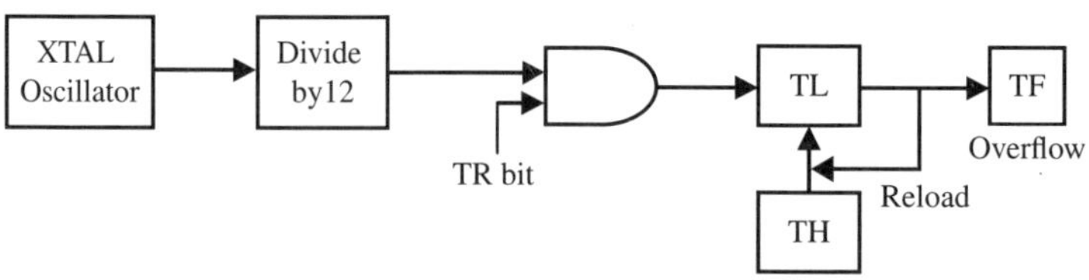

FIGURE 9.11 Timer in mode 2

EXAMPLE 9.22

Here we shall consider the problem of generating a square wave with 50% duty cycle and ON/OFF periods of 20 μS each. The delay required has been generated by Timer 0 in mode 2.

In this case, *n*, is computed as, 20 μS ÷1.085 μS = 19. Thus, the initial value needed for the timer is 237 (EDH). The TH0 register is thus loaded with the value EDH. The Timer 0 is programmed in mode 2. Hence, as soon as TH0 is loaded with the value, it gets automatically copied into TL0. The assembly language program to generate the square wave is as follows.

```
    ORG 0000H        ;
    CLR P2.5         ;  Clear port bit P2.5
    MOV TMOD, #02H   ; Set Timer 0 in Mode 2
    MOV TH0, #0EDH   ; Load TH0
L1: CPL P2.5         ; Complement P2.5
    SETB TR0         ; Start timer
L2: JNB TF0, L2      ; Check roll over of TF0 flag
    CLR TF0          ; Clear timer flag
    SJMP L1          ; Repeat
    END
```

The C language program for the same problem is as follows.

```
#include <reg51.h>
sbit p2_5 = P2^5;
void main (void)
{
    TMOD = 0x02;             // Timer 0, Mode 2
    TH0 = 0xED;              // Load TH0
    while (1) {
       p2_5 = ~p2_5;         // Toggle P2.5
       TR0 = 1;              // Start timer
       while (! TF0);        // Wait for overflow
       TF0 = 0;              // Reset timer flag
    }
}
```

9.5.2 Counter Programming

The timer modules available in 8051 can also be made to operate as counters to count external pulses. For this, the C/T bit for the corresponding timer in the TMOD register needs to be set to 1,

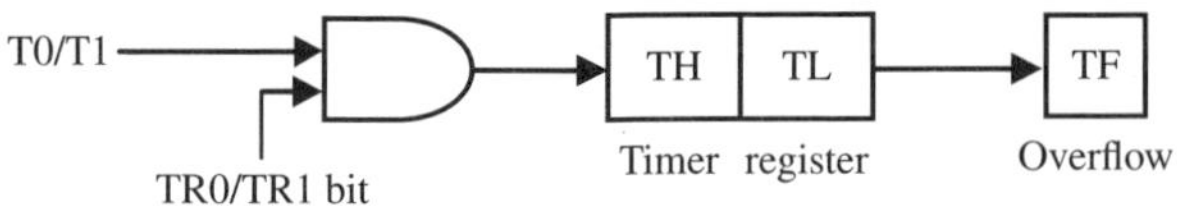

FIGURE 9.12 Counter in mode 1

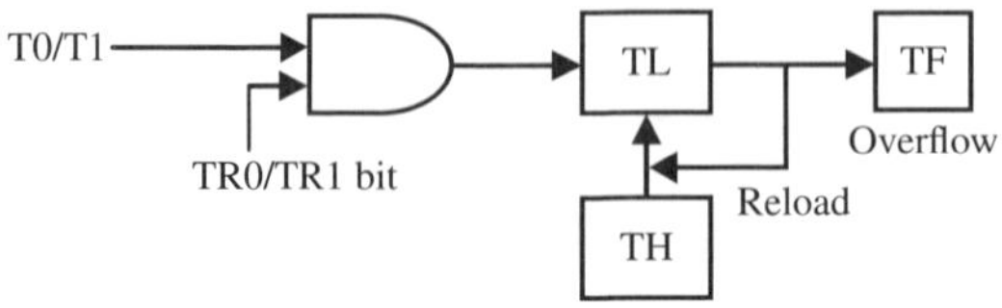

FIGURE 9.13 Counter in mode 2

indicating a counter mode of operation. Two external pins *T*0 (Pin 14) and *T*1 (Pin 15) of 8051 are used as pulse inputs to Counter 0 and Counter 1, respectively. On external pulses reaching those pins, the corresponding counters do upcounting. It may be noted that the pins 14 and 15 are multiplexed with Port 3 bits, P3.4 and P3.5, respectively. Hence, a program using the counters cannot utilize those two bits of Port 3 simultaneously for I/O operations. The operation of the counters in Mode 1 and Mode 2, counting external input pin events have been illustrated in Figures 9.12 and 9.13, respectively.

EXAMPLE 9.23

To illustrate the counter operation, let us look into the problem of counting pulses on T0 pin. The counter is made to operate in Mode 2. The counting proceeds from 0 to 255, on occurrence of further pulse on T0, the count value becomes 0. To realize this, the TH0 register has been cleared, so that, whenever the count in TL0 exceeds 255 and rolls over, the TL0 register will be reloaded with zero from TH0. On this roll over, the TF0 flag will also be set. In the program the current count available in TL0 has been output to port P2. The assembly language program for the problem is as follows.

```
    ORG 0000H       ;
    MOV TMOD, #06H  ; Set Counter 0 in Mode 2
    SETB P3.4       ; Make T0 input line
    MOV TH0, #0     ; Clear TH0
    SETB TR0        ; Start counter
L1: MOV A,TL0       ; Get TL0 content in A
    MOV P2, A       ; Output to Port 2
    JNB TF0, L1     ; Continue till roll over
    CLR TF0         ; Clear timer flag
    SJMP L1         ; Repeat
    END
```

The corresponding C language program is noted next.

```
#include <reg51.h>
sbit p3_4 = P3^4;
void main (void)
{
    TMOD = 0x06;                    // Counter 0, Mode 2
    TH0 = 0;                        // Clear TH0
    p3_4 = 1;                       // Configure T0 in input mode
    TR0 = 1;                        // Start counter
    while (1) {
       P2 = TL0;                    // Output TL0 count to Port 2
       if ( TF0 )  TF0 = 0;         // If roll over, reset TF0

    }

}
```

9.6 Interrupts in 8051

LO 6

There are six interrupts supported by the 8051 processor. Table 9.15 summarizes the types of interrupts, their vector address, pin number etc.

1. *Reset*: This interrupt is available at Pin 9 of the chip. If the pin is high, the 8051 gets a Reset interrupt. As a result, the control branches to ROM address 0000H, that is, the PC gets loaded with 0000H.
2. *Timers*: As noted previously, there are two timers in 8051 – *Timer 0* and *Timer 1*. Each timer generates an interrupt when it rolls over and the corresponding TF flag is set. The interrupts corresponding to the two are named as *Timer 0 interrupt* and *Timer 1 interrupt*. For the Timer 0 interrupt, the designated branch address is 000BH, while for the Timer 1 interrupt, the address is 001BH.
3. *INT0, INT1*: These are known as external interrupts. The physical pin corresponding to INT0 is Pin 12 which is multiplexed with P3.2 (Port 3, Bit 2). Similarly, the INT1 is applied via Pin 13

TABLE 9.15 Interrupts in the 8051

Interrupt	ROM Address	Pin No.
Reset	0000H	9
INT0	0003H	12 (P3.2)
Timer 0 interrupt	000BH	
INT1	0013H	13 (P3.3)
Timer 1 interrupt	001BH	
Serial communication	0023H	

(multiplexed with P3.3). The designated branch addresses for INT0 and INT1 are 0003H and 0013H, respectively.

4. *Serial communication*: The serial communication interrupt is activated on receiving/transmitting a character serially by the 8051. The same branching address 0023H is utilized for both the transmission and reception events.

9.6.1 Interrupt Enable Register

The enable/disable set up of the 8051 interrupts can be controlled via programming the Interrupt Enable (IE) register. By default, on reset, all interrupts are disabled. Hence, if not enabled explicitly by programming the IE register via software, the 8051 does not respond to any of the interrupts. The IE register structure has been shown in Figure 9.14. In the following, its constituent bits have been enumerated.

EA: This is the master control over all interrupts. Setting EA = 0 will disable all interrupts. On the other hand, if EA = 1, the enable/disable status of individual interrupts are controlled by other bits of the register as noted in the following.

Bit 6 is left for future extension.

ET2, ET1, ET0: These three bits are for controlling the timer interrupts. Since the 8051 has only two timers, Timer 0 and Timer 1, the bits ET1 and ET0 are only meaningful for it. The bit ET1 decides the enable status of Timer 1 while ET0 decides the status for Timer 0. The bit ET2 is useful for 8052. Setting these bits to 1 enable the corresponding timer while making any of them 0 disables the corresponding timer.

ES: This bit is devoted to enable/disable the serial communication interrupt.

EX1, EX0: The EX1 bit controls the enabling or disabling of the external interrupt INT1. Similarly, the bit EX0 controls the enable status of external interrupt INT0. Setting these bits to 1 enable the corresponding external interrupts and resetting the bits disable the interrupts.

The MOV instruction can be used to load the IE register with the desired enabling pattern of the interrupts. For example, if the programmer wishes to enable the two timer interrupts and INT0, keeping others disabled, the IE register should be set to the bit pattern "10001011" using an instruction like,

```
MOV IE, #10001011B
```

9.6.2 Timer Interrupts

To use the timer interrupts, the IE register has to be programmed to enable the generation of interrupts when a particular timer rolls over and the TF bit is set. This has the advantage as in this case, after starting a timer, the 8051 need not wait checking continually the status of the TF bit. On occurrence of roll back and TF flag getting set, an interrupt gets generated and the program control gets transferred to the

(MSB)							(LSB)
EA	–	ET2	ES	ET1	EX1	ET0	EX0

FIGURE 9.14 Interrupt enable (IE) register structure

corresponding service routine address (000BH for Timer 0 and 001BH for Timer 1). Further, the resetting of TF flags is also not necessary, as it is done automatically by the processor, once control jumps to the interrupt service routine.

EXAMPLE 9.24

Let us revisit the example of generating a square wave with 50% duty cycle and ON/OFF periods of 20 μS each. As in the previous case, Timer 0 in mode 2 will be used to generate the necessary delays in the program. However, instead of checking the TF flag continually, the program will use the facility to generate interrupt whenever Timer 0 rolls over. The interrupt service routine will complement port bit P2.5. Here is the assembly language program for the same.

```
        ORG 0000H        ;
        LJMP MAIN        ; Jump to MAIN, bypass IVT
        ORG 000BH        ; ISR for Timer 0
        CPL P2.5         ; Complement port bit P2.5
        RETI             ; Return from ISR
        ORG 0030H        ; Main routine starts here
  MAIN: MOV TMOD, #02H   ; Set Timer 0 in Mode 2
        MOV TH0, #0EDH   ; Load TH0
        MOV IE, #82H     ; Enable Timer 0
        SETB TR0         ; Start timer
    L1: SJMP L1          ; Stay here
        END
```

The last statement "SJMP L1" makes the program to loop infinitely. This has been done to demonstrate that the 8051 need not check the TF0 flag anymore. The same program written in C will be as follows.

```
#include <reg51.h>
sbit p2_5 = P2^5;
void timer0 (void) interrupt 1
{
     p2_5 = ~p2_5;          // Complement P2.5
}
void main (void)
{
     TMOD = 0x02;           // Timer 0, Mode 2
     TH0 = 0xED;            // Load TH0
     IE = 0x82H;            //  Enable Timer 0 interrupt
     while (1);             //  Infinite loop
}
```

9.6.3 External Interrupts

There are two external interrupts in 8051, called INT0 and INT1 that are multiplexed with port pins P3.2 and P3.3, respectively. An interfaced device can activate these pins to interrupt the program being executed by the 8051.

As far as the triggering modes are concerned, the 8051 can be programmed to support two different modes for INT0 and INT1 – *low-level* and *falling edge*. As noted in Figure 9.8, the TCON register contains the bits IT0 (TCON.0) and IT1 (TCON.2) controlling the triggering modes of the interrupt lines INT0 and INT1. The two bits can be set/reset via bit manipulation instructions of 8051. Setting the bit IT0 to 1 causes INT0 to be configured as a falling edge triggered interrupt. On the other hand, resetting IT0 configures INT0 as low level triggered. Similar convention follows for the bit IT1 and interrupt INT1. The TCON register contains the IE bits to remember the high-to-low pulse for falling edge triggering. For INT0, the edge is remembered by setting of the bit IE0 (TCON.1). For INT1, the falling edge is remembered in the bit IE1 (TCON.3). The flags IE0 and IE1 are reset on executing the corresponding RETI instruction in the interrupt service routine. Unlike falling-edge triggering, the IE0 and IE1 bits are not used to remember low-level triggering of external interrupts. However, it may be noted that the low-level of INT0 and INT1 are not remembered internally. For the low-level triggered interrupt to be sensed properly, it must be held low till control branches to the ISR. Also, the low-level trigger must be withdrawn before the RETI instruction in the ISR. Otherwise, the interrupt line may be sensed multiple times indicating multiple occurrences of the same interrupt. As a general rule, the low-level should be maintained for around four machine cycles, but not more. For the falling edge-triggered case, the interrupt pin should be kept high for one machine cycle followed by low for at least one machine cycle.

EXAMPLE 9.25

To illustrate the operation of external interrupts, let us consider a system that has a switch connected to the INT0 pin and an LED to port bit P2.1. The switch configuration is such that on closing it, a 0 value is sent to INT0, whereas when the switch is open, the pin INT0 will be at 1. Thus, the switch may be closed to generate a low-level trigger for the INT0 pin. The interrupt service routine will turn the LED ON. The LED remains ON for a short while, turns OFF after that. Following this, control returns from the service routine.

```
    ORG 0000H     ;
    LJMP MAIN     ; Jump to MAIN, bypass IVT
    ORG 0003H     ; ISR for INT0
    SETB P2.1     ; Set port bit P2.1, LED ON
    MOV R2, #0FFH ; R2 used as counter
L1: DJNZ R2, L1   ; Wait for some time
    CLR P2.1      ; LED OFF
    RETI          ; Return from ISR
    ORG 0030H     ; Main routine starts here
```

```
MAIN: MOV IE, #81H   ; Enable INT0
  L2: SJMP L2        ; Wait in loop
      END
```

The same program written in C will be as follows.

```
#include <reg51.h>
sbit p2_1 = P2^1;

void external0 (void) interrupt 0
{
     p2_1 = 1;              // Turn LED ON
     for (i = 0; i < 10000; i++);
     p2_1 = 0;
}

void main (void)
{
     IE = 0x81H;            //  Enable INT0 interrupt
     while (1);             //  Infinite loop
}
```

It may be noted that in order to follow the falling edge triggering of INT0, in the MAIN routine, IT0 (TCON.0) bit be set by executing the instruction "SETB TCON.0".

9.6.4 Interrupt Priorities

The interrupts in 8051 has certain priority defined between them. Upon reset, the 8051 interrupts are assigned the default priorities. As per this setting, INT0 has the highest priority, followed by Timer 0, INT1, Timer 1, Serial Communication, in sequence. That is,

INT0 > Timer 0 > INT1 > Timer 1 > Serial Communication.

The priority order among the interrupts, at any point of operation, is decided by the setting of bits in the Interrupt Priority (IP) register. The structure of the IP register has been shown in Figure 9.15. The register is bit-addressable. Among the eight bits in the register, the most significant two bits are reserved. The bit marked PT2 controls priority of Timer 2 interrupt, PS controls priority of Serial interrupt, PT1 and PT0 determine priorities of Timer 1 and Timer 0, while PX1 and PX0 correspond to the external interrupts INT1 and INT0, respectively. For example, in the IP register, if only the bit PT1 is set to 1 while all others are 0, Timer 1 accords the highest priority among all the interrupts. If more than one bits are set in the IP register, all of them will have higher priorities than other interrupts for which the bits are not set. Within this group of higher priority interrupts, the priority ordering will be as per the default priority rule.

(MSB)							(LSB)
–	–	PT2	PS	PT1	PX1	PT0	PX0

FIGURE 9.15 Interrupt priority (IP) register structure

For example, consider the situation in which the bits PX1 and PT0 are set to 1 while all other are 0 in the IP register. In this setting, the interrupts INT1 and Timer 0 will have higher priority than others, while between INT1 and Timer 0, Timer 0 interrupt will have priority over INT1. As far as the execution of interrupt service routines is concerned, if a higher priority interrupt occurs while the service routine of a lower priority interrupt is being executed, the execution of current service routine is suspended and the control gets transferred to the service routine of the higher priority interrupt. Once the service routine of the higher priority interrupt is over, control resumes the execution of the lower priority interrupt service routine.

9.7 Serial Communication Programming LO 7

While discussing on 8085, it has been observed that it contains two dedicated pins SID and SOD through which the serial transmission and reception of data are possible, in a bit-by-bit fashion. However, the basic 8085 does not support the full serial communication protocol. In contrast, in 8051 the complete serial communication protocol has been supported. Under such protocol, serial communication takes place between a transmitter and a receiver, both of which are devices having electronic interface supporting transmission and/or reception. There are three modes of communication between the sender and the receiver, as noted next.

- *Simplex Mode*: Unidirectional communication between a master and a slave device.
- *Duplex Mode*: Simultaneous bidirectional communication between two devices.
- *Half-duplex Mode*: Bidirectional communication. However, at any point of time, data is transferred only in one direction.

Serial communication can also be classified into two categories – *synchronous* and *asynchronous*. In synchronous communication, the transmitter and the receiver work on the same clock signal. Thus, the two devices must be physically close to each other and also able to work on a common clock signal. This restricts its usage in many applications where it is difficult to make the devices work on the same clock. On the other hand, if the transmitter and the receiver work on their own clock signals, the communication between them is termed as an asynchronous transmission. Due to its inherent facility of independence between the communicating devices, asynchronous transmission is often preferable to synchronous transmission. However, it necessitates following a fixed protocol for faithful communication. Each character to be transmitted is put into a *frame* structure. In the idle situation, when no data transmission is taking place, the line is at 1 (high). A 0-bit in the line, called the *start bit*, marks the beginning of a frame. The 8-bit ASCII code of the character being transmitted constitute the next eight bits in the frame. The end of frame is marked by one/two 1 (high) bit(s), called *stop* bit(s). Figure 9.16

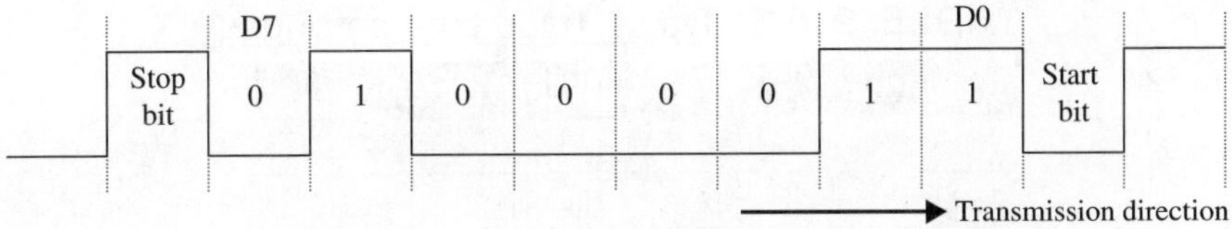

FIGURE 9.16 Frame structure for character 'C' (ASCII code 43H)

shows the frame structure. Sometimes, after the 8-bit character, a frame may hold a *parity* bit. The parity may be calculated as *even* or *odd*. For even parity system, the parity bit is set/reset such that the total number of 1's in the character code along with the parity bit is even. Similarly, the total number of 1's is odd in the odd-parity system. The parity bit may help in detecting some of the probable transmission errors. Upon receiving a frame, the receiver also computes the parity bit from the character bits. In case one or any odd number of bits in the character bits or the parity bit got corrupted in the transmission process, the recomputed parity bit at receiver will not match with the received parity bit, detecting an error. The data transfer rate between the sender and receiver is also preset. The transfer rate is stated as *bps* (*bits per second*). It is also called *baud rate*. Since there are several factors controlling the successful serial communication between sender and receiver, the two devices must agree upon the frame format, baud rate etc.

There are two hardware pins assigned the task of serial communication in 8051 –*TxD* (*Transmit Data*) and *RxD* (*Receive Data*). These pins have been multiplexed with port 3 bits, P3.0 and P3.1, respectively. The voltage levels supported are +5V for a 1 and 0V for a 0. This is a deviation from the standard serial communication protocol RS232C, that recognizes –3 to –25 V for logic 1 and +3 to +25 V for logic 0. Hence, to be fully compatible with the RS232C the 8051 has to use some additional interfacing chip (such as MAX232). Further discussion on this has been omitted for the sake of brevity.

9.7.1 Setting Baud Rate

As it may be necessary to communicate serially with different devices with varying parameters, the 8051 can also be programmed to support many different baud rates. To set the baud rate, the Timer 1 in Mode 2 (auto-reload) has to be used. To decide the clock frequency of the timer, the 8051 crystal frequency is first divided by 12. The serial communication circuitry within the 8051 further divides this frequency by 32 to drive the Timer 1. For auto-reload, the TH1 register has to be loaded with appropriate value. Once the timer starts operating in Mode 2, pulses (due to roll over) are generated at appropriate time instants. These pulses are used for bit timing. To illustrate further, let us assume that the crystal frequency is 11.0592 MHz, accordingly, the clock signal fed to the Timer 1 has a frequency of $((11.0592 \div 12) \div 32) = 28800$ Hz. Loading the TH1 register with 0FDH (–3 in decimal), will cause the timer to roll over after every three pulses, setting the baud rate to $28800 \div 3 = 9600$.

Some typical baud rates used by the devices in serial communication are 2400, 4800, 9600 etc. Table 9.16 notes the corresponding TH1 register initialization values. Here, the calculations assume the crystal frequency to be 11.0592 MHz.

TABLE 9.16 Typical baud rate settings

Baud Rate	TH1 (Hex)
9600	FD
4800	FA
2400	F4

9.7.2 Registers for Serial Communication

There are two important registers associated with serial communication in 8051 – SBUF and SCON. Both the registers are 8-bit wide.

- **SBUF:** This is the *Serial Buffer* register used to hold the character transmitted or received via serial communication. Program intending to transmit a character serially should load it into the SBUF register. Once the character to be transmitted is written onto SBUF, the start and the stop bits are added to make a frame. The serial transmission of the frame through TxD pin initiates immediately. Similarly, in the receiving process, when the bits are received serially via the RxD pin, the framing bits are removed and the character present in the frame is put into the SBUF register. The SBUF register content can now be read by the program to get the character. The following statements illustrate the process.

```
MOV SBUF, # 'N'  ; Load ASCII code of 'N' into SBUF
MOV SBUF, A      ; SBUF gets content of A register
MOV A, SBUF      ; SBUF content copied to A register
```

- **SCON:** The 8-bit *Serial Port Control* register decides the framing structure to be followed in serial communication. It also controls the enabling and disabling of transmission and reception. The structure of the register has been shown in Figure 9.17. It is an 8-bit bit-addressable register with the individual fields having the following meanings.
    - **SM0, SM1:** These are the mode control bits for serial communication. Out of the four alternative modes (Mode 0, Mode 1, Mode 2 and Mode 3), only the Mode 1 is used extensively. We shall concentrate on this mode only. To activate Mode 1, SM0 and SM1 bits are set to 0 and 1, respectively. As far as the frame structure is concerned, Mode 1 implies 8-bit data with one start bit and one stop bit in each frame, making the frame size 10 bits. The mode also allows various baud rate settings by programming the TH1 register.
    - **SM2:** This mode bit is used for multiprocessor communication. Thus, for most of the applications involving single processor, the bit will be set to 0 to disable multiprocessor mode.
    - **REN:** The *Receive Enable* bit enables data reception via the RxD pin in 8051. By default, the 8051 performs only data transmission operation via the TxD line. To receive data serially through RxD pin, the programmer must set the REN bit to 1. Thus, if the application does not wish to receive serial data, the REN bit may be reset by the programmer to block reception.
    - **TB8, RB8:** For most of the applications, these two bits are reset to 0, as these bits are used mostly in Mode 2 and Mode 3 of communication.

(MSB)							(LSB)
SM0	SM1	SM2	REN	TB8	RB8	TI	RI

FIGURE 9.17 Serial port control (SCON) register structure

- **TI:** The bit, called *Transmit Interrupt* gets set to 1 by the 8051 whenever the transmission of the 8-bit character available in the SBUF register has been completed. It indicates that the 8051 is ready to transfer another character. In fact, the TI flag bit is set just while transmitting the stop bit in the frame. A program requiring multibyte serial data transfer can check this bit to determine whether the SBUF register is ready or not to accept the next character to be transmitted.
- **RI:** This is the *Receive Interrupt* bit that gets set when the 8051 has placed a received byte into the SBUF register. The RI flag is set halfway through the reception of the stop bit. The RI flag should be checked by the user program to know whether any character is ready to be read from the SBUF register. Upon detecting the RI flag set, the user program should read the character from the SBUF register, before it is lost.

9.7.3 Serial Transmission

For serial transmission of characters, first the Timer 1 has to be programmed in mode 2 (auto-reload) and TH1 register be loaded appropriately for the selected baud rate. Assuming mode 1 data transmission, the SCON register is programmed in mode 1. The timer is started by setting the TR1 bit. The TI bit is cleared. To send a character, the program writes it into the SBUF register. The program then continually monitors the TI flag to check the completion of this character transmission. Once TI is found to be 1, the transfer of the character is complete. To transmit further characters, the TI flag should be cleared and the next character be copied into the SBUF register.

EXAMPLE 9.26

Let us look into the problem of transmitting the character string "1947" serially through the TxD line of 8051. The baud rate used is 9600 while the mode of transmission is Mode 1. The assembly language program for the problem has been noted next.

```
ORG 0000H         ;
MOV TMOD, #20H    ; Timer 1, Mode 2
MOV TH1, #0FDH    ; Baud rate 9600
MOV SCON, #40H    ; Mode 1, REN disabled
SETB TR1          ; Start Timer 1, auto-reload
MOV A, # '1'      ; Load '1' into A
ACALL SEND        ; Transfer '1'
MOV A, # '9'      ; Load '9' into A
```

```
        ACALL SEND          ; Transfer '9'
        MOV A, # '4'        ; Load '4' into A
        ACALL SEND          ; Transfer '4'
        MOV A, # '7'        ; Load '7' into A
        ACALL SEND          ; Transfer '7'
    L1: SJMP L1             ; Loop here continually
  SEND: MOV SBUF, A         ; Load SBUF
    L2: JNB TI, L2          ; Wait for last bit to be sent
        CLR TI              ; Get ready for next byte
        RET                 ; Return from subprogram
        END
```

The same program written in C will be as follows.

```
#include <reg51.h>
void Send_Ser (unsigned char);
void main (void)
{
     TMOD = 0x20;           // Timer 1, Mode 2
     TH1 = 0xFD;            // 9600 baud rate
     SCON = 0x40;           // Set SCON in Mode 1, REN disabled
     TR1 = 1;               // Start timer
     Send_Ser('1');         // Send '1'
     Send_Ser('9');         // Send '9'
     Send_Ser('4');         // Send '4'
     Send_Ser('7');         // Send '7'
     while (1);             // Wait here infinitely
}
void Send_Ser(unsigned char m)
{
     SBUF = m;              // Copy character to SBUF
     while (TI == 0);       // Wait for last bit to be sent
     TI = 0;                // Reset TI bit
}
```

9.7.4 Serial Reception

In order to receive characters serially, the Timer 1 needs to be programmed in mode 2 (auto-reload) and TH1 initialized appropriately to set the required baud rate. The SCON register should be programmed in mode 1 with REN enabled. The timer has to be started by setting the TR1 bit. The RI bit be cleared. The program should now continually monitor the RI flag to check the receipt of a character. When RI becomes 1, the SBUF will contain the received character. The character now be copied from the SBUF register. The RI flag should next be cleared to enable reception of the next character, if any.

EXAMPLE 9.27

The following example receives characters serially and the received characters are sent to port P2 for display. The assembly language program for the problem is as follows.

```
       ORG 0000H        ;
       MOV TMOD, #20H   ; Timer 1, Mode 2
       MOV TH1, #0FDH   ; Baud rate 9600
       MOV SCON, #50H   ; Mode 1, REN enabled
       SETB TR1         ; Start Timer 1, auto-reload
   L1: ACALL RECV       ; Get next character serially
       MOV P2, A        ; Copy A to port P2
       SJMP L1          ; Continue
 RECV: JNB RI, RECV     ; Wait for the character
       MOV A, SBUF      ; Wait for last bit to be sent
       CLR RI           ; Get ready for next byte
       RET              ; Return
       END
```

The C language program for the same problem will be as follows.

```
#include <reg51.h>
void main (void)
{
    unsigned char my_char;
    TMOD = 0x20;           // Timer 1, Mode 2
    TH1 = 0xFD;            // 9600 baud rate
    SCON = 0x40;           // Set SCON in Mode 1, REN disabled
    TR1 = 1;               // Start timer
    while (1) {            // Loop
         while (RI == 0);  // Wait for RI to become 1
         my_char = SBUF;   // Read character from SBUF
         P2 = my_char;     // Put character into port P2
         RI = 0;           // Reset RI flag
    }
}
```

9.7.5 Serial Communication Interrupt

The serial data transmission and reception techniques discussed so far put the program into a busy-loop to check whether the transmission/reception of the previous character is over or not. This makes the programs inefficient. The busy wait loop can be avoided by utilizing the serial interrupt. As mentioned previously, in 8051, only one interrupt has been kept for both transmission and reception. As a result, assuming that the serial interrupt has been enabled in the IE register, whenever the TI flag or the RI flag is set, it generates an interrupt to the 8051 and the control branches to the address 0023H. It is the responsibility of the service routine to check the TI and RI flags to determine the actual source of interrupt – successful transmission or successful reception and take the appropriate action based upon that.

EXAMPLE 9.28

To illustrate the usage of interrupts in serial communication, let us consider an application that reads port P1 and transmits the character read serially. Simultaneously, it receives characters serially and outputs it to port P2. The assembly language program accomplishing the same has been noted next.

```
        ORG 0000H
        LJMP MAIN             ; Jump to main routine
        ORG 0023H
        LJMP SERIAL           ; Jump to Serial Int. routine
        ORG 0030H
  MAIN: MOV P1, #0FFH         ; Make P1 input port
        MOV TMOD, #20H        ; Timer 1, Mode 2
        MOV TH1, #0FDH        ; Baud rate 9600
        MOV SCON, #50H        ; Mode 1, REN enabled
        MOV IE, #10010000B    ; Enable serial interrupt
        SETB TR1              ; Start Timer 1, auto-reload
    L1: MOV A, P1             ; Read port P1
        MOV SBUF, A           ; Serially transmit
        SJMP L1               ; Loop continually
        ORG 100H              ;
SERIAL: JB TI, L2             ; Jump if TI is set
        MOV A, SBUF           ; Else character received
        MOV P2, A             ; Copy to P2
        CLR RI                ; Get ready for next byte
        RETI                  ; Return
    L2: CLR TI                ; Clear TI
        RETI                  ; Return
        END
```

The corresponding C language program will be as follows.

```
#include <reg51.h>
void serial0 () interrupt 4
{
   if (TI == 1) TI = 0;    // Clear interrupt
```

```
    else {
        P2 = SBUF;          // Put serial data onto P2
        RI = 0;             // Clear interrupt
    }
}

void main (void)
{
    unsigned char my_char;
    P1 = 0xFF;              // Make P1 input port
    TMOD = 0x20;            // Timer 1, Mode 2
    TH1 = 0xFD;             // 9600 baud rate
    SCON = 0x50;            // Set SCON in Mode 1, REN enabled
    IE = 0x90;              // Enable serial interrupt
    TR1 = 1;                // Start timer
    while (1) {             // Loop
          my_char = P1;     // Read character from port P1
          SBUF = my_char;   // Put character into SBUF
    }
}
```

SUMMARY

A large number of machine-level instructions are supported by the microcontroller 8051. The groups of instructions are data transfer, arithmetic and logic operations, branching etc. The instructions will have operands which can be a constant value, a register, a memory location etc. The 8051 supports a number of ways to specify the operands. Though assembly language instructions can be used to express the program logic of a specific problem to be solved, for problems with moderate to high logical complexities, it may be difficult to express the logic neatly in the assembly language. Hence, such programs are often developed in the C language. The embedded C has the augmented data types signed/unsigned char/int, bits etc. To design precision control applications, timers and counters are often very important. There are two modules in 8051, each of which can be configured both as timers and counters to operate in different modes. Several interrupts are supported by 8051, such as timers, reset, external interrupt pins and serial communication related one. The 8051 supports quite powerful serial communication that is helpful in situations where parallel communication is not possible due to pin/wire limitations. The chapter includes many example programs to illustrate the issues.

EXERCISES

Multiple Choice Questions

1. An example of direct addressing is
   (a) MOV A, @R0 (b) MOV A, #45H
   (c) MOV A, 45H (d) MOV A, R1
2. Which of the following is a relative jump?
   (a) JC (b) LJMP
   (c) AJMP (d) None of the above
3. Flags(s) affected by "INC R5" is/are
   (a) CY (b) AC
   (c) OV (d) None of the above
4. The target address range for AJMP instruction is
   (a) 2 kB (b) 4 kB
   (c) 8 kB (d) 16 kB
5. Bit width of operands in arithmetic operations in 8051 can be
   (a) 8 (b) 4
   (c) 16 (d) All of the above
6. The maximum nonzero number for which "DA A" instruction works correctly is
   (a) 255 (b) 99
   (c) 100 (d) None of the above
7. NOP instruction can be useful to introduce
   (a) Address (b) Delay
   (c) Memory location (d) None of the above
8. While executing the CJNE instruction, the carry flag will be set to 1 if
   (a) destination = source
   (b) destination < source
   (c) destination > source
   (d) destination ≤ source
9. Assuming R3 to be equal to 1, the instruction "DJNZ R3, L1" will take control to
   (a) L1
   (b) Next instruction
   (c) Depends on assembler
   (d) None of the above
10. The jump instruction that clears the operand bit is
    (a) JB (b) JBC
    (c) JNB (d) None of the above
11. MOV instruction in the 8051 can handle data of size (in bits)
    (a) 1 (b) 8
    (c) 16 (d) All of the above
12. Identify the option in which the data to be moved is located very close to the current instruction.
    (a) MOVC A, @A+DPTR
    (b) MOVC A, @A+PC
    (c) Both (a) and (b)
    (d) None of the above
13. The value of CY flag after executing MUL instruction is
    (a) 0
    (b) 1
    (c) Depends on result
    (d) Not affected
14. Identify the instruction not valid for 8051.
    (a) PUSH A (b) PUSH R4
    (c) POP R5 (d) All of the above
15. The minimum number of instructions from the interrupted program that is executed after a RETI is
    (a) 1 (b) 2
    (c) 0 (d) 3
16. The instruction to be used to bring LSB of A to CY is
    (a) RL (b) RR
    (c) RLC (d) RRC
17. Suppose the 8051 has executed a SWAP instruction. In the result, the original bit 3 will go to the bit position
    (a) 0 (b) 7
    (c) 5 (d) CY

18. After a DIV instruction, remainder is left in
(a) A (b) B
(c) R1 (d) R7

19. Identify the instruction not valid for 8051.
(a) INC DPTR (b) DEC DPTR
(c) INC DPL (d) DEC DPH

20. In embedded C, the default data type for byte is
(a) Unsigned char (b) Signed char
(c) Unsigned byte (d) Signed byte

21. Identify the nibble definitely exchanged in 8051 the instructions XCH and XCHD.
(a) Lower (b) Upper
(c) Both (a) and (b) (d) None of the above

22. Identify the byte comparison instruction(s) in
(a) CMP (b) CJNE
(c) Both (a) and (b) (d) None of the above

23. The hexadecimal equivalent for the BCD number 76 is
(a) 76H (b) 4CH
(c) 0C4H (d) None of the above

24. Assume that the accumulator A contains 25H. If the instruction "RR A" is executed, the content of A will become
(a) 12H (b) 4AH
(c) 25H (d) None of the above

25. Assume that the accumulator A contains 25H. After the instruction "RL A" has been executed once, the content of A will become
(a) 12H (b) 4AH
(c) 25H (d) None of the above

26. In order to operate in timer mode, the C/T bit of the 8051 should be set to
(a) 0 (b) 1
(c) Either 0 or 1 (d) Tristate

27. For auto-reload mode, the reload value is stored in register(s)
(a) TH
(b) TL
(c) Both (a) and (b)
(d) Neither (a) nor (b)

28. Assuming that the registers TH and TL contain FFH and 00H respectively and that the timer has been programmed in Mode 1 for the 8051 processor connected with crystal frequency 11.0592 MHz, the delay produced will be
(a) 277.00 μS (b) 277.76 μS
(c) 278.00 μS (d) 278.76 μS

29. The instruction "MOV IP, #00011000B" has been used to set priorities of interrupts. The resultant priority set up will be
(a) RI+TI > TF1 > INT0 > TF0 >INT1
(b) TF1 > RI+TI > INT0 > TF0 > INT1
(c) INT0 > TF0 > INT1 > TF1 > RI+TI
(d) INT0 > RI+TI > TF1 > TF0 > INT1

30. Among all interrupts in 8051, the default priority of INT0 is
(a) More than INT1 (b) Less than INT1
(c) Same as INT1 (d) Double of INT1

31. The bit controlling the run of Timer 1 is
(a) TR1 (b) TF1
(c) IE1 (d) IT1

32. Assuming that the GATE of TMOD register is set, the timer run can be controlled by
(a) INT
(b) TR
(c) Both (A) and (B)
(d) Neither (A) nor (B)

33. If a timer is run in Mode 1 in the 8051 processor with a crystal frequency of 12 MHz, the maximum delay that can be produced is
(a) 1 μS (b) 65536 μS
(c) 65535 μS (d) None of the above

34. The total number of pins for interrupts in 8051 is
(a) 1 (b) 2
(c) 3 (d) 4

35. Which parity scheme is better to detect transmission errors?
(a) Even
(b) Odd
(c) Mixed Even-Odd
(d) None of the above

Answers to Multiple Choice Questions

1. c	2. a	3. d	4. a	5. a	6. b	7. b	8. b	9. a	10. b
11. d	12. b	13. a	14. d	15. a	16. d	17. b	18. b	19. b	20. b
21. a	22. b	23. b	24. a	25. b	26. a	27. a	28. b	29. b	30. a
31. a	32. c	33. b	34. c	35. d					

Long Answer Type Questions

1. State the strategy followed by the assemblers to differentiate between immediate and direct operands in 8051.
2. In a MOV instructions, it is not allowed to have registers as both source and destination operands. Explain why this is not counted as a serious restriction.
3. Distinguish between the instructions XCH and XCHD.
4. The MOVC instruction can use either the DPTR or the PC as the base register. Distinguish between the two usages with their relative advantages and disadvantages.
5. Distinguish between the LJMP and AJMP instructions in 8051.
6. How can the DJNZ instruction be used to implement loop statements available in high-level languages?
7. Distinguish between the operations of RET and RETI instructions.
8. Explain the necessity of clearing the carry flag before the subtraction operation in 8051.
9. It is required to perform logical NAND of two bytes stored in the on-chip RAM. Write an assembly language code fragment to implement it.
10. Describe the utility of a NOP instruction in 8051 programs.
11. Describe the procedure to calculate the initial value for timers in 8051.
12. How are the priorities of interrupts set in 8051?
13. Explain the procedure to set baud rate for serial communication in 8051 with an example.
14. Explain how the timer modules of 8051 can be used as counters.
15. How can one determine whether a serial communication interrupt marks successful transmission and reception of a character?
16. Explain how the interrupts in 8051 can be enabled and disabled.
17. An array of N numbers have been stored at external RAM location 2000H onwards. Write an assembly language program for 8051 to determine the minimum value and store it at external RAM location 1000H.
18. An array of N numbers stored at RAM locations 2000H onwards. Assume the resulting numbers to be within 255. Write a program in the assembly language of 8051 to compute the squares of each of them. Write the corresponding C program as well.
19. It is required to check the primality of a number stored at RAM location 40H. If the number is a prime, store 1 at 41H, else store 0 at 41H. Write the corresponding 8051 assembly language program. Also write the corresponding C program.

20. Two arrays of N numbers are stored from memory locations 1000H and 2000H, respectively. Write an assembly language program for 8051 to check if all the corresponding elements in two arrays are same or not. Write the corresponding C language program as well.
21. Write a program in the assembly language of 8051 to determine if a character string stored from memory location 40H is a palindrome or not. Assume that the number of characters in the string is available at location 55H. The program should store 1 at location 30H if the string is a palindrome and a 0 otherwise.
22. Design an 8051 assembly language program to generate a square wave with 25% duty cycle and 50 μS time period on port bit P1.5. Also write the corresponding C program.
23. It is required to count the number of pulses on the line T1 of the 8051 microcontroller. The current count value should be available at port P1. Write the corresponding program in the assembly language of 8051 and also in C.
24. Consider a system with two switches connected to the INT0 and INT1 pins of 8051. The switches are such that, when a switch is closed, it gives a 0 value to the corresponding pin. An LED has been connected to port bit P2.5. The LED is ON when the switch to INT0 is closed. Closing the switch of INT1 turns the LED OFF. Design an interrupt based assembly language program for this system. Also write the corresponding C language program.
25. The string "MICROCONTROLLER" has to be transmitted serially by the 8051. Develop an assembly language program for the same. The program should also read port bit P0.1 and the transmission should repeat as long as the bit is kept at 1. Write the corresponding C language program as well.

10

I/O Interfacing with the 8051

LEARNING OUTCOMES

After reading this chapter, students will be able to:

LO 1 Interface switches and LEDs with 8051

LO 2 Interface a hex keyboard with 8051

LO 3 Interface multiplexed seven-segment display with 8051

LO 4 Interface LCD with 8051

LO 5 Interface ADC with 8051

LO 6 Interface DAC with 8051

LO 7 Interface 8255 with 8051

LO 8 Interface a stepper motor with 8051

10.1 Introduction

The microcontroller 8051 has been used extensively in control applications. It is used to sense the environment/plant for various parameters and actuate the devices attached to control it. The type of device to be connected naturally depends upon the activity being controlled. As discussed in Chapter 5, devices can be broadly classified into two categories – the input devices (mostly sensors) and the output devices (mostly actuators). In Chapter 5, several such device interfacings have been discussed. For this, the 8085 needs to connect via I/O ports. The I/O ports are to be designed separately. In the case of 8051, the microcontroller has its own I/O ports that could be used for interfacing. This chapter will primarily show such interfacings with the 8051. The hardware part remains mostly the same as that in Chapter 5, however, the software can be in the assembly language of 8051 or in embedded C.

10.2 Switch and LED Interfacing LO 1

Figure 10.1 shows an interfacing of switches and LEDs to the 8051. Eight switches have been connected to Port 1 while eight LEDs have been connected to Port 0. The control program reads the value from Port 1 and displays the same in the LEDs connected to Port 0. The corresponding program has been noted next.

```
    ORG 0
    MOV P1, #0FFH      ; Set P1 in input mode
L1: MOV A, P1          ; Read Port P1
    MOV P0, A          ; Write to Port P0
    SJMP L1            ; Continue
    END
```

The C language program is as follows:

```
#include <reg51.h>
void main (void)
{
      while (1) P0 = P1;
}
```

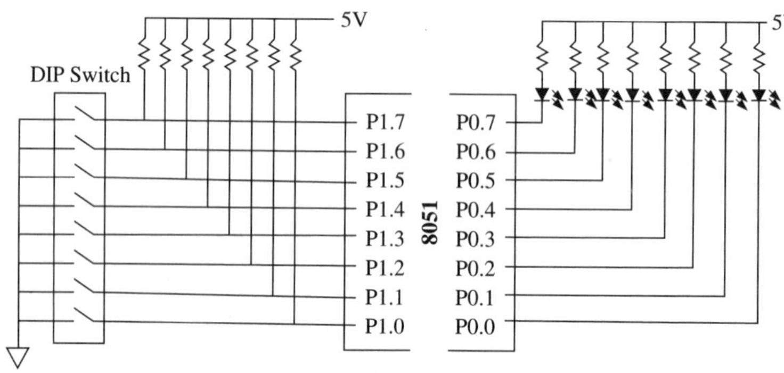

FIGURE 10.1 Interfacing eight switches and eight LEDs

10.3 Hex Keyboard Interfacing LO 2

A schematic to interface a hexadecimal matrix keyboard has been shown in Figure 10.2. The rows and columns of the 4×4 keyboard are driven by four bits of ports P1 and P2, respectively. Software-based debouncing has been used. After debouncing, the individual rows are set to 0 by turns. It aids in

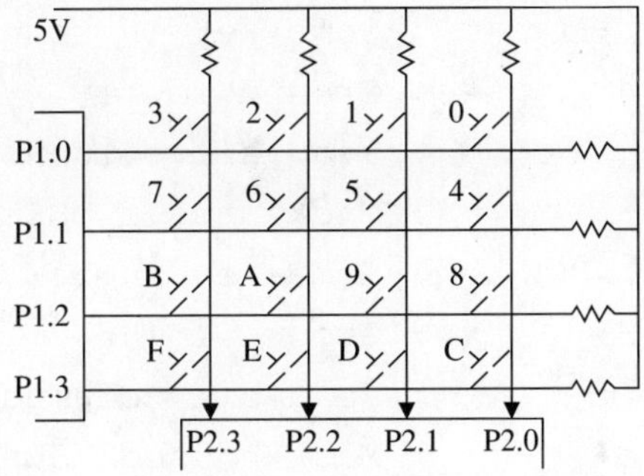

FIGURE 10.2 Interfacing a 4×4 hex keyboard

detecting if a key has been pressed. Once a row bit has been set to 0, the columns are read. The column bit read as 0 in the port identifies the key. A code table has been utilized to get the key code. For example, the co-ordinate of row-1 and column-1 corresponds to the key code 5. In the following, the assembly language program has been listed that scans the keyboard and gets the code of the key pressed (if any) at RAM location 55H.

```
    ORG 1000H
    MOV P2, #0FFH          ; Set P2 in input mode
L1: MOV P1, #00H           ; Ground all rows
L2: MOV A, P2              ; Read all columns
    ANL A, #0FH            ; Mask unused MSBs
    CJNE A, #0FH, L2       ; Check for all keys open
    ACALL DELAY            ; Some key may be pressed, wait for 20 ms
    MOV A, P2              ; Read all columns
    ANL A, #0FH            ; Mask unused MSBs
    CJNE A, #0FH, L2       ; No key pressed, go back
    ACALL DELAY            ; Key pressed, wait for 20ms
    MOV P1, #0FEH          ; Ground row 0
    MOV A, P2              ; Read all columns
    ANL A, #0FH            ; Mask unused MSBs
    CJNE A, #0FH, ROW_0    ; Some key in row 0 pressed
    MOV P1, #0FDH          ; Ground row 1
    MOV A, P2              ; Read all columns
    ANL A, #0FH            ; Mask unused MSBs
```

```
          CJNE A, #0FH, ROW_0  ; Some key in row 0 pressed
          MOV P1, #0FDH        ; Ground row 1
          MOV A, P2            ; Read all columns
          ANL A, #0FH          ; Mask unused MSBs
          CJNE A, #0FH, ROW_1  ; Some key in row 1 pressed
          MOV P1, #0FBH        ; Ground row 2
          MOV A, P2            ; Read all columns
          ANL A, #0FH          ; Mask unused MSBs
          CJNE A, #0FH, ROW_2  ; Some key in row 2 pressed
          MOV P1, #0F7H        ; Ground row 3
          MOV A, P2            ; Read all columns
          ANL A, #0FH          ; Mask unused MSBs
          CJNE A, #0FH, ROW_3  ; Some key in row 3 pressed
          AJMP L1              ; No key pressed, repeat scanning
   ROW_0: MOV DPTR, #KEY_ROW_0 ; DPTR = Start of Row 0
          SJMP FIND_KEY        ; Find column for the key
   ROW_1: MOV DPTR, #KEY_ROW_1 ; DPTR = Start of Row 1
          SJMP FIND_KEY        ; Find column for the key
   ROW_2: MOV DPTR, #KEY_ROW_2 ; DPTR = Start of Row 2
          SJMP FIND_KEY        ; Find column for the key
   ROW_3: MOV DPTR, #KEY_ROW_3 ; DPTR = Start of Row 3
FIND_KEY: RRC A                ; Rotate A and get column status to CY
          JNC MATCHED          ; Column found, if CY=0
          INC DPTR             ; DPTR = Next column address
          SJMP FIND_KEY        ; Check for next column
 MATCHED: CLR A                ; Clear accumulator
          MOVC A, @A+DPTR      ; Get code from the table
          MOV 55H, A           ; Get key code to RAM[55]
          AJMP L1              ; Repeat scanning

   DELAY: MOV TMOD #01         ; Timer 0, Mode 1
          MOV TL0, #0E0H       ; Load TH and TL
          MOV TH0, #0B1H       ; 0B1E0H for 20ms delay, clock 12MHz
          SETB TR0             ; Start timer
   AGAIN: JNB TF0, AGAIN       ; Monitor Timer 0
```

```
            CLR TR0                         ; Stop Timer 0
            CLR TF0                         ; Clear Timer 0 flag
            RET

            ORG 2000H
KEY_ROW_0: DB '0', '1', '2', '3'      ; Row 0
KEY_ROW_1: DB '4', '5', '6', '7'      ; Row 1
KEY_ROW_2: DB '8', '9', 'A', 'B'      ; Row 2
KEY_ROW_3: DB 'C', 'D', 'E', 'F'      ; Row 3
            END
```

The embedded C program for the matrix keyboard interface has been noted next. The key code will be stored in a variable key_pressed.

```
# include <reg51.h>
# define COLUMN   P2   // P2 will read the columns
# define ROW      P1   // P1 will select rows
void delay_rtn( unsigned int del_val );
unsigned char keys[4][4] = { '0', '1', '2', '3',
                             '4', '5', '6', '7',
                             '8', '9', 'A', 'B'
                             'C', 'D', 'E', 'F' };

unsigned char key_pressed;
void main()
{
     unsigned char col_id, row_id;
     COLUMN = 0xFF;  // Configure P2 in input mode
     while (1) do {
            do {
                ROW = 0x00;              // Ground all rows
                col_id = COLUMN;         // Read all columns
                col_id &= 0x0F;          // Mask unused bits
            } while (col_id != 0x0F); // All keys open now
            do{
              do{
                col_id = COLUMN;         // Read all columns
                col_id &= 0x0F;          // Mask unused bits
              }while (col_id == 0x0F); // Wait till some key pressed
```

```
    delay_rtn( 20 );           // Wait for 20 ms
    col_id = COLUMN;           // Read all columns
    col_id &= 0x0F;            // Mask unused bits
  } while (col_id == 0x0F);

  // Some key definitely pressed - identify it
  while (1) do {

      ROW = 0xFE;              // Ground Row 0
      col_id = COLUMN;         // Read columns
      col_id &= 0x0F;          // Mask unused bits
      if (col_id != 0x0F) {
              row_id = 0;      // Save row index
              break;           // Exit while-loop
      }

      ROW = 0xFD;              // Ground Row 1
      col_id = COLUMN;         // Read columns
      col_id &= 0x0F;          // Mask unused bits
      if (col_id != 0x0F) {
              row_id = 1;      // Save row index
              break;           // Exit while-loop
      }

      ROW = 0xFB;              // Ground Row 1
      col_id = COLUMN;         // Read columns
      col_id &= 0x0F;          // Mask unused bits
      if (col_id != 0x0F){
              row_id = 2;      // Save row index
              break;           // Exit while-loop
      }

      ROW = 0xF7;              // Ground Row 1
      col_id = COLUMN;         // Read columns
      col_id &= 0x0F;          // Mask unused bits
      if (col_id != 0x0F){
              row_id = 3;      // Save row index
              break;           // Exit while-loop
      }
  }
```

```
            if (col_id == 0x0E)
                    key_pressed = keys[ row_loc ][ 0 ];
            else if (col_id == 0x0D)
                    key_pressed = keys[ row_loc ][ 1 ];
            else if (col_id == 0x0B)
                    key_pressed = keys[ row_loc ][ 2 ];
            else
                    key_pressed = keys[ row_loc ][ 3 ];
        }
}
void delay_rtn(unsigned int del_val)
{
        unsigned int x, y;
        for (x = 0; x < 1000; x++)
                for (y = 0; y < del_val; y++);
}
```

10.4 Multiplexed Seven-Segment Display Interfacing

LO 3

The hardware connection to the 8051, multiplexing four seven-segment display modules has been shown in Figure 10.3. The least significant four bits of port P2 have been used for module selection.

The bit patterns corresponding to the digit to display are sent via port P1. The 8051 assembly language program for this interfacing has been noted next.

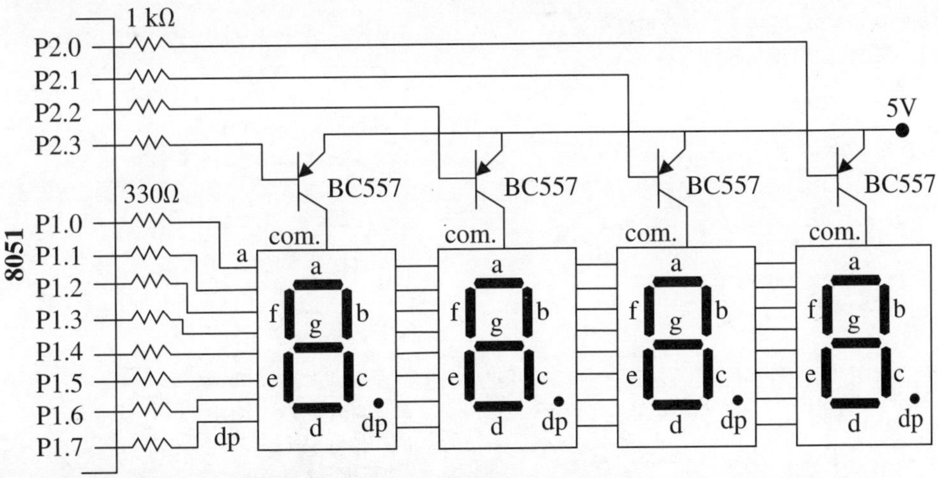

FIGURE 10.3 Four multiplexed seven-segment displays

```
            ORG 1000H
 DGT_PATT:  DB 0C0H, 0CFH, 0A4H,         ; Bit patterns for digits
                  0B0H, 99H, 92H, 82H,
                  0F8H, 80H, 90H
                                         ; Display most significant digit
       L1:  MOV R1, #0                   ; Set R1
            ACALL BIT_PATT               ; Get bit pattern in A
            MOV P1, A                    ; Send pattern to glow segments
            MOV P2, #11110111            ; Select module
            ACALL DELAY_RTN              ; Put a small delay
                                         ; Display second digit
            MOV R1, #1
            ACALL BIT_PATT
            MOV P1, A
            MOV P2, #11111011
            ACALL DELAY_RTN
                                         ; Display third digit
            MOV R1, #2
            ACALL BIT_PATT
            MOV P1, A
            MOV P2, #11111101
            ACALL DELAY_RTN
                                         ; Display fourth digit
            MOV R1, #3                   ; Read all columns
            ACALL BIT_PATT               ; Mask unused MSBs
            MOV P1, A                    ; Some key in row 2 pressed
            MOV P2, #11111110            ; Ground row 3
            ACALL DELAY_RTN
            AJMP L1                      ; Repeat display

 BIT_PATT:  MOV DPTR, #STRING            ; Get start address of string
            MOV A, R1                    ; Get digit index in A
            MOVX A, @A+DPTR              ; Get the digit (character
                                           format) in A
            CLR C                        ; Clear carry bit
            SUBB A, #'0'                 ; Subtract ASCII code of '0'
            MOV DPTR, #DGT_PATT          ; DPTR = Start of digit display
                                           array
            MOVX A, @A+DPTR              ; Get bit pattern into A
            RET                          ; Return

DELAY_RTN:  MOV TMOD #01                 ; Timer 0, Mode 1
            MOV TL0, #0E0H               ; Load TH and TL
            MOV TH0, #0B1H               ; 0B1E0H for 20ms delay, clock
                                           12MHz
            SETB TR0                     ; Start timer
```

```
AGAIN: JNB TF0, AGAIN                  ; Monitor Timer 0
       CLR TR0                         ; Stop Timer 0
       CLR TF0                         ; Clear Timer 0 flag
       RET

       ORG 2000H
STRING: DB "1579"                      ; String to be displayed
       END
```

The corresponding C language program has been noted next.

```
# include <reg51.h>
void delay_rtn( unsigned int del_val );
unsigned char DGT_PATT[10] = { 0xC0, 0xCF, 0xA4, 0xB0, 0x99,
                               0x92, 0x82, 0xF8, 0x80, 0x90 };
unsigned char STRING[4] = "1579";
void main()
{
       while (1) do {
              P1 = DGT_PATT[ STRING[ 0 ] - '0'];  P2 = 0xF7;
              delay_rtn( 20 );
              P1 = DGT_PATT[ STRING[ 1 ] - '0'];  P2 = 0xFB;
              delay_rtn( 20 );
              P1 = DGT_PATT[ STRING[ 2 ] - '0'];  P2 = 0xFD;
              delay_rtn( 20 );
              P1 = DGT_PATT[ STRING[ 3 ] - '0'];  P2 = 0xFE;
              delay_rtn( 20 );
       }
}
void delay_rtn( unsigned int del_val )
{
       unsigned int x, y;
       for (x = 0; x < 1000; x++)
              for (y = 0; y < del_val; y++);
}
```

10.5 Liquid Crystal Display (LCD) Interfacing

LO 4

An interfacing of the Liquid Crystal Display (LCD) module with 8051 has been shown in Figure 10.4. Port P1 of 8051 has been used to drive the data pins DB0 through DB7 of the LCD module. Other important pins RS, R/W and E are connected via port bits P3.2, P3.1 and P3.0, respectively. The program has been noted next.

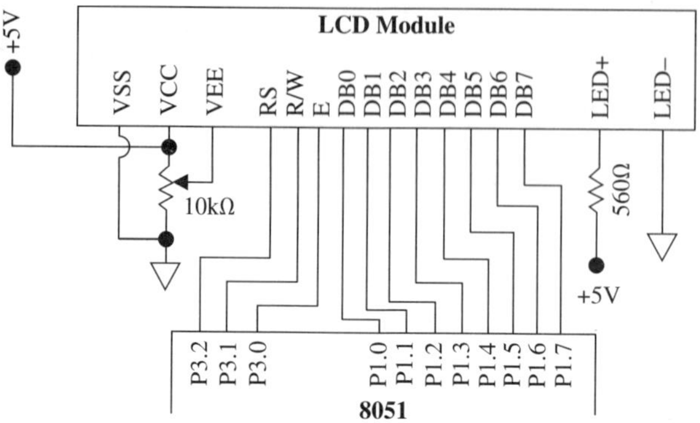

FIGURE 10.4 Interfacing LCD module with the 8051

```
           ORG 1000H
           MOV A, #38H          ; Initialize 2 line LCD, 5×7 matrix
           ACALL ISS_CMD        ; Issue the command
           MOV A, #0EH          ; LCD on, cursor on
           ACALL ISS_CMD        ; Issue the command
           MOV A, #01H          ; Clear LCD
           ACALL ISS_CMD        ; Issue the command
           MOV A, #92H          ; Cursor at line 1, position 18
           ACALL ISS_CMD        ; Issue the command
           MOV A, #'8'          ; Display letter '8'
           ACALL DSP_DATA
           MOV A, #'0'          ; Display letter '0'
           ACALL DSP_DATA
           MOV A, #'5'          ; Display letter '5'
           ACALL DSP_DATA
           MOV A, #'1'          ; Display letter '1'
       L1: SJMP L1              ; Remain at this position

 ISS_CMD:  ACALL CHK_READY      ; Check if LCD ready
           MOV P1, A            ; Issue the command code
           CLR P3.2             ; Make RS = 0
           CLR P3.1             ; Set R/W to 0 for a write
```

```
          SETB P3.0          ; Set E = 1 for high-to-low pulse
          ACALL DELAY_RTN    ; Give a delay
          CLR P3.0           ; Set E = 0 to latch the command
          RET                ; Return

DSP_DATA: ACALL CHK_READY    ; Check if LCD ready
          MOV P1, A          ; Send data
          SETB P3.2          ; Make RS = 1
          CLR P3.1           ; Set R/W to 0 for a write
          SETB P3.0          ; Set E = 1 for a high-to-low pulse
          ACALL DELAY_RTN    ; Give a delay
          CLR P3.0           ; Set E = 0  to latch data
          RET                ; Return

CHK_READY: SETB P1.7         ; Make P1.7 input to read DB7
          CLR P3.2           ; Make RS = 0
          SETB P3.1          ; R/W = 1 for a read operation
    BACK: CLR P3.0           ; Set E = 0 for low-to-high pulse
          ACALL DELAY_RTN    ; Put some delay
          SETB P3.0          ; Set E = 1 for low-to-high pulse
          JB P1.7, BACK      ; Loop DB7 is clear
          RET                ; Return

DELAY_RTN: MOV TMOD #01      ; Timer 0, Mode 1
          MOV TL0, #0E0H     ; Load TH and TL
          MOV TH0, #0B1H     ; 0B1E0H for 20ms delay, clock 12MHz
          SETB TR0           ; Start timer
   AGAIN: JNB TF0, AGAIN     ; Monitor Timer 0
          CLR TR0            ; Stop Timer 0
          CLR TF0            ; Clear Timer 0 flag
          RET
          END
```

The C language program can be as follows:

```
# include <reg51.h>
sfr LCD_DB_Pins = P1; // LCD data pins connected to P1
sbit RS = P3^2;       // RS connected to Port 3, pin 2
sbit RW = P3^1;       // Read/Write to Port 3, pin 1
sbit E = P3^0;        // Enable to Port 3, pin 0
sbit BUSY = P1^7;     // DB7 gives the busy status

void delay_rtn( unsigned int del_val );
void LCD_Command( unsigned char value );
void LCD_Data( unsigned char value );
void LCD_Ready();
void main()
{
       LCD_Command( 0×38 );   // Initialize 2 line LCD, 5×7 matrix
       LCD_Command( 0×0E );   // LCD on, cursor on
       LCD_Command( 0×01 );   // Clear LCD
       LCD_Command( 0×92 );   // Line 1, position 18
       LCD_Data( '8' );       // Display '8'
       LCD_Data( '0' );       // Display '0'
       LCD_Data( '5' );       // Display '5'
       LCD_Data( '1' );       // Display '1'
}
void LCD_Command( unsigned char value )
{
       LCD_Ready();             // Check LCD busy flag DB7
       LCD_DB_Pins = value;     // Put value on DB pins
       RS = 0; RW = 0; E = 1;
       delay_rtn( 1 );
       E = 0;                   // High-to-Low pulse on E
}
void LCD_Data( unsigned char value )
{
       LCD_Ready();             // Check LCD busy flag DB7
       LCD_DB_Pins = value;     // Put value on DB pins
       RS = 1; RW = 0; E = 1;
       delay_rtn( 1 );
       E = 0;                   // High-to-Low pulse on E
```

```
}
void LCD_Ready()
{
        BUSY = 1;                   // Make DB7 input
        RS = 0; RW = 1;
        while (BUSY == 1) { E = 0; delay_rtn( 1 ); E = 1; }
        // Wait for BUSY = 0
}
void delay_rtn( unsigned int del_val )
{
        unsigned int x, y;
        for (x = 0; x < 1000; x++)
                for (y = 0; y < del_val; y++);
}
```

10.6 Analog-to-Digital Converter (ADC) Interfacing

LO 5

Figure 10.5 enumerates the connection for interfacing ADC0808/0809 to the 8051. The assembly language program for the interfacing has been noted in the following.

```
ORG 1000H
MOV P1, #0FFH       ; Configure P1 as an input port
SETB P2.6           ; Make EOC an input line
```

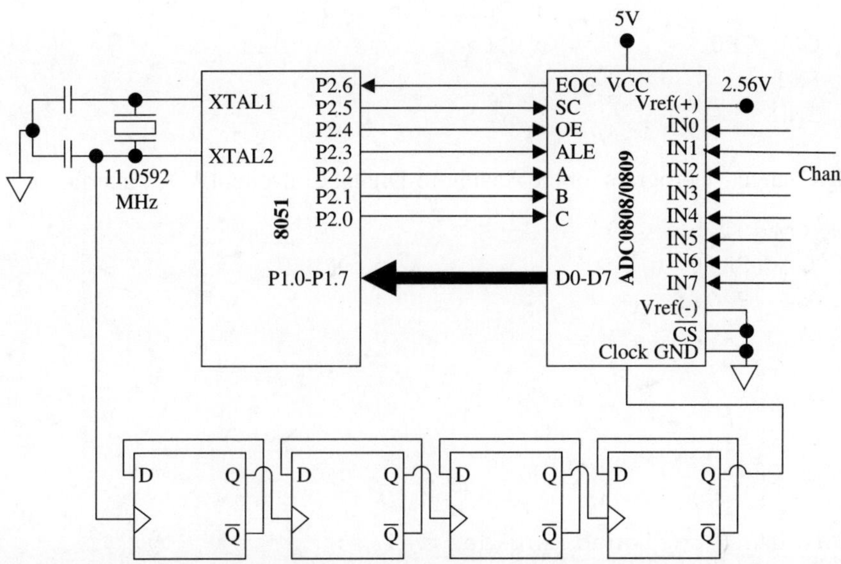

FIGURE 10.5 Interfacing ADC0808/0809 with the 8051

```
            CLR P2.3          ; Clear ALE
            CLR P2.4          ; Clear OE
            CLR P2.5          ; Clear SC
        L1: CLR P2.0          ; Select channel 1, Set C=0, B=0, A=1
            CLR P2.1
            SETB P2.2
            ACALL DELAY_RTN   ; Put small delay
            SETB P2.3         ; Latch selection, low-to-high for ALE
            ACALL DELAY_RTN
            CLR P2.3          ; Clear ALE
            CLR P2.5          ; Clear SC
        L2: JB P2.6, L2       ; Wait for EOC to become low
        L3: JNB P2.6, L3      ; Wait for EOC to become high
            SETB P2.4         ; Set OE to high
            ACALL DELAY_RTN   ; Put delay
            MOV A, P1         ; Read the converted digital value
            CLR P2.4          ; Clear OE
            SJMP L1           ; Proceed to convert next analog input
 DELAY_RTN: MOV TMOD #01      ; Timer 0, Mode 1
            MOV TL0, #0E0H    ; Load TH and TL
            MOV TH0, #0B1H    ; 0B1E0H for 20ms delay, clock 12MHz
            SETB TR0          ; Start timer
     AGAIN: JNB TF0, AGAIN    ; Monitor Timer 0
            CLR TR0           ; Stop Timer 0
            CLR TF0           ; Clear Timer 0 flag
            RET
            END
```

The corresponding C language program for the Analog-to-Digital Converter (ADC) interface is as follows:

```
#include <reg51.h>
sbit SEL_C = P2^0;
sbit SEL_B = P2^1;
sbit SEL_A = P2^2;
sbit ALE = P2^3;
sbit OE = P2^4;
sbit SC = P2^5;
sbit EOC = P2^6;
sfr DIG_VAL = P1;
void delay_rtn( unsigned int del_val );
void main()
```

```
{
    DIG_VAL = 0xFF;          // Configure P1 as input
    EOC = 1;                 // Configure EOC line as input
    ALE = 0;                 // Clear ALE
    SC = 0;                  // Clear SC
    OE = 0;                  // Clear OE
    while (1) {
           SEL_C = 0; SEL_B = 0; SEL_A = 1;  // Choose channel 1
           delay_rtn(1);
           ALE = 1;
           delay_rtn(1);
           SC = 1;
           delay_rtn(1);
           ALE = 0;
           SC = 0;                  // Start conversion
          while (EOC == 1);         // Wait for conversion to complete
          while (EOC == 0);
          OE = 1;                   // Enable output to be read
          delay_rtn(1);
          A = DIG_VAL;
          OE = 0;
          }
  }
  void delay_rtn( unsigned int del_val )
  {
          unsigned int x, y;
          for (x = 0; x < 1000; x++)
                for (y = 0; y < del_val; y++);
  }
```

10.7 Digital-to-Analog Converter (DAC) Interfacing

LO 6

Figure 10.6 shows an interfacing of DAC0808 with the 8051. The 8051 source code to produce a triangular waveform on this interface has been presented next.

```
              ORG 0
              CLR A              ; Clear accumulator
       L1:  MOV P1, A            ; Send data to DAC
```

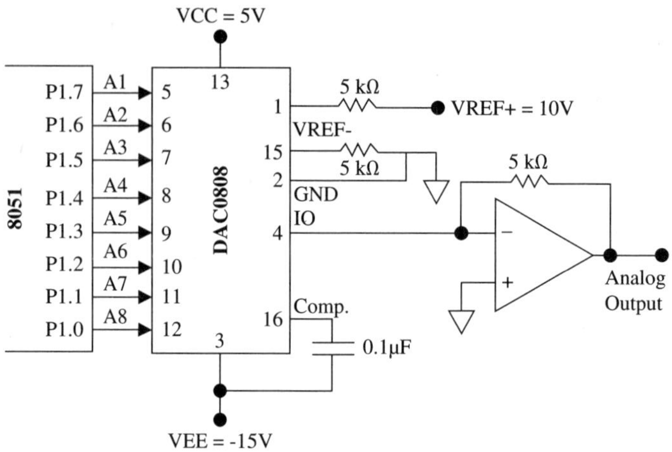

FIGURE 10.6 Interfacing DAC0808 with 8051

```
        ACALL DLA          ; Insert delay
        CJNE A, #0FFH, L2  ; Check if 255 reached
        SJMP L3            ; Jump to decrement
L2:     INC A              ; Increment A
        SJMP L1            ; Repeat increment
L3:     DEC A              ; Continue
        MOV P1, A          ; Send to DAC
        ACALL DLA          ; Insert delay
        CJNE A, #01, L4    ; Check if 0 reached
        CLR A              ; Clear accumulator
        SJMP L1            ; Repeat increment
L4:     SJMP L3            ; Repeat decrement
DLA:    NOP                ; Small delay routine
        NOP
        RET
        END
```

The corresponding C language program is as follows:

```
#include <reg51.h>
     void main (void)
     {
     unsigned char i;
     while (1)  {
```

```
            for (i = 0; i < 255; i++) { P1 = i; delay_rtn(2); }
            for (i = 255; i > 0; i --) { P1 = i; delay_rtn(2); }
        }
    }
void delay_rtn( unsigned int del_val )
{ unsigned int x, y;
  for (x = 0; x < 1000; x++)
          for (y = 0; y < del_val; y++);
}
```

10.8 8255 Interfacing

LO 7

Figure 10.7 shows an interfacing of the 8255 chip with 8051. Though 8051 itself has got four I/O ports, it is often advisable to have an 8255 chip also in the system. This is primarily because of the fact that the 8051 port bits are mostly multiplexed with other functionalities and thus may not be available for I/O. The electrical parameters of the ports may also be restrictive for certain applications. Having an 8255 eliminates many of these shortcomings. It is assumed that the 8255 will get selected only for addresses with A15 = A14 = 1. As per the requirements of 8255, Port A gets selected for A1 = 0, A0 = 0. Thus, in order to perform data transfer through Port A, the address should be "11xxxxxxxxxxxx00". For ports B, C and control register, the addresses should be "11xxxxxxxxxxxx01", "11xxxxxxxxxxxx10" and "11xxxxxxxxxxxx11", respectively. Assuming the don't care bits to be all zeros, one set of addresses for ports A, B and C and the control register is, C000H, C001H, C002H and C003H, respectively. Many more addresses are possible with different settings of the don't care bits.

In the following, an assembly language program has been noted that outputs patterns to ports A, B and C in such a fashion that the 8-bit outputs in each port are "01010101" (55H) followed by "10101010" (AAH) continually. In order to configure the ports in output mode, the control word 80H is written to the control register. The program then outputs the desired patterns in an infinite loop.

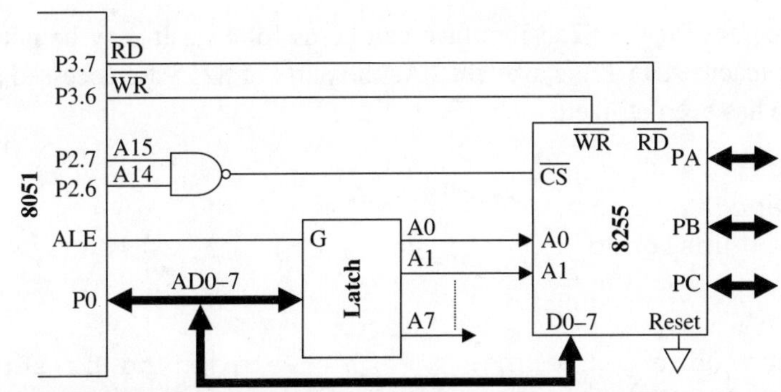

FIGURE 10.7 Interfacing 8255 with the 8051

As 8255 has been configured as an external memory device, the DPTR register is used to provide 16 bit addresses.

```
            ORG 0
            MOV DPTR, #0C003H    ; DPTR points to control register
            MOV A, #80H          ; Load control word
            MOVX @DPTR, A        ; Issue control word
            MOV A, 55H           ; Load 55H to A
     L1:    MOV DPTR, #0C000H    ; DPTR points to Port A
            MOV @DPTR, A         ; Output to Port A
            INC DPTR             ; DPTR points to Port B
            MOV @DPTR, A         ; Output to Port B
            INC DPTR             ; DPTR points to Port C
            MOVX @DPTR, A        ; Output to Port C
            CPL A                ; Complement A
            ACALL DL_RTN         ; Call Delay routine
            SJMP L1              ; Continue
DL_RTN:     MOV TMOD, #01        ; Timer 0, Mode 1
            MOV TL0, #0E0H       ; Load TH and TL
            MOV TH0, #0B1H       ; 0B1E0H for 20ms delay
            SETB TR0             ; Start timer
 AGAIN:     JNB TF0, AGAIN       ; Monitor Timer 0
            CLR TR0              ; Stop Timer 0
            CLR TF0              ; Clear Timer 0 flag
            RET
            END
```

The C language program for the 8255 interface can be as follows. It may be noted that for external memory access, the macro XBYTE is available. As the ports of 8255 are accessed as external memory locations, the macro has been utilized.

```
#include <reg51.h>
#include <absacc.h>
void delay_rtn( unsigned int del_val );
void main()
{
       unsigned char pattern;          // Pattern to be sent to ports
       XBYTE[0xC003] = 0x80;           // Configure all ports as output
```

```
        while (1) {
                pattern = 0x55;
                XBYTE[0xC000] = pattern;     // Send 55H to all ports
                XBYTE[0xC001] = pattern;
                XBYTE[0xC002] = pattern;
                delay_rtn( 100 );            // Put a delay
                pattern = 0xAA;
                XBYTE[0xC000] = pattern;     // Send 0AAH to all ports
                XBYTE[0xC001] = pattern;
                XBYTE[0xC002] = pattern;
                delay_rtn( 100 );            // Put a delay
        }
}
void delay_rtn( unsigned int del_val )
{
        unsigned int x, y;
        for (x = 0; x < 1000; x++)
                for (y = 0; y < del_val; y++);
}
```

10.9 Stepper Motor Interface

LO 8

Similar to the interfacing of the stepper motor, shown in Section 7.3.5, in the following a program has been presented to interface it with the microcontroller 8051. It is assumed that the motor has been driven by port P1.

```
        ORG 0
L1:     MOV A, #01H   ; Get 01H in A
        MOV P1, A     ; Energize winding A
        CALL DLYRTN   ; Call delay routine
        MOV A, #02H   ; Get 02H in A
        MOV P1, A     ; Energize winding B
        CALL DLYRTN   ; Call delay routine
        MOV A, #04H   ; Get 04H in A
        MOV P1, A     ; Energize winding C
        CALL DLYRTN   ; Call delay routine
        MOV A,#08H    ; Get 08H in A
```

```
            MOV P1, A      ; Energize winding D
            CALL DLYRTN    ; Call delay routine
            SJMP L1        ; Repeat from beginning
DLYRTN:     NOP            ; Small delay routine
            NOP
            RET
            END
```

The corresponding C language program is as follows:

```
#include <reg51.h>
void main (void)
{
     while (1)  {
            P1 = 0x01;
            delay_rtn(2);
            P1 = 0x02;
            delay_rtn(2);
            P1 = 0x04;
            delay_rtn(2);
            P1 = 0x08;
            delay_rtn(2);
     }
}
void delay_rtn( unsigned int del_val )
{     unsigned int x, y;
     for (x = 0; x < 1000; x++)
              for (y = 0; y < del_val; y++);
}
```

SUMMARY

To interact with the real world, a microcontroller-based system needs to interface with different input-output devices. The simplest input device to interface with 8051 is a switch, while the simplest output device is an LED. A matrix organization of switches can lead to a matrix keyboard. A hexadecimal matrix keyboard can be interfaced using two ports of the 8051. The interface program repeatedly scans the rows and columns of the keyboard to detect any keypress. An LCD can be interfaced with 8051 and can be operated by sending the required set of commands from the 8051. To interact with the analog world, it is necessary to interface analog-to-digital and digital-to-analog converters. To augment the number of I/O ports in an 8051-based system, an 8255 chip can be connected to get three additional 8-bit programmable ports. Interfacing of a stepper motor with the 8051 has been illustrated as well.

EXERCISES

Multiple Choice Questions

1. If a current limiting resistance is not used while driving an LED through the output port of 8051
   (a) LED intensity may be low
   (b) LED may be damaged
   (c) Port may be damaged
   (d) None of the above

2. For interfacing an LED with 8051 port, which of the following is/are safe ways?

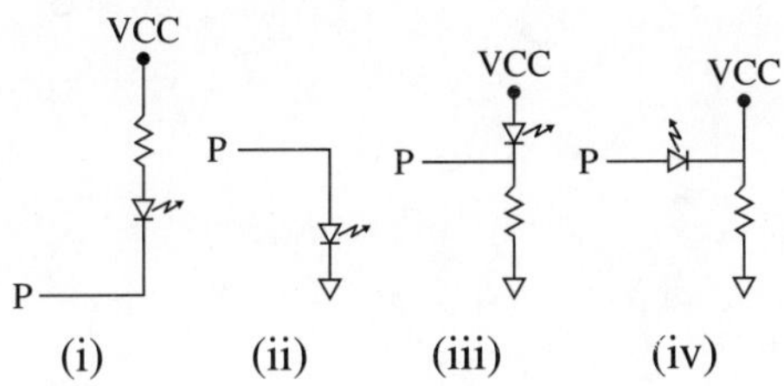

   (a) (i) (b) (i), (ii)
   (c) (ii), (iv) (d) (ii)

3. For interfacing an LED with 8051 port, if P = 1, which configuration will glow the LED?

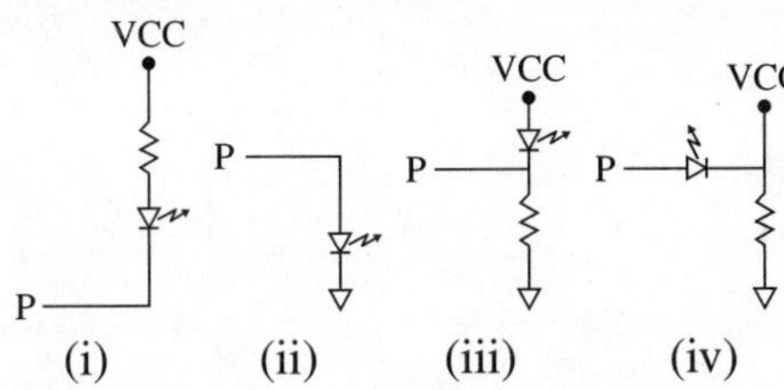

   (a) (i) (b) (i), (ii)
   (c) (ii), (iv) (d) (ii)

4. If ports P1 and P2 are used to drive the rows and columns of a matrix keyboard, number of keys can be upto
   (a) 4 (b) 16
   (c) 32 (d) 64

5. If ports P1 and P2 are used to drive the multiplexed seven-segment displays, the number of display modules interfaced can be upto
   (a) 4 (b) 8
   (c) 16 (d) 32

6. The minimum number of port bits needed for an LCD module interface is
   (a) 8 (b) 9
   (c) 10 (d) 11

7. To interface an 8-bit ADC with the 8051, without any interrupt support being used, the minimum number of port bits needed is
   (a) 8 (b) 10
   (c) 14 (d) 16

8. To interface an 8-bit DAC, the minimum number of 8051 port pins is
   (a) 8 (b) 10
   (c) 12 (d) None of the above

9. In 8255 interfacing, the lines D0 through D7 of 8255 should be connected to the 8051 port
   (a) 0 (b) 1
   (c) 2 (d) None of the above

10. In 8051 C-programming, XBYTE macro is used for
    (a) External memory access
    (b) Extended byte access
    (c) External bit access
    (d) None of the above

Answers to Multiple Choice Questions

1. b 2. a 3. d 4. d 5. b 6. d 7. c 8. a 9. a 10. a

Long Answer Type Questions

1. Modify the program corresponding to Figure 10.1 to turn ON and OFF LEDs, in group of four.
2. Modify the Hex Keyboard interfacing routine to read all keys if multiple keys are pressed simultaneously.
3. Modify hardware and software corresponding to Figure 10.3 to interface 8, seven-segment display modules and display the pattern "01234567" continually there.
4. Write an assembly language program to generate a Sine-wave via DAC interfacing through 8051. Also, write the corresponding C program.
5. Study the other programming modes of 8255 and develop corresponding interfacing programs.

11

External Communication Interface

LEARNING OUTCOMES

After reading this chapter, students will be able to:

LO 1 Interface processors using RS-232

LO 2 Establish device communication using SPI

LO 3 Interface devices using I^2C

LO 4 Design device interfacing using IEEE-488 (GPIB)

LO 5 Utilize Current Loop for data transfer

LO 6 Interface devices using Bluetooth

LO 7 Connect devices with Zigbee interfacing

LO 8 Interface processors with the Internet using Wi-Fi

11.1 Introduction

A system design problem consists of two parts – the computation part and the communication part. Once the data has been collected from the environment using devices following the strategies discussed in previous chapters, certain parameters can be computed by the microprocessor/microcontroller. The computed values may need to be communicated to nearby devices or faraway places over the internet. For example, in an Internet-of-Thing (IoT) environment, the home appliances may need to be controlled remotely by the house owner from his/her workplace. For this, the parameters of the appliances may be sensed and communicated to a local host device in the house which may in turn communicate the collected information over the internet to the house owner at the workplace. Control commands issued remotely by the owner may be similarly communicated to the individual devices at home. Thus, the processor chosen for a system design should not only have good computational features but also efficient communication mechanisms should be there. As there are a plethora of devices from various manufacturers, certain communication standards have evolved over the time to interface with them. Some such standards are RS-232, SPI, I^2C, GPIB, Current Loops, Bluetooth, Zigbee, Wi-Fi etc. Processor designers have also included many of these standards into their microprocessors and microcontrollers. In this chapter, we shall look into these standards and their usage in system design. The communication standards are being used more and more in IoT-based designs. We shall start with the oldest of these standards, RS-232 and gradually introduce the advanced standards, covering up to Wi-Fi.

11.2 RS-232 (LO 1)

RS-232 is one of the oldest communication interfacing standards existing from way back in the 1960s. The maximum cable length for the interface is 25 m with a data rate up to 38.4 kbps. It is an *unbalanced* interface in which the data bits are transmitted at voltage levels with reference to local grounds only. Because of this simplicity, it is still under use, though several serial communication protocols have come up. To transmit a logic 0, the driver outputs a signal in the range +5 V to +15 V while for logic 1, the corresponding range is –5 V to –15 V. At the receiver end, a signal in the range +3 V to +15 V is recognized as logic 0, while –3 V to –15 V is considered logic 1. As per the RS-232 terminology, a logic 1 is called a *space* while a logic 0 is called a *mark*. The standard was originally designed to connect between a *Data Terminal Equipment* (*DTE*) and a *Data Communication Equipment* (*DCE*). For example, a computer may be a DTE while a modem is a DCE.

11.2.1 Handshaking

In the absence of synchronizing clock signal, the flow control of information becomes a concern for RS-232 communication. Flow control prevents data overloading at the receiver. Handshaking methods are used to ensure the same. There are three types of handshaking available in RS-232, as noted next.

1. *Software handshaking*: Software-based handshaking saves the hardware area and additional signal lines. A typical example of this is data transmission over telephone data lines. The software protocol is called *Xon/Xoff*. In this, control characters are sent along the data lines. The command *Xon* starts the transmission while *Xoff* stops it. *Xon* is ASCII character 17 and *Xoff* is 19. When the receiver is ready to receive further data, it sends *Xon* character to start transmission. Transmission can be suspended by the receiver sending an *Xoff* character.
2. *Hardware handshaking*: This is superior to software-based handshaking. The connection has been shown in Figure 11.1. Here, the *Request To Send* (*RTS*) of *Data Terminal Equipment* (*DTE*) is connected to the *Clear To Send* (*CTS*) pin of *Data Communication Equipment* (*DCE*). The lines *Data Terminal Ready* (*DTR*) and *Data Set Ready* (*DSR*) are also connected in a similar fashion. To initiate

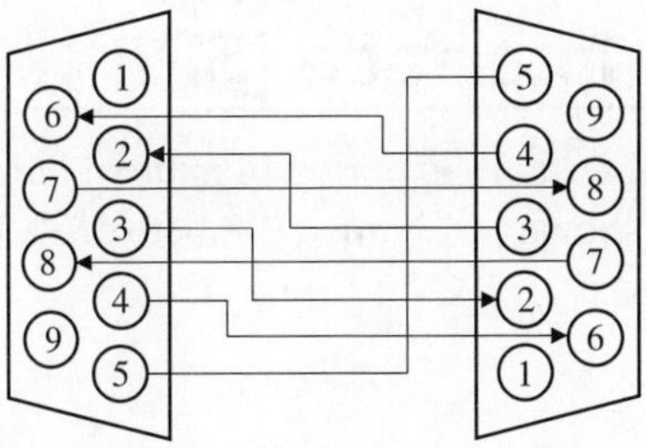

Pin details:
1: Data Carrier Detect (DCD)
2: Received Data (Rx)
3: Transmitted Data (Tx)
4: Data Terminal Ready (DTR)
5: Signal Ground
6: Data Set Ready (DSR)
7: Request To Send (RTS)
8: Clear To Send (CTS)
9: Ring Indicator (RI)

FIGURE 11.1 Nine-pin connector in RS-232

the transmission, *DTE* sets *RTS* ON. Once *DCE* becomes ready, it puts *CTS* ON. The *DTE* responds by making *DTR* ON and the line remains ON as long as data is being transmitted.

3. *Combined hardware software handshaking*: This is implemented by having the hardware connectivity as well as software-controlled *Xon/Xoff* character transmission, for flow control.

11.2.2 Interfacing with RS-232

Many of the microcontrollers (including 8051, PIC and AVR) work using TTL signals at voltage levels 0–5 V. The built-in UART of these processors can transmit and receive data as TTL waves only. For communication using proper RS-232, the signal levels need to be converted from TTL level to RS-232 compatible signals. The chip MAX-232 can be used for this purpose. It can convert a TTL level signal to an RS-232 level and an RS-232 level signal to a TTL level.

MAX-232 can operate up to a speed of 120 Kbps. It has two driver channels and two receiver channels. The maximum distance to which data can be sent is about 12–15 m. The pin layout of the chip has been shown in Figure 11.2(a). It requires an additional set of four capacitors of 1 μF each for its operation. Figure 11.2(b) shows the connection of capacitors to the chip.

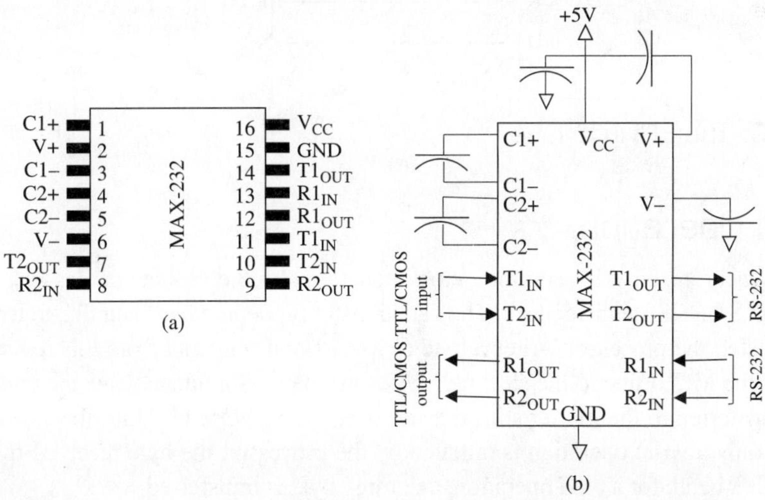

FIGURE 11.2 (a) MAX-232 pin layout (b) MAX-232 functional layout

11.3 Serial Peripheral Interface (SPI)

LO 2

SPI is a very simple, synchronous, four-wire interface supported in many microcontrollers and peripheral devices. The interface structure has been shown in Figure 11.3, between a processor and a typical peripheral device. The signal lines are as follows:

- MOSI – Master Out Slave In
- MISO – Master In Slave Out
- SCLK – Serial Clock
- CS – Chip Select

Some examples of peripheral devices with SPI capability are as follows:

- *LM74* temperature sensor working in the range of –55°C to +150°C with a resolution of 0.0625°C.
- *SCP1000* pressure sensor with 17-bit resolution, capable of detecting pressure difference within a 9-m^3 column of air.
- *LTC2452* ultra-tiny, differential, 16-bit delta-sigma ADC.
- *SX8652* low-power, high-reliability controller for resistive touch screens.
- *Microchip 25XXX* serial EEPROM with densities 128 bits to 512 kilobits, 3 to 20 MHz bus speed and low power with built-in write protection.
- *DS3234* low-cost, extremely accurate real-time clock.
- Memory cards – Multimedia Cards (MMC) and Secure Digital (SD) cards.

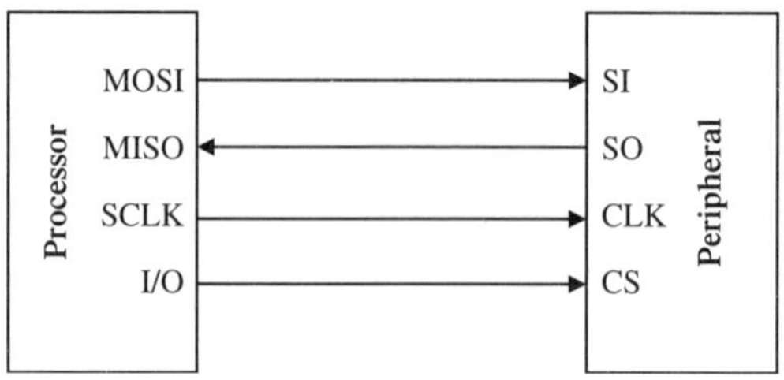

FIGURE 11.3 The SPI interface

11.3.1 SPI Operation

SPI operation is synchronized via a common clock generated by the master (processor). The master and the slave contain an 8-bit shift register each. The data transfer procedure has been illustrated in Figure 11.4. To initiate the transfer, the processor writes a byte to its SPI register. Data from this register are transmitted bit-by-bit over the MOSI line to the shift register of the slave. Simultaneously, the content of the slave shift register is transferred to the master shift register over the MISO line. Thus, the two register contents get exchanged. If only a write operation is initiated by the processor, the byte received from the device is ignored. On the other hand, for a read operation, a dummy byte is transferred.

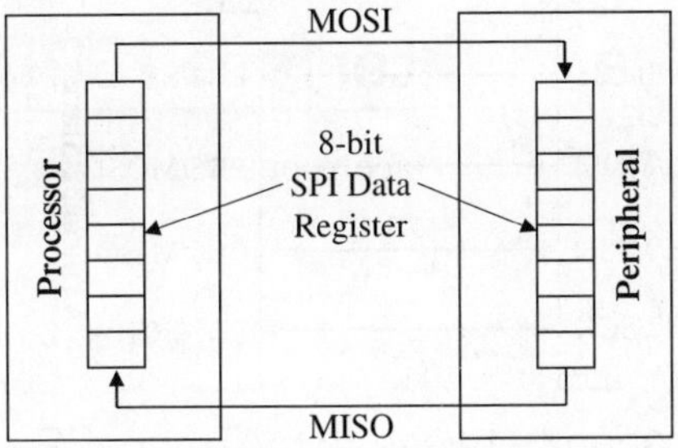

FIGURE 11.4 Data transfer through SPI interface

11.3.2 SPI – An Example

In this section, we shall look into an example interface between two AVR ATmega microcontrollers using the SPI interface. The AVR microcontroller details have been presented in Chapter 12. For SPI operation, the AVR microcontroller has the following three registers.

- *SPI Control Register* (*SPCR*): This is the master register that contains bits to initialize the SPI operation and control it.
- *SPI Status Register* (*SPSR*): Bits of this register can be used to read the status of SPI bus lines.
- *SPI Data Register* (*SPDR*): This is the read/write register through which the actual data transfer takes place. The register is 8-bit wide.

The SPCR register can be used to enable SPI operation, set up clock speed, configure master/slave modes etc. Various bits of the register are as follows:

- *Bit 7*: *SPI Interrupt Enable* (*SPIE*): The bit has to be set to 1 to enable interrupts in the SPI.
- *Bit 6*: *SPI Enable* (*SPE*): The bit is set to enable SPI operation.
- *Bit 5*: *Data Order* (*DORD*): The bit should be set to 1 to transmit LSB first and 0 to transmit MSB first.
- *Bit 4*: *Master/Slave Select* (*MSTR*): The bit needs to be set to 1 for master mode operation and 0 for slave mode operation of the processor/device.
- *Bit 3*: *Clock Polarity* (*CPOL*): The bit selects the clock polarity. If the bit is set to 1, in idle mode of the SPI bus, SCK is kept high by the master. On the other hand, if the bit is 0, SCK is kept low by the master. SPI data transfer takes place at SCK edges. If CPOL = 0, the rising edge of SCK is taken as the leading edge. Similarly, for CPOL = 1, the falling edge of SCK is considered the leading edge.
- *Bit 2*: *Clock Phase* (*CPHA*): Setting the bit to 1 samples the data at the leading (first) edge of SCK. Setting to 0, samples data at the trailing (second) edge of SCK.
- *Bits 1, 0*: *SPI Clock Rate Select* (*SPR0, SPR1*): Used to choose the oscillator frequency divider for the internal clock or externally connected crystal oscillator.

The SPSR register contains status bits for the SPI operation. The individual bits are as follows:

- *Bit 7*: *SPI Interrupt Flag* (*SPIF*): The flag bit is set whenever a serial transfer is complete. If the SPIE bit of SPCR is set, an interrupt is also generated. The SPIF flag is cleared when the interrupt service routine is executed.

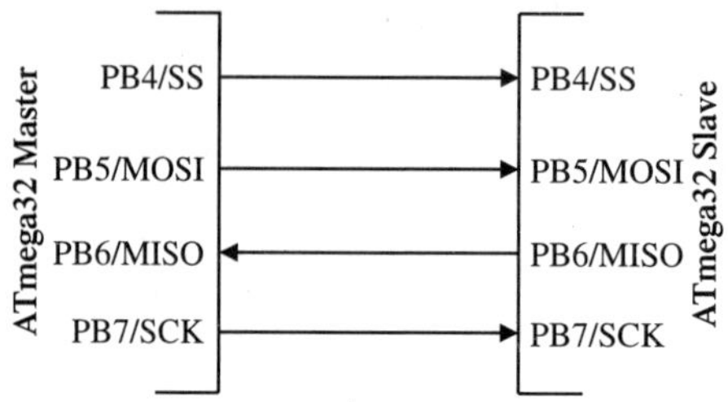

FIGURE 11.5 SPI communication between two ATmega32 processors

- *Bit 6*: *Write Collision Flag* (*WCOL*): The flag is set when data is written to the SPDR register, there is an impending transfer or the data lines are busy. Thus, it is advisable to first read the SPDR register so that the WCOL gets reset, before writing onto the register.
- *Bits 5–1*: Reserved.
- *Bit 0*: *SPI Double Speed Mode* (*SPI2x*): Though not needed in general, the bit can be set in conjunction with the bits 1 and 0 of the SPCR register, for very-high-speed data transfer.

Figure 11.5 shows the signal line connection between two ATmega32 processors for SPI communication. It may be noted that the SPI signal lines are multiplexed with digital I/O pins of Port B. The I/O pins are configured as input by default. To use some of them as output, the corresponding bits in the DDRB (Data Direction Register Port B) register should be set to 1. The following code fragment illustrates the software for the SPI transfer process. It consists of three functions – spi_master_init() to initialize the master mode operation, spi_slave_init() to initialize the slave mode operation and spi_transceiver() to transmit/receive data.

```
void spi_master_init ( void )
{
        // Set MOSI, SCK as output
        DDRB = (1 << 5) | (1 << 7);

        // Enable SPI, Set as master
        SPCR = (1 << SPE) | (1 << MSTR);

        // Enable interrupts
        sei();          // Function to enable SPIE
}

void spi_slave_init ( void )
{
        // Set MISO as output
        DDRB = (1 << 4);
```

```
        // Enable SPI
        SPCR = (1 << SPE);
}

unsigned char spi_transceiver ( unsigned char item)
{
        // Copy data item to SPI buffer
        SPDR = item;

        // Wait till transmission complete
        while (!( SPSR & (1 << SPIF)));

        // Return received data
        return (SPDR);
}
```

11.4 Inter-Integrated Circuit (IIC, I²C)

LO 3

I²C, also known as *Two Wire Interface* (TWI), is another very simple and effective strategy to interface peripheral devices to processors. Multiple devices can be connected via a multidrop bus which is bidirectional, low speed and synchronous to a common clock. The data rate varies between 100 and 400 kbps. The wires are named as follows:

- Serial Data (SDA)
- Serial Clock (SCL)

The wires are both open-drain and bidirectional. A typical interface has been shown in Figure 11.6.

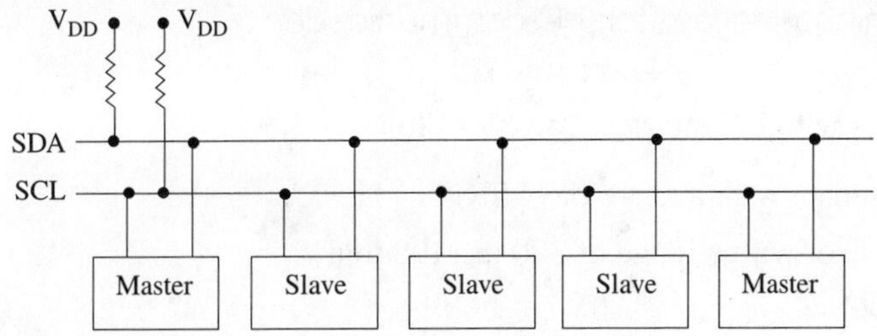

FIGURE 11.6 I²C wires

11.4.1 I²C Operation

In idle conditions, both SDA and SCL are high. The *start* condition is indicated by SDA going low followed by the SCL going low. The SDA wire now goes for a transition corresponding to the first valid data bit. The SDA line transits for the next bit when the SCL is low again, rising edge of SCL is used to sample the data bit. The bit must remain valid on the SDA line till SCL goes low once more. The process continues till all bits are transmitted. The *stop* condition is indicated by SCL returning to high, followed by SDA becoming high. Any number of bytes can be transmitted in an I²C packet. For acknowledgement, after transmitting the 8th bit of a byte, the master releases SDA and gives a pulse on SCL. The receiver acknowledges the byte transfer by pulling SDA low. For a slower slave device interface, whenever the slave is unable to receive further bytes, it can abort transmission by making SCL low. The transmitter then waits till SCL is released by the receiver. As there can be multiple slave devices on the bus, each device possesses a 7-bit unique address. The first transmitted byte contains a 7-bit address corresponding to the slave device and a *direction bit.* A '1' in the direction bit indicates a read operation to get data from the slave device. The bit being '0' indicates a write operation by the master. The timing diagram for I²C transmission has been shown in Figure 11.7.

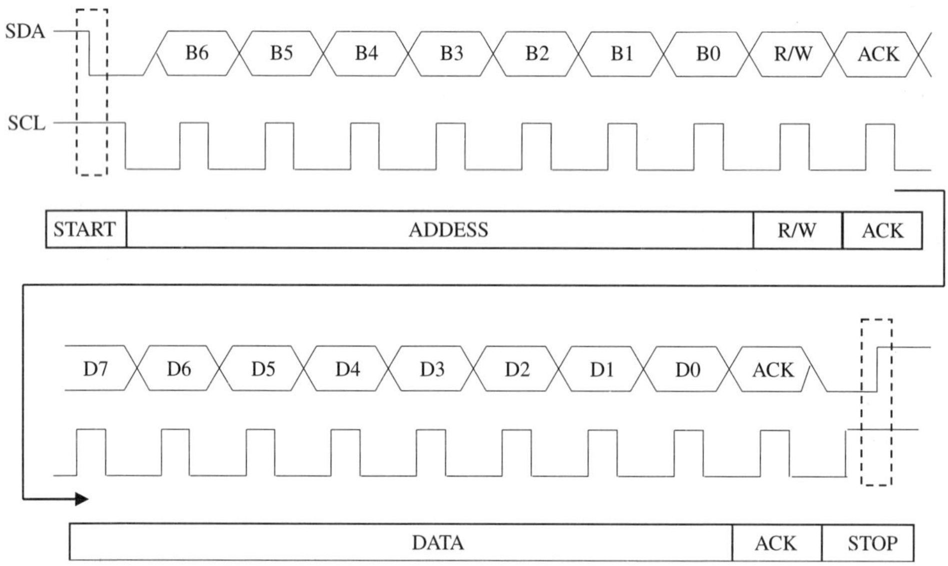

FIGURE 11.7 Timing diagram for I²C communication

Some example devices with I²C interfaces are as follows:

- *LM75*: Temperature sensor with a range of –55°C to +125°C.
- *BMP085*: Pressure sensor with a range of 300 to 1100 hPa.
- *ADC7992*: 12-bit ADC.
- *PCA94S08*: 8 kbit EEPROM.
- *PCA8565*: Real-time clock.

11.4.2 I²C – An Example

In this section, we shall look into the operation of I²C protocol in AVR ATmega processor. It has two hardware pins – Pin 22 and Pin 23, which act as the SCL and SDA lines respectively. The pins have the multiplexed function of Port C bits 0 (PC0) and 1 (PC1), respectively. It may be noted that I^2C is also called a *Two-Wire Interface* (*TWI*). Accordingly, names of the registers in ATmega, related to I^2C operation, start with the prefix "TW". The following are the important registers in AVR, related to I^2C.

1. *TWI Control Register* (*TWCR*): This register is used to control all events related to I^2C communication. Figure 11.8 shows the structure of the register. The bits in the register are as follows:
   - *TWINT* (*Bit 7*): It stands for *TWI Interrupt*. The bit gets set on completion of the current TWI event (such as start, stop, transmit and receive). The bit must be cleared by software by writing 1 to it.
   - *TWEA* (*Bit 6*): *TWI Enable Acknowledgement* bit should be set to 1 in receiver mode, so that, the receiver generates acknowledgement. For transmission, the bit should be cleared.
   - *TWSTA* (*Bit 5*): *TWI START Condition* bit is set by the master device to generate START condition by monitoring the free bus status to take control of the bus. As shown in Figure 11.7, for START, SDA goes low followed by SCL going low.
   - *TWSTO* (*Bit 4*): *TWI STOP Condition* bit is set by the master device to generate a stop condition with SCL going high, followed by SDA going high. This leads to the release of the TWI bus by the master.
   - *TWWC* (*Bit 3*): *TWI Write Collision* bit gets set when an attempt is made to write to the TWI Data Register before the current transmission is complete (TWINT is low).
   - *TWEN* (*Bit 2*): *TWI Enable* bit enables the I^2C interface so that pins 22 and 23 are assigned to it.
   - *Bit 1*: Reserved.
   - *TWIE* (*Bit 0*): *TWI Interrupt Enable* bit enanbles the TWI interrupt.
2. *TWI Status Register* (*TWSR*): This register holds the status information regarding I^2C operation. Figure 11.9 shows the structure of the register. The bits of the register have the following purposes.
   - *TWS7-TWS3* (*Bits 7-3*): These bits capture the status of TWI control and bus.
   - *TWPS1-TWPS0* (*Bits 1-0*): These are called prescalar bits and are used to calculate SCL frequency (explained at point no. 5). Table 11.1 shows the settings in more detail.
3. *TWI Data Register* (*TWDR*): This is an 8-bit register holding the data to be transmitted or received. While the communication is going on, the content is not writable. As long as the TWINT bit remains set, the data in TWDR remains stable.

7	6	5	4	3	2	1	0
TWINT	TWEA	TWSTA	TWSTO	TWWC	TWEN	–	TWIE

FIGURE 11.8 TWI control register

7	6	5	4	3	2	1	0
TWS7	TWS6	TWS5	TWS4	TWS3	–	TWPS1	TWPS0

FIGURE 11.9 TWI status register

TABLE 11.1 Prescalar value settings

TWPS1	TWPS0	Exponent	Prescalar Value
0	0	0	1
0	1	1	4
1	0	2	16
1	1	3	64

7	6	5	4	3	2	1	0
TWA6	TWA5	TWA4	TWA3	TWA2	TWA1	TWA0	TWGCE

FIGURE 11.10 TWI address register

4. *TWI Address Register* (*TWAR*): It contains the address of the TWI unit in slave mode of operation. Figure 11.10 shows the structure of the register. The individual bits of the register are as follows:
   - *TWA6-TWA0* (*Bits 7-1*): These bits constitute the 7-bit address of the slave device.
   - *TWGCE* (*Bit 0*): The *TWI General Call Enable* bit enables recognition of general call over the TWI bus.
5. *TWI Bit Rate Register* (*TWBR*): This is an 8-bit register used to generate the SCL frequency in the master mode of operation. The SCL frequency is computed by the formula,

$$SCL\ frequency = \frac{CPU\ clock\ frequency}{16 + 2 \times (TWBR) \times (4^{TWPS})}$$

   Here, 4^{TWPS} is the pre-scalar value determined by the settings of the bits *TWPS1* and *TWPS0* in *TWSR*. For example, setting both the bits to 1 results in the prescalar value 4^{TWPS} to $4^3 = 64$, as noted in Table 11.1.

In ATmega, I²C communication may take place in four modes, as follows:

1. MASTER as a transmitter
2. MASTER as a receiver
3. SLAVE as a receiver
4. SLAVE as a transmitter

To operate the transmitter and receiver as a pair, normally the modes 1 and 3 and the modes 2 and 4 are used together. In the following, we shall look into C-code snippets for different modes of operation.

1. *Code snippets for MASTER controller*

   The operation of the MASTER controller can be refined into a number of steps as noted next.

   ***Step 1:** Initialization of MASTER*

   This is accomplished by setting the bit rate in the TWBR and the prescalar bits in TWSR. The code is as follows:

```
void I2C_init_Master (void)
{
        TWBR = 0x01;            // Bit rate
        TWSR = (0 << TWPS1) | (0 << TWPS0)// Setting prescalar bits
        // Sets SCL frequency to Freq_CPU/(16 + 2(TWBR)×4^TWPS
}
```

Step 2: *Send START condition*

This is done by writing 1 to the bits TWINT, TWSTA and TWEN. The status of the TWINT flag is monitored for the START condition to get transmitted. The TWSR is monitored to check the ACK byte.

```
void I2C_START (void)
{
   TWCR = (1<< TWINT) | (1 << TWSTA) | (1 << TWEN);
   while (!( TWCR & (1 << TWINT))); // Wait for START condition
                                       to get transmitted
   while (( TWSR & 0xF8) != 0x08); // Check for acknowledgment
  // Refer to AVR manual for different acknowledgement values
}
```

Step 3: *Send SLAVE address, data direction bit and wait for ACK*

This is done by first putting a 7-bit slave address and direction control bit in TWDR. The TWINT bit is cleared to 1 and TWEN is set to 1. The TWINT flag is monitored. Once the flag has become 0, the TWDR byte has been transmitted. The MASTER next checks for the correct acknowledgement.

```
void I2C_write_address ( unsigned char data_addr_dir )
{
        TWDR = data_addr_dir;  // Address and direction put in TWDR
        TWCR = (1 << TWINT) | (1 << TWEN);
        while (!(TWCR & (1 << TWINT)));
        while ((TWSR & 0xF8) != 0x40);   // Check for correct ACK
}
```

Step 4: *Send 8-bit data and wait for ACK*

The step first puts the 8-bit data in TWDR and the bits TWINT and TWEN are made 1. The TWINT flag is monitored for completion of data transmission and the acknowledgement is checked.

```
void I2C_write_data ( unsigned char data )
{
TWDR = data;
        TWCR = ( 1<< TWINT ) | ( 1 << TWEN );
        while ( !( TWCR & ( 1 << TWINT )));
        while (( TWSR & 0xF8) != 0x28);
}
```

Step 5: Send the STOP condition

To send the STOP condition, the TWSTO bit of TWCR is set along with TWINT and TWEN. Now, the TWSTO bit is continually monitored to become 0 as that will indicate the situation that the STOP condition has been transmitted.

```
void I2C_STOP ( void )
{
        TWCR = ( 1 << TWINT ) | ( 1 << TWEN ) | ( 1 << TWSTO );
        while (!( TWCR & ( 1 << TWSTO )));
}
```

2. *Code snippets for the SLAVE controller*

The operation of the SLAVE controller consists of the following steps.

Step1: Initialization of SLAVE

The slave controller is initialized by assigning an address to the slave. The slave address is held in the seven most significant bits of TWAR. The LSB and TWGCE are used to acknowledge general call addresses. Assuming that the slave address is 20H, the code fragment to initialize the slave is as follows:

```
void I2C_init_slave ( void )
{
        TWAR = 0x20;
}
```

Step 2: Check the status of TWSR

The slave can now wait in a loop till TWSR content becomes 60H. The value 60H in TWSR indicates that the data sent by the master is to be read by this slave only. The code is as follows:

```
void I2C_match_read_slave ( void )
{
    while (( TWSR & 0xF8 ) != 0x60 )
    {
        TWCR = ( 1 << TWEA ) | ( 1 << TWEN ) | ( 1 << TWINT );
        while ( ! ( TWCR & ( 1 << TWINT )));
    }
}
```

Step 3: Read data

This step reads the data sent by the master. The code for the same is as follows:

```
void I2C_read_data ( unsigned char *data )
{
  TWCR = ( 1 << TWINT ) | ( 1 << TWEA ) | ( 1 << TWEN );
           // Clear interrupt flag, get acknowledgement, enable TWI
  while ( ! ( TWCR & ( 1 << TWINT ))); // Wait for TWINT
  while (( TWSR & 0xF8 ) != 0x80 ); // Wait for acknowledgement
  *data = TWDR;             // Read data from TWDR
}
```

These routines can be used by the devices connected to the I²C bus to carry out the data transfer activities efficiently. For details on other modes of operation (i.e., modes 2 and 4), readers may refer to the AVR ATmega manual pages.

11.5 IEEE-488

LO 4

This is a bus specification developed by Hewlett-Packard in the 1960s. Also known as *Hewlett-Packard Interface Bus* (*HP-IB*) and *General Purpose Interface Bus* (*GPIB*), *IEEE-488* is a short-range communication standard supporting an 8-bit, parallel, multi-master interface. With licensing from HP, GPIB became a de facto standard for automated industrial instrumentation. In 1975, IEEE standardized the bus as *Standard Digital Interface for Programmable Instrumentation – IEEE-488*. The standard has been subsequently revised to *IEEE-488.1* in 1987 and *IEEE-488.2* in 1992. The pin distribution of IEEE-488 has been shown in Figure 11.11. Table 11.2 enumerates the functionalities of individual pins.

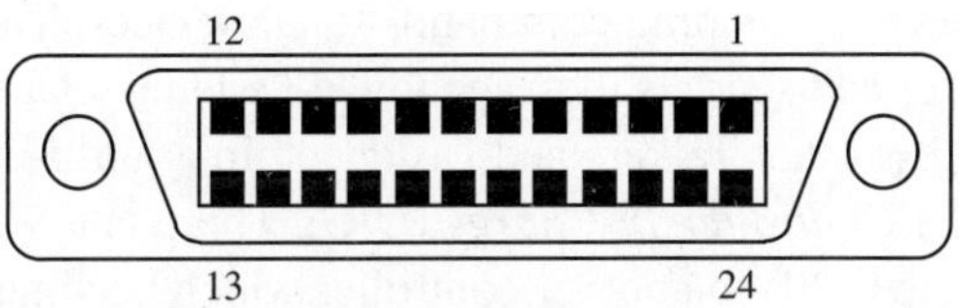

FIGURE 11.11 Pin distribution in IEEE-488

TABLE 11.2 Pin functionalities in IEEE-488

Pin No.	Name	Functionality
Pin 1–4	DIO1–4	Data input/output bits 1–4
Pin 5	EOI	End of identity
Pin 6	DAV	Data valid
Pin 7	NRFD	Not ready for data
Pin 8	NDAC	Not data accept
Pin 9	IFC	Interface clear
Pin 10	SRQ	Service request
Pin 11	ATN	Attention
Pin 12	SHIELD	Shield
Pin 13–16	DIO5–8	Data input/output bits 5–8
Pin 17	REN	Remote enable
Pin 18–23	GND	Ground wire twisted with DAV, NRFD, NDAC, IFC, SRQ and ATN, respectively
Pin 24	Logic ground	Logic ground

GPIB instrumentation bus structure has been shown in Figure 11.12. The bus supports 31 primary devices (numbered 0–30) using a 5-bit device address. However, up to 15 devices can share a physical bus of cable length up to 20 m. To connect more devices, active extenders need to be attached to the bus. As each device has its own asynchronous handshaking signals, both slow and fast devices can be mixed on the bus. The typical data transfer rate is about 1 MB per second, which can go up to 8 MB per second. At this rate, transmission line issues like impedance matching and line termination can be ignored.
A GPIB/IEEE-488 device communicates with other GPIB devices via *device-dependent* messages and *interface* messages, as follows:

- *Device-dependent messages* are also known as *data* or *data messages*. They contain device-specific information, like programming instructions, measurement results, machine status and data files.
- *Interface messages* are used to manage the bus, such as initializing the bus, addressing and unaddressing devices and setting device modes. These messages are also known as *commands* or *command messages*.

The devices in a GPIB network can be of one of the three types – *Talkers*, *Listeners* and *Controllers*. A Talker device sends data messages to one/more Listeners which receive data. The Controller manages the flow of information on GPIB by sending commands to all devices. The Controller enables a Talker and a Listener before the Talker can send its message to the Listener. Once the message transmission is over, the Controller may address other Talkers and Listeners. In a multi-controller environment, at any time only one Controller is the *Controller-In-Charge* (*CIC*). There is a *System Controller* that can take control or give up control of the GPIB to another controller which becomes the active controller.

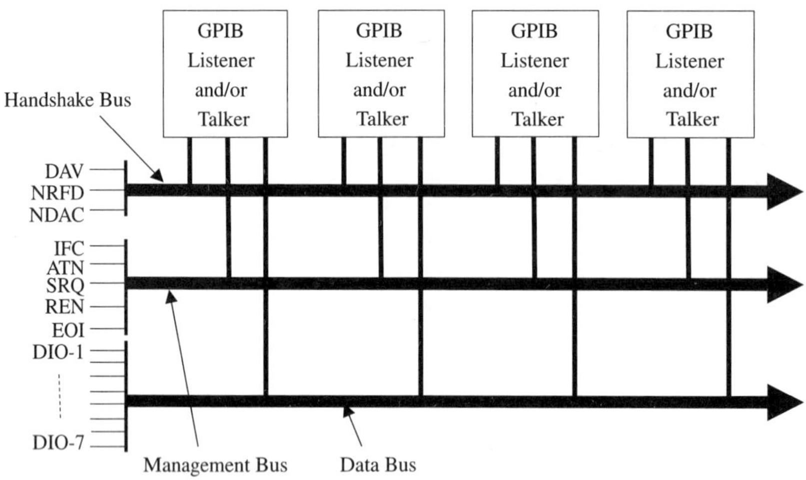

FIGURE 11.12 GPIB instrumentation bus structure

11.5.1 Interfacing GPIB Controller

In this section, we shall look into a brief overview of *μPD7210*, an intelligent GPIB controller from *NEC Electronics* that can be integrated into a system to meet the functional requirements of Talker, Listener and Controller. Placed between the processor bus and the GPIB, it provides high-level management of the GPIB to unburden the processor and simplify system design. The *μPD7210* is a 40-pin DIP chip, fully compatible

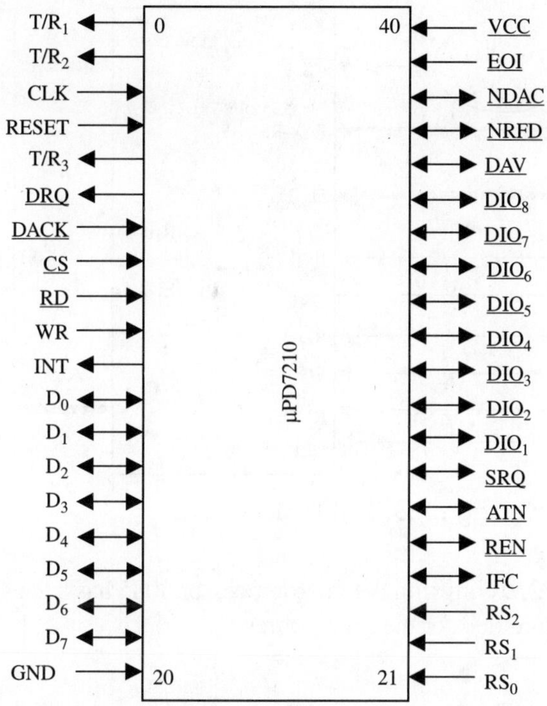

FIGURE 11.13 Pin layout of μPD7210

with most of the processor architectures and requires only bus driver/receiver components to implement GPIB. Figure 11.13 shows a pin layout of the chip. While most of the pins correspond to the IEEE-488 connector pins (noted in Table 11.2), the remaining ones are for interfacing with a processor, as noted next.

- *T/R_1 to T/R_3 (Pins 1, 2, 5)*: These are input/output control signals for the GPIB transceivers, controlled by internal registers.
- *CLK* (*Pin 3*): This is a 1 to 8 MHz reference clock.
- *RESET* (*Pin 4*): Reset input.
- *DRQ* (*Pin 6*): DMA requests output to ask processor to release data bus.
- *DACK* (*Pin 7*): DMA acknowledgement from processor, once data bus is released. The data bus now gets connected to the *data registers* of μPD7210.
- *Chip Select* (*Pin 8*): It selects the chip for internal register access.
- *RS_0–RS_2* (*Pins 21, 22, 23*): Register select inputs. One of the eight internal registers in *μPD7210* gets selected for read/write operation.
- *RD*, *WR* (*Pins 9, 10*): Performs a read/write operation with the selected register.
- *INT* (*Pin 11*): Interrupts the processor on the occurrence of any of the 13 internal interrupt factors.
- *D0 to D7* (*Pins 12 to 19*): An 8-bit bidirectional data bus for interface to the processor.
- *VCC, GND* (*Pins 40, 20*): +5 V power supply and ground.

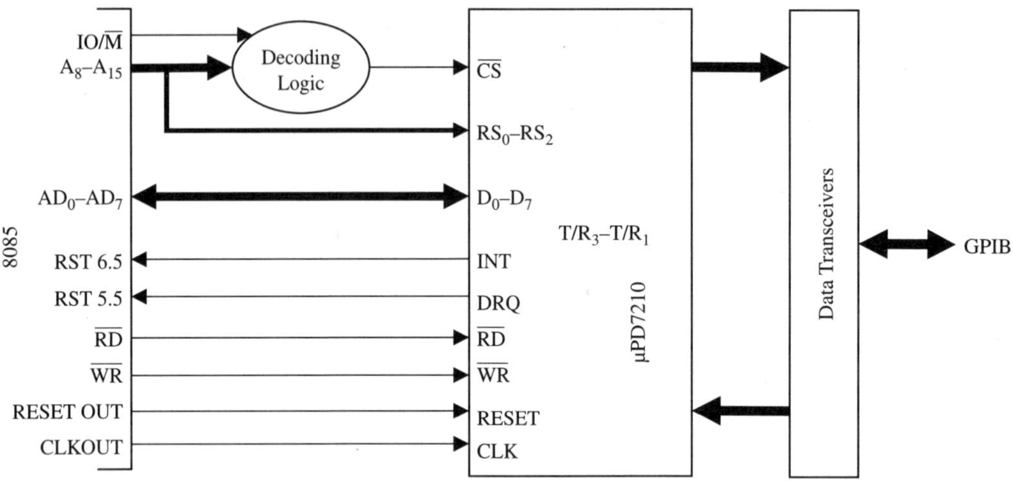

FIGURE 11.14 Interfacing µPD7210 with 8085

An interfacing of the *µPD7210* chip with the microprocessor 8085 has been shown in Figure 11.14. The detailed program has been omitted for the sake of brevity.

11.6 4–20 mA Current Loop

LO 5

Typically used for electronic signalling in industrial process control, the 4–20 mA current loops work with two current values of 4 and 20 mA to represent 0 to 100% of the range of measurement and control. The loops can be utilized for carrying sensor information from the field to the controller and also for carrying control signals (from the controller) to the on-field actuators. That way, current loops are ways of transmitting information. The situation has been shown in Figure 11.15. The information is carried as varying current levels, depending on the value of the sensor output. Its operation is thus similar to an ADC. However, an ADC produces equivalent voltage as output, whereas a current loop transmitter produces current as output.

In a 4–20 mA current loop, 4 mA represents a logic low level, while 20 mA represents a logic high. The current loop structure, shown in Figure 11.15 contains a sensor (for pressure, temperature, level, pH, flow or other process variables), transmitter, receiver and current source (or power supply). Depending upon the

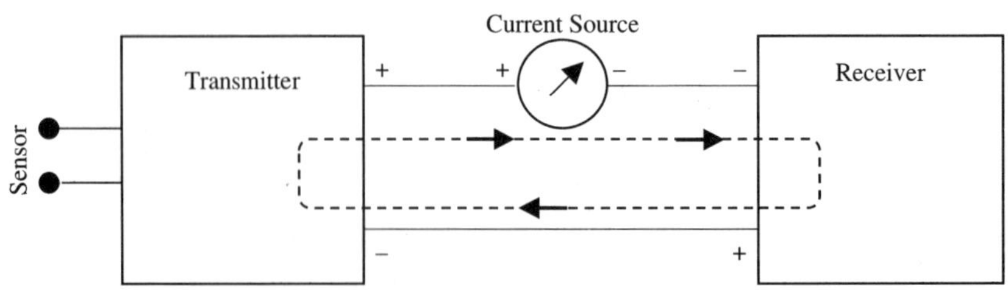

FIGURE 11.15 Current loop

value of the physical parameter being sensed, the sensor produces an equivalent output voltage. The transmitter converts this voltage to a proportional 4–20 mA DC current. On the other hand, the receiver converts the 4–20 mA current into equivalent voltage for further processing. The current source supplies power to the system. The current loop interface provides a number of key advantages, as noted next.

- As there is no voltage drop issue in interconnecting wiring, the accuracy of the signal is not affected.
- The implementation is often via twisted-pair conductors, providing high noise immunity.
- Has the facility of fault self-monitoring – currents less than 3.8 mA or more than 20.5 mA may be counted as a fault.
- Long cables can be used.
- Can be easily converted to voltage by using a resistor.
- The loop wire carries current (4 mA even when process output is low), that can be used to power remote devices (often called *loop-powered* device).

11.6.1 Example System

Figure 11.16 shows an example system depicting the interfacing of a sensor with a processor. From the sensor, a current loop has been used to transfer the current signal to an ADC (analog-to-digital converter) connected to the processor. Similarly, the digital output from the processor has been converted to analog voltage via a DAC (digital-to-analog converter). The analog voltage is next converted into current signal and transmitted over another current loop to the load for actuation. It may be noted that the following two modules have been used in the interface that need specific mention.

- *XTR110*: Precision voltage-to-current converter/transmitter. It has two input voltage ranges – 0 to +5 V and 0 to +10 V. The outputs can be 4–20 mA, 5–25 mA and some more commonly used ranges.
- *RCV420*: Precision current-loop receiver. It can convert a 4–20 mA input signal into a 0 to 5 V output signal.

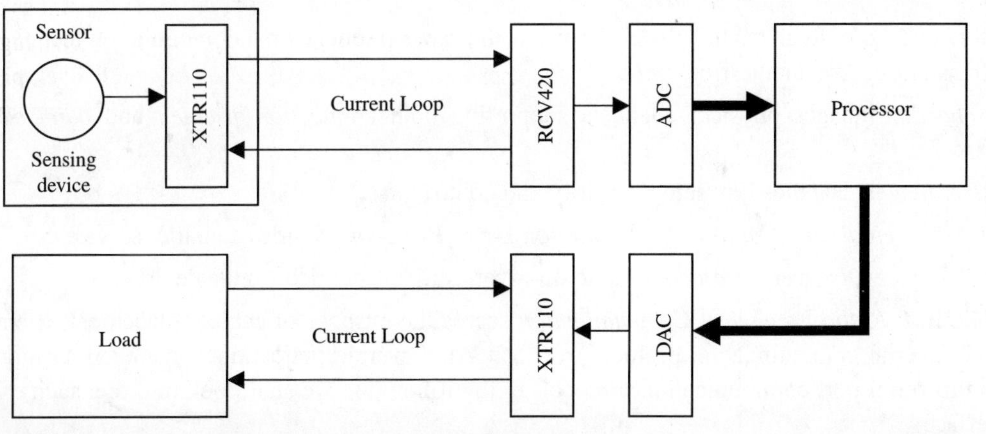

FIGURE 11.16 Current loop example system

As we have already seen ADC and DAC interfacings in previous chapters, further details in the current loop interface have been omitted.

11.7 Bluetooth

LO 6

Bluetooth is a short-range wireless technology initiated by *Ericsson* in the year 1989. It has been subsequently adopted by myriads of companies to be plugged into electronic devices like computers, printers, mobile phones, speakers, headsets, keyboards and mice, to name a few. Truly speaking, it has come up as a substitute for all other interfacings used for short-range networking like RS-232C, USB and Infrared. It provides a one-to-one communication technology, transmitting data through low-power radio waves (about 1 mW) at a frequency of 2.45 GHz (between 2.40 GHz and 2.48 GHz). This frequency band has been reserved globally for unlicensed (regulated) communication for industrial, scientific and medical applications. The Bluetooth transmission is restricted to 10 m but can penetrate through opaque medium. Thus, the communication can span across several rooms in a building. Up to eight devices can be connected simultaneously. Though the devices may be very closely spaced, signals do not interfere with each other due to the usage of *spread spectrum frequency hopping* in which a device uses 79 individual randomly chosen frequencies with the transmitter changing frequencies on a regular basis. The transmitters change frequency 1600 times per second.

The devices in Bluetooth form a *Personal Area Network* (*PAN*), called *piconet*. A piconet may have one major device and up to seven slave devices that are active. There can be up to 255 further devices that are inactive or *parked*. When two Bluetooth-enabled devices come close to each other, an electronic conversation takes place to decide whether they have any data to share or if a device needs to control the other. No user intervention is necessary for the process. This establishment of communication between two devices is called *pairing* or *bonding*. The pairing process may be secured by introducing a *passkey* entry, in which at least one device should have a display and at least one device a numeric keypad. A six-digit numeric code is displayed, and the user has to enter the code through the keypad to establish connection.

Bluetooth communication follows the Bluetooth protocol stack. The stack consists of the following layers from bottom to top.

1. *Bluetooth Radio*: Realizes the physical layer with proper frequency band, modulation technique etc. to transmit/receive data to/from peers.
2. *Baseband*: Manages physical channels supporting both *connection-oriented* and *connectionless* services.
3. *Link Manager*: Handles link setup, security and control.
4. *L2CAP*: *Logical Link Control and Adaptation Layer Protocol* provides data link services.
5. *SDP*: *Service Discovery Protocol* used to discover available services in the device.
6. *RFCOMM*: *Radio Frequency Communication* serves the purpose of cable replacement. It emulates RS-232 serial communication protocol over L2CAP. Several applications have been designed that use this serial port communication protocol. In the following, we shall look into one such example interface.

11.7.1 Bluetooth HC-05 Interfacing with 8051

HC-05 is a Bluetooth module that can be used in both master and slave configurations. It has been incorporated into many applications such as wireless headsets, game controllers, wireless mice and keyboards. The module uses serial communication to communicate between devices. It talks to the microcontroller (such as 8051) using serial port (USART). HC-05 has six pins, as noted in the following.

1. *Key/EN*: HC-05 can operate in two modes – *Data mode* and *Command mode*. If the pin is low, the module works in data mode, else it works in command mode. The default operation is in data mode with a baud rate of 9600. The command mode may be used to change the settings of HC-05. Data transfer rate can vary up to 1 Mbps in the 10 m range.
2. *VCC*: +5 V or +3.3 V may be used.
3. *GND*: Ground connection.
4. *TXD*: Transmit Serial Data. The data received by the wireless Bluetooth module is transmitted out serially on the TXD pin.
5. *RXD*: Receive Serial Data. The received serial data on the RXD pin is transmitted by the wireless Bluetooth module.
6. *State*: Indicates whether the module is connected or not.

Detailed discussion on the command mode is beyond the scope of the book. Figure 11.17 shows an interfacing of HC-05 with the 8051 microcontroller. The serial data interface of 8051 has been used to connect to the HC-05 module. The USART programming used in Section 9.7 can be utilized to send/receive characters to/from HC-05.

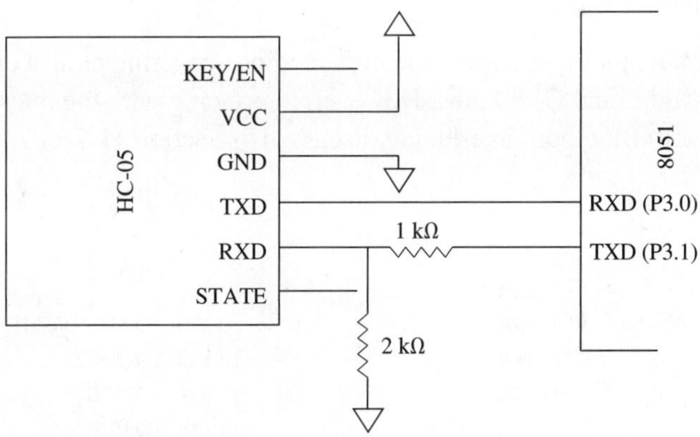

FIGURE 11.17 HC-05 interfacing with 8051

11.8 Zigbee

LO 7

This is a low-cost, low-power wireless technology for IoT data networks. Zigbee is primarily used for wireless communication, monitoring and control of battery-operated devices and sensor networks, particularly for extremely limited power situations. It supports several network topologies, such as point-to-point, point-to-multipoint and mesh. Zigbee devices have *sleep mode*, facilitating low power consumption. The data rate supported is up to 250 kbps. Security has been integrated into Zigbee with the usage of 128-bit AES encryption. Over-the-Air (OTA) software upgrade feature allows a single device to update firmware on all network devices.

Among the topologies supported by the protocol, *point-to-point* is a network with just two nodes. It is not very much used in IoT applications. On the other hand, in a *point-to-multipoint* topology (also known as *star* network), a Zigbee device acts as the central hub that connects with several other Zigbee devices acting as end nodes. A *tree* topology is also supported that contains some router nodes connected with the central hub or other router nodes in a tree structure. The end nodes are connected to routers. Zigbee mesh topology provides redundancy and robustness with multiple interconnect devices. The network is decentralized with each node having the self-discovery capability. Adhoc routing in mesh creates better stability with events like nodes joining the network, leaving the network or developing faults.

Zigbee operates in a 2.4 to 2.4835 GHz frequency band with a distance range of 10–100 m between devices. However, using mesh topology, data can be passed over longer distances as individual nodes can function as repeaters. Before starting communication, a Zigbee device checks for channel clearance using *Carrier Sense Multiple Access with Collision Avoidance* (*CSMA/CA*). If the channel is clear, data is transmitted, else the device waits for a backoff period before retrying.

11.8.1 XBee Interfacing

The XBee RF module offers multiple communication methods covering almost all the microcontrollers. It also has a number of additional GPIO pins. However, interfacing with the microcontroller is done via the USART pins (similar to Bluetooth interfacing discussed in Section 11.7.1). A pin layout of the XBee

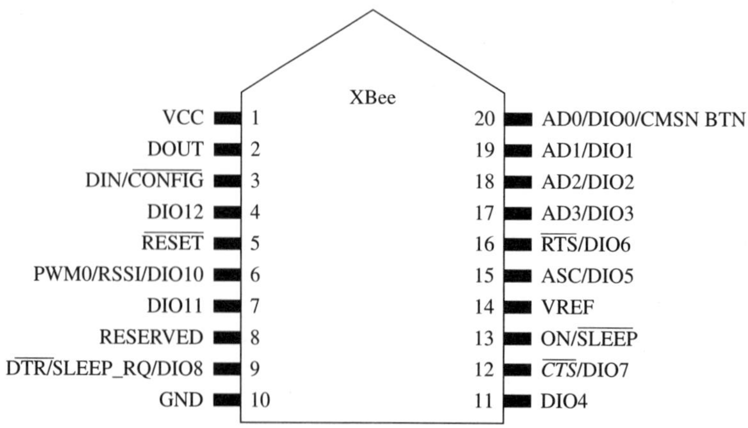

FIGURE 11.18 XBee pin layout

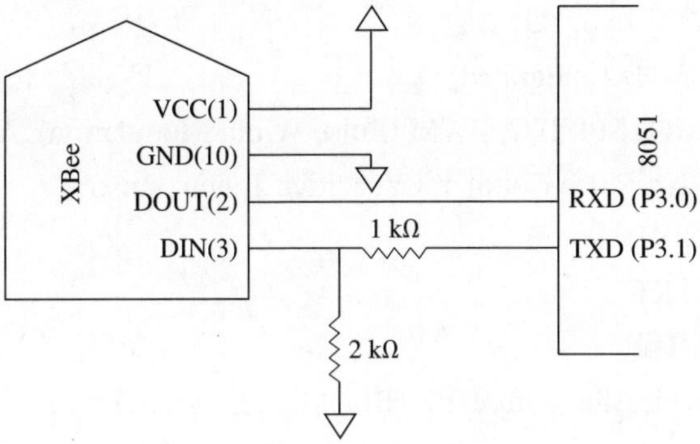

FIGURE 11.19 XBee interfacing with 8051

devices has been shown in Figure 11.18. Detailed discussion on the pins is skipped for the sake of brevity. However, the pins important for USART interfacing with microcontrollers are as follows:

- *DOUT* (*Pin 2*): It acts as a USART serial receive data line for the microcontroller.
- *DIN* (*Pin 3*): The pin acts as a UART serial transmit data for the microcontroller.

An interface with the 8051 microcontroller has been shown in Figure 11.19.

11.9 Wi-Fi

LO 8

Based on the IEEE 802.11 family of standards, Wi-Fi is a wireless networking technology allowing devices to connect to the internet and/or communicate with each other without any physical cables. Electromagnetic waves are used for transmitting information. The frequencies are further divided into channels. The most commonly used frequency bands are 2.4 GHz UHF and 5 GHz SHF radio bands with the 6 GHz SHF band being used in new generations. Wi-Fi devices connect to a Wireless Access Point (WAP). The communication works best for line-of-sight operation. An access point may have a range of about 20 m; however, it may extend up to 150 m.

Several microcontrollers have been designed to have Wi-Fi capabilities, making them very much suitable for IoT applications. The following are some such microcontrollers.

- ESP8255, ESP8266 and ESP32 from *Espressif*.
- CYW43340 from *Cypress*.
- MT7681, MT7687 from *MediaTek*.
- WFI32E01PC, ATSAMW25 from *Microchip*.
- CC3200, CC3220 from *Texas Instruments*.

11.9.1 Wi-Fi Microcontroller ESP32

To understand the features that are available with the Wi-Fi microcontrollers, we shall look into ESP32. This is a System-on-Chip (SoC) microcontroller used extensively in IoT applications. A specification of the microcontroller is as follows:

- Operating voltage: 3.3 V
- Operating current: 80 mA (average)

- Operating range: –40°C to 85°C
- Crystal: 40 MHz (integrated)
- Interfaces: UART, SPI, I2C, PWM (Pulse Width Modulation), ADC, DAC, GPIO, Pulse Counter, Capacitive Touch Sensor
- ROM: 448 KB
- RAM: 520 KB
- Integrated Flash: 4 MB
- Connectivity: Wi-Fi, Bluetooth, BLE (Bluetooth Low Energy)
- On-chip sensor: Hall sensor
- RF components: Power amplifier, Low-noise receive amplifier, Antenna switch, RF Balun

The Wi-Fi interface supports protocols like TCP/IP, HTTP and even the secured HTTPS. The footprint of the chip is quite small – 25 mm × 18 mm. The chip is low cost and can be programmed in high-level languages using its IDE.

11.9.2 Wireless Network Processor CC3100

Unlike Wi-Fi microcontrollers, traditional microprocessors and microcontrollers do not possess facilities to connect to the internet. The problem can be resolved by interfacing a wireless network processor with them. One such example processor is the *Wireless Network Processor CC3100* from *Texas Instruments*. It can be interfaced with traditional non-Wi-Fi microcontrollers to connect them to the internet. It minimizes the host processor software requirements. Similar to ESP32, CC3100 supports all Wi-Fi features, while the interface with the host processor is done via USART and SPI. Figure 11.20 shows such interfacing.

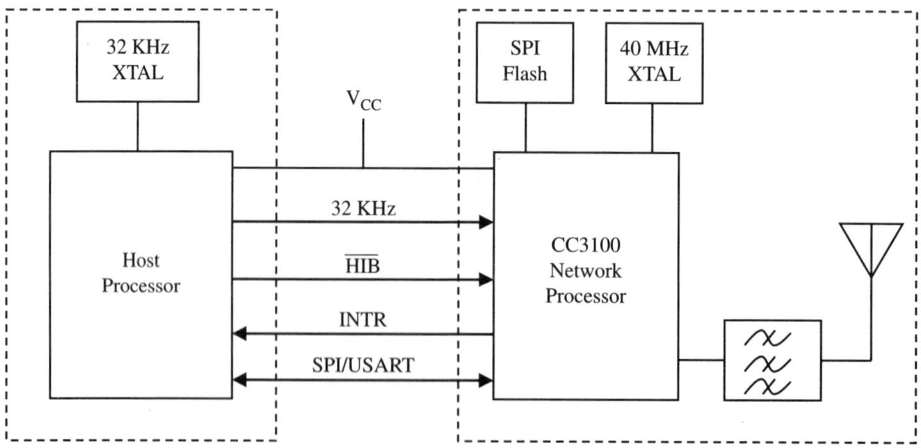

FIGURE 11.20 Interfacing CC3100 with host processor

SUMMARY

LO 8

Microprocessors and microcontrollers need to communicate with other electronic devices which may be spread over geographically distributed locations. Thus, apart from being computationally efficient, these processors must support communication infrastructures. Depending upon the distances between the devices, different types of communication backbones may be necessary. For closely located devices, communication over RS-232, SPI, I^2C, Current Loop etc. may be sufficient for wired connections. Whereas, for wireless connection, Bluetooth, Zigbee etc. can be used for short-length communication. For communicating over the internet in an IoT environment, the processor must have a built-in Wi-Fi facility. For older processors (without Wi-Fi facility), a wireless network processor may be introduced. This chapter has provided a detailed overview of all these communication platforms. Elaborate discussion on each theme is beyond the scope of the book. The readers may refer to the manual pages of the corresponding devices to learn more about their interfacing requirements and modes of operation.

EXERCISES

Multiple Choice Questions

1. RS-232 communication is
   (a) Balanced
   (b) Unbalanced
   (c) Can be balanced
   (d) None of the above or unbalanced
2. With respect to RS-232, a computer is a
   (a) DCE (b) DTE
   (c) DCE or DTE (d) None of the above
3. SPI protocol is
   (a) Synchronous
   (b) Asynchronous
   (c) Both synchronous or asynchronous
   (d) None of the above
4. An SPI operation can be
   (a) Read (b) Write
   (c) Exchange (d) None of the above
5. Number of wires in I^2C is
   (a) 2 (b) 4
   (c) 8 (d) None of the above
6. Maximum number of devices in I^2C is
   (a) 4 (b) 16
   (c) 64 (d) 128
7. Number of possible prescalar values in I^2C is
   (a) 4 (b) 16
   (c) 64 (d) 128
8. Number of data bits in IEEE-488 is
   (a) 4
   (b) 8
   (c) 16
   (d) None of the above
9. Number of primary devices in GPIB is
   (a) 7 (b) 15
   (c) 31 (d) 63
10. Which of the following is not a valid GPIB device
    (a) Talker
    (b) Listener
    (c) Controller
    (d) None of the above
11. A current loop has logic level transmitted as
    (a) Voltage
    (b) Current
    (c) Both voltage and current
    (d) None of the above

12. To be considered a fault, in current loop, the current value should be
    (a) Less than 3.8 mA
    (b) More than 20.5 mA
    (c) Beyond 3.8 to 20.5 mA
    (d) All of the above
13. Frequency of Bluetooth communication is
    (a) 2.45 GHz (b) 5 GHz
    (c) 3 GHz (d) None of the above
14. Size of the passkey in Bluetooth is
    (a) 4 (b) 2
    (c) 6 (d) 3
15. Baud rate of HC-05 is
    (a) 2400 (b) 4800
    (c) 9600 (d) None of the above
16. Data rate of Zigbee in kbps is
    (a) 100 (b) 150
    (c) 200 (d) 250
17. Zigbee topology with redundancy is
    (a) Tree (b) Star
    (c) Mesh (d) None of the above
18. UHF band of Wi-Fi is
    (a) 2.4 GHz (b) 5 GHz
    (c) 6 GHz (d) None of the above
19. Protocol(s) supported by Wi-Fi is/are
    (a) TCP/IP
    (b) HTTP
    (c) HTTPS
    (d) All of the above
20. CC3100 can be connected to a host processor via
    (a) SPI
    (b) USART
    (c) Both SPI and USART
    (d) None of the above

Answers of Multiple Choice Questions

1. b	2. b	3. a	4. c	5. a	6. d	7. a	8. b	9. c	10. d
11. b	12. d	13. a	14. c	15. c	16. d	17. c	18. a	19. d	20. c

Long Answer Type Questions

1. Explain the interfacing of RS-232 with 8051 via MAX-232.
2. State the advantages of SPI and I²C over RS-232.
3. Explain the data transfer process in SPI with respect to ATmega.
4. Explain the data transfer process in I²C.
5. Enumerate the data transfer process in GPIB.
6. How can we interface a GPIB controller with 8085?
7. State the advantages of current loop.
8. How can current loops be interfaced with a processor to control a remote load?
9. State the advantages of Bluetooth.
10. Show the interface of HC-05 with 8051.
11. State the advantages of Zigbee interfacing.
12. How can we interface XBee with 8051?
13. State the advantages of Wi-Fi.
14. State the major specification of ESP32.
15. Explain the usage of CC3100 in enabling internet access in traditional microcontrollers.

12

Advanced Processors

LEARNING OUTCOMES

After reading this chapter, students will be able to:

LO 1 Enumerate the features of CISC architecture

LO 2 List the features of RISC architecture

LO 3 Define memory hierarchy in a computer system

LO 4 State the Cache memory organization

LO 5 Enumerate the operation of virtual memory

LO 6 State the features of 80286, 80386, 80486 and Pentium microprocessors

LO 7 Enumerate the architecture of PIC, MSP and AVR microcontrollers

LO 8 List the features of ARM processors

LO 9 State the Internet of Things architecture and platforms like Arduino UNO and Raspberry Pi

12.1 Introduction

In a computing system, program instructions and data are stored in the memory. The processor fetches instructions from memory and executes them. As the fetched instructions are first decoded by the processor, the instruction must be meaningful to it. The instructions acceptable to a certain processor are fixed by the processor designer, while deciding upon its *instruction set architecture* (*ISA*). At the time of designing the ISA, the processor designer makes crucial decisions about making the processor powerful enough to handle the individual instructions included there. Naturally, a complex instruction set increases the complexity of the processor design and its associated logic circuitry. The other approach by the designer could be to keep the individual machine instructions simple, making the associated processor design less complex. In that case, the complex operations of high-level languages need to be realized by a group of machine instructions. An optimizing compiler can transform complex high-level operations into a sequence of simple processor instructions. Thus, the processor may be designed following two approaches. In the first approach, complex operations are directly supported as machine instructions. In the second approach, the processor may support only simple operations with compilers doing the job of translating complex operations into a sequence of simple machine instructions. Both the approaches are effective. However, special care has to be taken at the hardware and/or software level for efficient utilization of the processor.

To understand the concept better, let us consider a high-level multiplication operation "$a = a \times b$", where a, b are memory locations. In the light of above discussion, one of the following two approaches may be taken in the processor design.

1. The hardware may be designed to be capable of multiplying two memory operands via a dedicated machine instruction,

   $$\text{MULT } a, b \quad ; \quad a = a \times b$$

   However, to carry out the multiplication, the following substeps are to be performed.

   (i) Bring the operands a and b into CPU registers.

   (ii) Multiply the contents of two registers and store the result in a third register.

   (iii) Copy the content of the third register to the memory location designated for a.

   In this case, the language compiler can translate the multiplication statement in a high-level language program into a MULT instruction for the processor. The design of the compiler becomes easy.

2. The other option may be to design the processor to support multiplication of CPU registers only. It may have separate instructions for memory read and write operations to transfer data between a CPU register and a memory location. In this case, the processor hardware becomes simple, However, the compiler needs to do extra work to determine the best possible instruction subsequence for realizing the multiplication operation of the high-level language program.

The first approach of introducing reasonably complex operations in the instruction set itself is commonly known as *Complex Instruction Set Computer* (*CISC*). The second one is known as *Reduced Instruction Set Computer* (*RISC*). *Performance ability* of a computer system is measured in terms of time needed per program. The parameter is often captured by the following equation.

$$\frac{time}{program} = \frac{time}{cycle} \times \frac{cycles}{instruction} \times \frac{instructions}{program}$$

Both CISC and RISC instruction sets want to augment this parameter. With complex instructions in its repertoire, the CISC approach attempts to minimize the number of instructions per program. On the other hand, the RISC approach attempts to reduce cycles per instruction. In the process, CISC sacrifices the number of cycles per instruction, while RISC sacrifices the required number of instructions per program. Example 12.1 illustrates the concept further.

EXAMPLE 12.1

Let a program contain 80% simple and 20% complex instructions. Assume that for CISC implementation, each simple instruction takes 4 cycles and each complex instruction 10 cycles, with cycle time being 100 ns. On the other hand, for RISC implementation, all instructions are simple and uniform in nature. Let us assume that all such simple instructions take one clock cycle each. In RISC, any complex operation in the program is converted into a series of simple instructions. Let any complex operation take on an average of 12 simple instructions. Cycle time for RISC is assumed to be 75 ns.

If the program contains 10^6 instructions, the estimated time requirement for CISC implementation will be,

$$(10^6 \times 0.80 \times 4 + 10^6 \times 0.20 \times 10) \times 100 \times 10^{-9} = 0.52 \text{ sec.}$$

The time needed by the RISC processor for the same program will be about,

$$(10^6 \times 0.80 \times 1 + 10^6 \times 0.20 \times 12 \times 1) \times 75 \times 10^{-9} = 0.24 \text{ sec.}$$

Thus, in this example, the RISC processor is expected to take half of the time needed by the CISC processor to accomplish the given program. However, this cannot be generalized, it is not always true that RISC will perform better than CISC. Depending upon the class of the application and the operations involved in the program, the RISC and CISC architectures exhibit different time requirements. Moreover, the success of the RISC CPU depends heavily upon the optimization carried out by the compiler to determine the best possible sequence of machine instructions while translating high-level programs to their corresponding machine code.

12.2 Features of Complex Instruction Set Computer (CISC) LO 1

The following are the important features of a CISC processor.

1. The number of instructions supported by a CISC processor is high, typically more than 200. The instructions are quite complex in the sense that the amount of computation done by any individual instruction is large. For example, the operations multiplication and addition may be clubbed into one instruction to compute "$s = s + a_i \times b_i$" – a multiply-accumulate operation. Such an instruction is very much useful for signal processing applications. The instruction includes the simple operations – a multiplication and an addition. The processor may also support various data types for the operands, such as, *byte*, *word, and double word.*
2. Apart from the regular addressing modes such as direct, indirect and immediate, a CISC processor may support additional complex addressing modes. Some such examples are auto-increment and auto-decrement. These two modes are similar to the indirect mode. However, additionally, the index

register gets incremented (for auto-increment) or decremented (for auto-decrement) automatically, after accessing the memory location. These modes are particularly useful for operations on arrays that modify each element of it. However, the instruction format becomes complex and requires additional time/hardware to decode the instruction and modify the index register.

3. CISC architectures often possess hardware implementation of some high-level language constructs. The constructs like *if-then-else*, *switch-case*, *loop,* etc. are generally supported. In the absence of hardware support, the machine language realizations of these constructs are cumbersome, requiring additional instructions as overhead to ensure the correct flow-control. Hardware realization of the constructs makes the software avoid these overheads. However, the process increases the hardware complexity.
4. The controller for a CISC processor is generally implemented using microprograming techniques. This helps in maintaining the upward compatibility of the processors. Through microcoding, the instruction set of a new processor can be made to be a superset of the instructions of the earlier processor. It may be noted that microprogrammed control units are slower than complete hardwired units. However, with the increased complexity of instructions, a full hardware realization of the controller also becomes an issue in terms of area, delay and power consumption.
5. Small machine code size for programs. Due to their complex nature, each instruction in a CISC processor can accomplish more computations, reducing the size of the translated machine code for user programs.
6. Instruction decoding becomes complex. A CISC processor supports a large number of instructions having possibly non-uniform formats. Availability of additional addressing modes makes the problem of decoding further difficult.
7. Number of general purpose registers is small in CISC. As a result, many operations allow memory locations as operands. The result of the operation is also stored in memory. Usage of memory operands reduces the requirements of general purpose registers. Some additional special purpose registers are often included to support special instructions. Example 12.2 illustrates a CISC processor.

EXAMPLE 12.2

Intel i486 is an example of a typical CISC processor. This is a 32-bit integer processor. To perform real arithmetic, an on-chip floating-point unit is included. There are 235 instructions, with the size of individual instructions varying from 1 byte to 12 bytes. In line with CISC features, the processor has eleven different addressing modes. Number of general-purpose registers is only eight. Different execution models are supported: *register-register*, *register-memory* and *memory-memory*. The complexity of instructions increases the difficulty in their decoding. To improve throughput, instruction execution is divided into five pipelined stages: *Fetch*, *Decode1*, *Decode2*, *Execute* and *Write-Back.*

The CISC processors reduce the complexity of compiler design. The generated machine code is short. The hardware solution for complex operations is expected to aid in faster program execution. There are several disadvantages also. Some such points have been noted next.

1. Due to the inclusion of a large number of complex instructions, the processor hardware complexity also increases significantly, escalating the area, delay and power requirements of the chip. For example,

in the *i486* number of decode stages has been made to two instead of one – a clear indication of increased decoding complexity.

2. The motivation behind introducing complex machine instructions in CISC is to match with the instructions commonly available in high-level languages. However, it is also very much likely that there is a small mismatch between a high-level language instruction and its closest counterpart in the machine language. This may turn the complex machine instruction to be mostly useless. With the computer supporting a number of high-level languages, the problem aggravates further. A complex instruction of CISC may not necessarily match with similar instruction variants across all supported high-level languages. This has the potential to make the compiled program inefficient.
3. Though CISC architecture advocates compiler simplification, this may not always be achievable. While trying to match the intended computation of a high-level language program, the optimizer may find it difficult to determine the machine instruction sequence for the same, as the instructions themselves are not very simple and uniform in nature.
4. With rapid decrease in the cost of memory chips, small program size in CISC may not be considered to be much beneficial. Further, with the increase in instruction sizes (for example, upto 12 bytes in *i486*), significant code size reduction may not be obtained.
5. Most importantly, the instructions are highly non-uniform in terms of size and execution time. This makes instruction pipelining an issue.

12.3 Features of Reduced Instruction Set Computer (RISC)

LO 2

The CISC approach of processor designing solves the problem of efficient execution of high-level language programs by supporting a large number of complex machine instructions directly by the processor. However, the approach has several drawbacks, as well. The diametrically opposite approach is to support a simple instruction set with less number of instructions, in which all instructions are of uniform size and take the same amount of time (typically, one cycle) for execution. This approach is known as *RISC* processors. The main features of a RISC processor are as follows:

1. Only a small number of simple instructions are supported by the processor. Both the size of the instructions and their execution times get reduced, compared to CISC. Instruction size is often made equal to the memory word size, so that an instruction can be fetched in a single memory access.
2. Unlike in CISC architecture, in RISC processors, the controller is often hardwired. A hardwired control unit makes the processor fast. CPU takes less silicon area. Program execution may also require possibly less power/energy.
3. The execution model is register-register. Operations could be completed in a single cycle. The LOAD and STORE are the only instructions that refer data in memory. Such an architecture is aptly called *load-store* architecture.
4. Addressing modes are made simple, reducing the instruction complexity significantly.

5. A RISC processor possesses a large number of general-purpose registers that are made available to the programmer. The availability of a large number of registers makes it possible for the variables and intermediate results of a computation to be stored in registers only. This avoids the necessity of repeated loading from memory. The compiler attempts to accommodate the local variables and parameters of a procedure in CPU registers only. Example 12.3 illustrates a RISC processor.

EXAMPLE 12.3

A typical RISC processor is SPARC. This is a 32-bit processor having an integer unit for integer operations and a floating point co-processor for real arithmetic. The instruction set contains only 69 basic instructions. In line with RISC philosophy, the processor may have 40 to 520 general-purpose 32-bit registers. It can have 2 to 32 overlapping register windows created via the grouping of registers. While invoking a new procedure, a register window is assigned to it. This ensures that the called procedure does not inadvertently modify the registers of the calling procedure. To facilitate the parameter passing and getting the return values, the register windows of calling and called procedures are overlapped. Figure 12.1 illustrates the idea of overlapping. Here, a level m procedure has called another one (at level $m + 1$). Via overlapping of register windows, the temporary registers of level m procedure have been made accessible as the parameter registers of the called procedure at level $m + 1$. As a result, the parameters need not be passed to the called procedure separately, and these are already available in its register window.

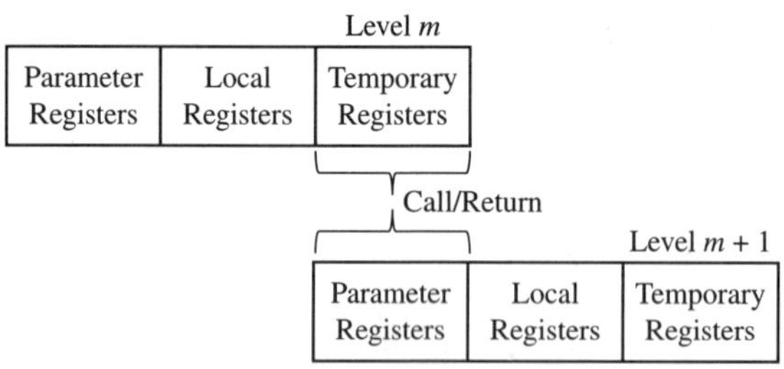

FIGURE 12.1 Overlapping register windows

It may be noted that RISC processors also have some disadvantages that need to be taken care of to make the resulting architecture suitable for classes of applications. The following issues need to be looked into.

1. Due to the simplicity of machine instructions, multiple such instructions will be needed to accomplish the intended task of a high-level language operation. This has the potential to effectively increase the number of memory accesses for certain classes of applications.
2. Programmers often develop codes in assembly language to make their programs highly efficient and optimized to exploit the hardware resources. With individual instructions becoming less powerful, it may be cumbersome to develop code to realize an operation of a high-level language.

TABLE 12.1 CISC vs. RISC

Feature	CISC	RISC
Instruction Set	Large, 120 to 350	Small, less than 100
Supported Data Types	More	Less
Addressing Modes	Large	Few
Instruction format	Variable	Fixed
Operands	Register, Memory	Register
Clocks per instruction	Variable 1–20	Fixed to 1 due to pipelining
General purpose registers	8 to 32	Large
Control Unit	Microprogrammed	Hardwired

3. Though hardware design is simplified, RISC chips require software support. The minimum requirement is a good optimizing compiler that can utilize the instruction set effectively.
4. Programming community need some important CISC features, such as, hardware support for loop, facility to copy a chunk of data from a source memory block to a destination memory block etc. Non-availability of these features in RISC may be treated as a difficulty to the users.

Major differences between the CISC and the RISC architectures have been summarized in Table 12.1. As it can be observed from the table, none of the alternatives between RISC and CISC has come up as the better one. The recent trend in processor designs is to have the better features of both the alternatives incorporated into them.

12.4 Memory Hierarchy

LO 3

A computer system consists of different types of storage elements with varying capacities and speed of operation. The fastest among them is the set of CPU registers, as they are inside the processor chip. However, the number of CPU registers is limited, and many of them have special functions, thus limiting their usage by the programmer. As far as the memory is concerned, some parts of it can be directly accessed by the processor, while others need some special interfacing hardware. The part of memory that a processor can directly access is called its *internal memory*, while the memory modules requiring special access mechanism constitute the *external memory*. The processor can access instruction/data elements from internal memory quite fast, almost at the rate of its own operation. On the other hand, data from the external memory (also called secondary memory) needs to be transferred to the internal memory before being available to the processor. A full memory hierarchy is shown in Figure 12.2, depicting the types of memory modules present in a computer system – their size, speed and price comparison.

As far as the access methods are concerned, the processor has direct access to the registers within it. The time needed to access a register is fixed for a processor design. The Cache memory has a policy of *associative access*. In this, selected bit locations within a word to match, are compared simultaneously across all the available words in the Cache. Word is retrieved from the Cache based on a portion of its content, rather than the address. Since all locations are checked simultaneously, the access time to the

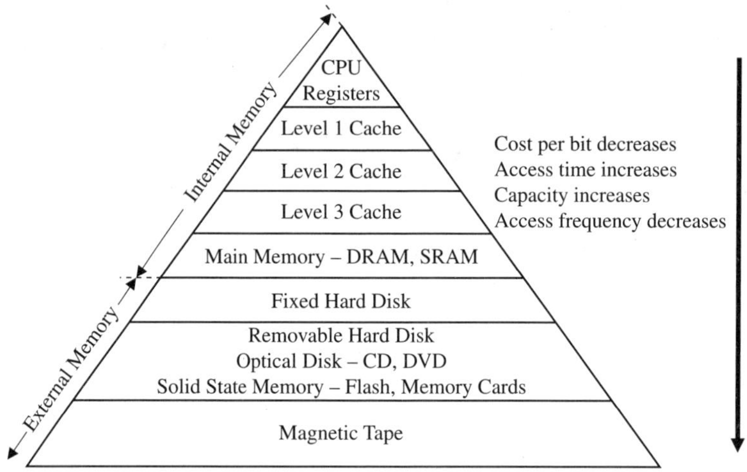

FIGURE 12.2 Memory hierarchy

Cache is constant. There are different levels of Cache memory – Level 1, Level 2 and Level 3. The Level 1 (also called L1) Cache is often put in the same chip as the processor. Thus the access time is comparable to the register access time by the processor. Level 2 and Level 3 (L2 and L3) Cache memories are present in systems between the CPU and the main memory. A frequently accessed part of the main memory is copied into the Cache memory, so that the content could be made available to the processor fast. In case a desired instruction/data item is not available in the Cache, it is searched for in the main memory.

Locality of Reference: The principle of locality of reference states that there is a high probability that future memory accesses are in the near vicinity of recent memory accesses. This is particularly true for program instructions due to the sequential nature of program execution and also the presence of loops within a program. For data accesses, once an array element has been accessed by the program, it is very much likely that the neighbouring elements will also be accessed in the near future, by the program.

The main memory is accessed in terms of bytes or words using the address generated by the CPU – data transfer is in units of a number of bits read/written during a single event. However, data transfers between memory and cache, between main memory and external memory, are in terms of blocks. A block contains a fixed number of bytes, as decided by the operating system. The block transfer is advisable as the secondary memory is slow, making it unsuitable for frequent access by the processor. Further, by the principle of locality, transferring a block of content will ensure that future CPU accesses can be served by the memory modules at higher levels closer to the CPU, thus saving time. Among the external memory modules, the disk drives follow a *direct access* principle. In this, a specified block number translates to a *track number* and a *sector number*. The read/write head gets transferred to the specified sector, and a full sector of data is transferred. A block generally corresponds to a sector-full of data. As the read/write head can be positioned at any random sector, the overall access time for the disk is variable but within strict limits. On the other hand, magnetic tapes are completely sequential in organization. Thus, access time to a tape is highly variable. While the main memory and secondary memory devices are supported by most of the operating systems, the advanced processors contain different types of Cache memories to improve the throughput (number of instructions completed per unit time). The Cache memory organization has been discussed in detail in the next section.

12.5 Cache Memory Organization

LO 4

Cache contains a part of the content of main memory. While looking for a word in the main memory, the CPU attempts to retrieve that data from the Cache. Two situations can occur. The word may be available in the Cache, called a *Cache hit*, or it may not be available in Cache leading to a *Cache miss*. In the case of a hit, the data is transferred to/from CPU in almost no time. However, for a miss, the word has to be transferred from the memory. Assuming that the main memory access time is 0.1 μS and the Cache memory access time is 0.01μS, if the *hit ratio* (fraction of total memory accesses that could be found in Cache itself) is 0.95, then the average access time for the overall memory architecture is given by,

$$0.95 \times 0.01 + 0.05 \times (0.01 + 0.1) = 0.015\ \mu\text{S}.$$

Figure 12.3 shows two organizations for the Cache memory with respect to the main memory and the CPU. When a Cache miss occurs, in the first case, the word from main memory is transferred simultaneously to the Cache and the CPU. In the other case, the memory transfers the word to the Cache only, from where it is further transferred to the CPU.

A part of the main memory is loaded into the Cache for quick reference. For this, the Cache memory is divided into *lines*. A Cache line may hold a number of words of the main memory. For this, the main memory is divided into *blocks* of words. For example, if the main memory block size is K words, in the Cache memory, each Cache line is also made to store K words, that is one block. A Cache line, apart from storing the data word, also needs to identify the memory block currently stored in the line. This is done by augmenting each Cache line with a few *tag* bits. Also some control bits are needed for each line to indicate whether the line has been modified by the CPU recently, etc. The details have been explained in Figure 12.4.

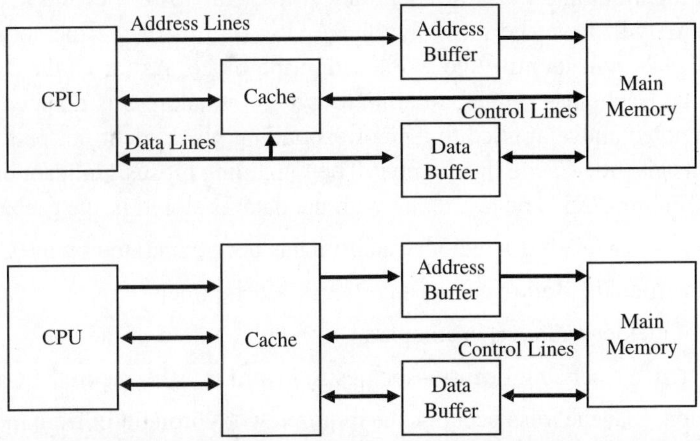

FIGURE 12.3 Cache memory hierarchy

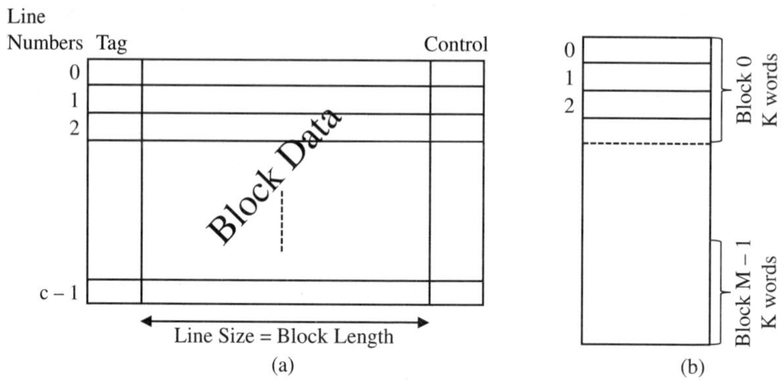

FIGURE 12.4 (a) Cache with c lines, each line with K word block; (b) Main memory with M number of K word blocks $M = 2^n/K$, n being number of address bits

12.5.1 Cache Mapping Function

The total space available in the Cache is much smaller than the main memory. In other words, the number of Cache lines is much smaller than the number of main memory blocks. Thus, when a memory block is identified to be put in Cache, a Cache line needs to be selected for holding the block. Based upon the policy of Cache line selection, the Cache memory organization can be divided into three classes – direct mapping, associative mapping and set associative mapping.

1. **Direct Mapping:** This is the simplest possible mapping of memory blocks onto Cache lines. Each memory block can get mapped onto a single Cache line only. For a Cache with c Cache lines, the memory block j gets mapped onto Cache line i, using the relationship,

$$i = j \text{ modulo } c$$

For the sake of Cache access, the address of a main memory word may be considered to contain two parts – the block identifier and the word identifier. If there are M blocks and K words per block, the block identifier part will have the most significant $\lceil \log M \rceil$ bits of the memory address, while the remaining $\lceil \log K \rceil$ bits will identify the word within the block. As far as the direct mapped Cache is concerned, if the Cache has c lines, the line to which the address is mapped is identified by the modulo function noted above applied to the corresponding block identifier part. Thus, the least significant $\lceil \log c \rceil$ bits give the Cache line number. The remaining most significant bits constitute the *tag* part, as shown in Figure 12.5. The tag, along with the data, is stored in the cache line.

The steps to access a Cache after a memory address has been generated by the CPU are as follows.

- The line number field is used to access a particular Cache line.
- The tag field of memory address is compared with the tag field of the line.
- If two tags match, a Cache hit occurs – the desired word is available in the Cache.
- If tags mismatch, a Cache miss occurs – the required word brought in from memory and placed in Cache with new tag replacing the old one.

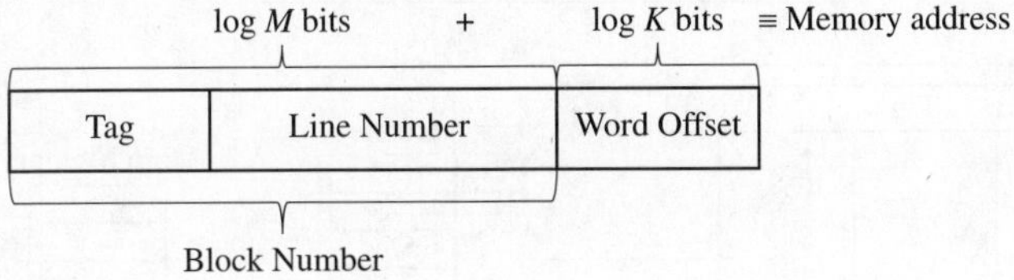

FIGURE 12.5 Memory address to Cache address translation

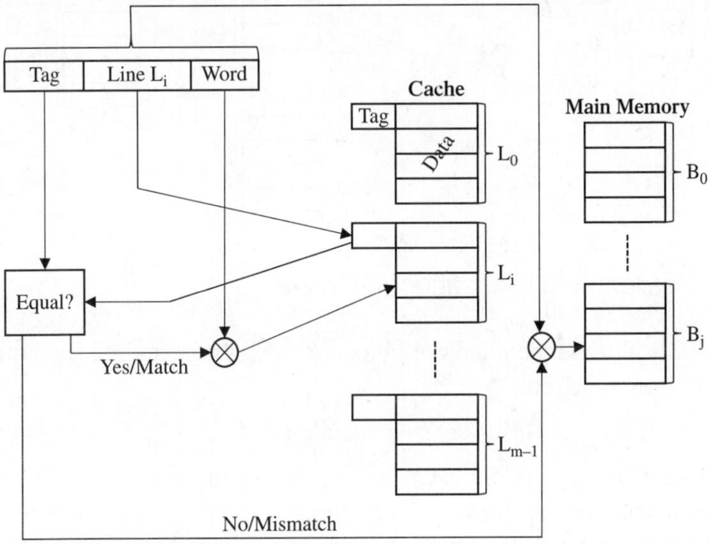

FIGURE 12.6 Search in Cache for word in Block B_j mapped to Cache line L_i

The matching process has been enumerated in Figure 12.6.

2. **Fully Associative Mapping:** The direct mapping policy is a bit restrictive, since a memory block can be stored into a fixed Cache line only. The restriction has been relaxed in the associative mapping policies. In *fully associative mapping*, any memory block may be stored in any of the Cache lines. Thus, the memory address does not identify the line number. Rather, the memory address is divided into two parts only – tag and word. Whenever a search is issued for a word in the Cache, a comparator compares tags of all Cache lines with the tag specified in the address. In case a match is found, the corresponding word is retrieved from the Cache line and returned to the CPU. In case of no match, the main memory is accessed, the block is brought into a Cache line, and its tag field is updated. The situation has is in Figure 12.7. It may be noted that the implementation of a fully associative Cache is costly as the control logic must simultaneously examine every line's tag for a match with the tag part of the specified block address.
3. **Set Associative Mapping:** This is an intermediary between the direct mapping policy and the fully associative mapping policy. For a Cache, a *Set* consists of a group of Cache lines. Similarly, a Cache may be taken to be a group of Sets. If a set is made up of k lines, the organization is called a *k-way set associative Cache*. If there are m lines in the Cache, a k-way set associative Cache will have v sets in

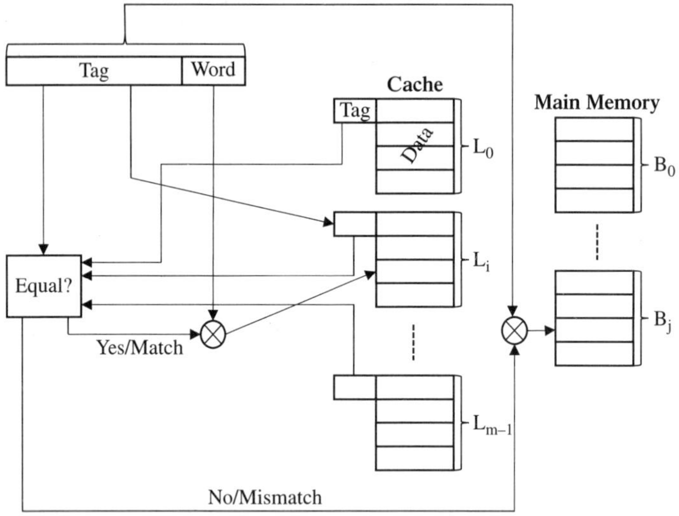

FIGURE 12.7 Search in fully associative Cache

it, such that, $m = v \times k$. For the main memory block B_j, the corresponding set number is identified by the associative mapping function,

$$i = j \textit{ modulo } v$$

The block B_j may be placed in any of the Cache lines belonging to the set i. To locate the final Cache line, the hardware must compare the tags of all k lines belonging to the set with the tag specified in the memory address. The addressing mechanism has been illustrated in Figure 12.8. The values v and k provide different variants of the set associativity.

- ($v = m$ and $k = 1$) $\rightarrow$ Direct mapping
- ($v = 1$ and $k = m$) $\rightarrow$ Full associative mapping
- ($v = m/2$ and $k = 2$) $\rightarrow$ 2-way set associative mapping, observed to have significant improvement in hit ratio over direct mapping
- ($v = m/4$ and $k = 4$) $\rightarrow$ 4-way set associative mapping, observed to provide modest improvement over 2-way set associativity

12.5.2 Replacement Policies

The size of the Cache memory is much smaller than the main memory. As a result, only a few blocks of the main memory can be made available in the Cache lines, at any point of time. In the case of a Cache miss, the missing block is read from the main memory. This block is also copied to the Cache. However, depending upon the mapping function, a suitable victim Cache line needs to be selected, onto which the missing block content be written and its tag be updated.

- **Direct Mapping:** In this case, a block can only be mapped to a fixed Cache line. Hence, there is no choice regarding the victim line.

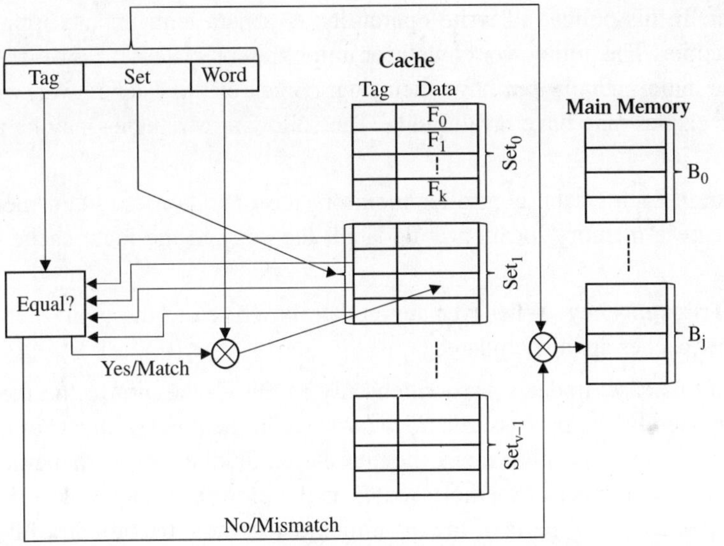

FIGURE 12.8 Search in set associative Cache

- **Fully and Set Associative Mappings:** For these two mappings, more than one Cache lines are available to choose from. For fully associative mapping, the choice is from all the Cache lines. For set associative mapping, replacement must be made from the Cache lines belonging to the set identified by the missing block address. The replacement algorithms commonly followed are noted next.
  - **Random choice:** A Cache line from the qualified set is selected randomly.
  - **First-In-First-Out (FIFO):** In this case, the block which entered first into the Cache, is selected as victim. The justification may be that since the block is the oldest one, perhaps it is not going to be referred to in the near future. However, this may not always be true as the block may be a very useful one for a code segment.
  - **Least Recently Used (LRU):** The block that has not been referred to for the longest amount of time is selected as the victim. The policy works well for many cases, however, extra hardware is needed to keep track of last usage time for each block.
  - **Least Frequently Used (LFU):** In this case, a counter is maintained for each Cache line. Whenever a line is referred to, it is incremented by one. The counter is initialized to 0 when a block is brought into the Cache line. For replacement, the line with the minimum count is chosen. The policy also has hardware overhead.

12.5.3 Write Policies

The replacement policy selects a victim Cache line to be loaded with the new block. However, the old block might have been modified by the processor, hence, needs to be written back to the main memory before overwriting with the new block. The following are the different policies followed to resolve the issue.

- **Write-through:** In this policy, all write operations to a data item update both the Cache and the main memory copies. The policy works well for uniprocessor systems. However, if multiple devices access the same main memory but have their own copies of the Cache, even with a write-through policy, multiple caches may have invalid data. The following strategies may be used to ensure data consistency.
    - **Bus Watching:** Each Cache controller monitors the operations to main memory by other bus masters. In case a memory location is updated, the copy in the local cache of the processor is invalidated.
    - **Hardware Transparency:** Whenever any update is written to the main memory, any matching words in other caches are also updated.
- **Write-Back:** In this case, updates are written only to the Cache, not to the memory. Whenever a Cache line is updated by the processor, a *dirty bit* is set in the *control part* (Figure 12.4) of the line. When the line is selected for replacement, the line should first be written to memory, only if the dirty bit is set. This saves a number of memory writes as the processor may update the data item several times in the Cache. However, parts of the main memory content become invalid – the valid content is in the Cache. Hence, all access to I/O modules should be through Cache. This increases the complexity of the circuitry and leads to potential data consistency problems.

12.6 Virtual Memory (LO 5)

While executing a program, the processor interacts with the main memory (or Cache, which is also a copy of parts of the main memory) to get the program instructions and data. In this situation, the largest program that could be executed by the processor must fit into the available memory. Moreover, in a multiuser system, there may be several user processes existing simultaneously in the main memory. This further limits the amount of main memory available to a program/process. The concept of *virtual memory* alleviates this problem by utilizing a part of the disk memory to hold the full program/process. Only a small part of the program is loaded into the main memory to be executed by the processor. If the CPU generates any reference to some instruction/data address that is not present in the main memory, it is loaded from the disk into the main memory, and the execution resumes only after the referred address is available in the main memory. As the disk space is very large, compared to the main memory, the virtual memory space can be considered as infinite, for any practical purpose.

The concept of virtual memory is similar to the Cache memory, with the roles changed as follows. The main memory now acts as the Cache, which is loaded from the disk on a demand basis. The disk takes up the role of the main memory. The concepts of replacement policies and writing policies are applicable to virtual memory environments as well. Similar to blocks in main memory, the virtual memory space of a process is divided into fixed-sized *pages*. The main memory is also divided into *frames* of size equal to the page size. Transfers between the main memory and the disk are in terms of pages. Any CPU-generated address for a process can be considered as a tuple <*page number*, *offset*>. For example, if the page size is 4KB, the least significant 12 bits of the CPU address give the offset of the instruction/data within the page, while the higher order bits give the page number. For a CPU generated address, if the corresponding page is not present in the

TABLE 12.2 Analogy between Cache and virtual memory

Cache	Virtual Memory
Block/Cache line	Page
Block size	Page size
Word offset	Offset
Miss	Page fault
Tag	Virtual page number

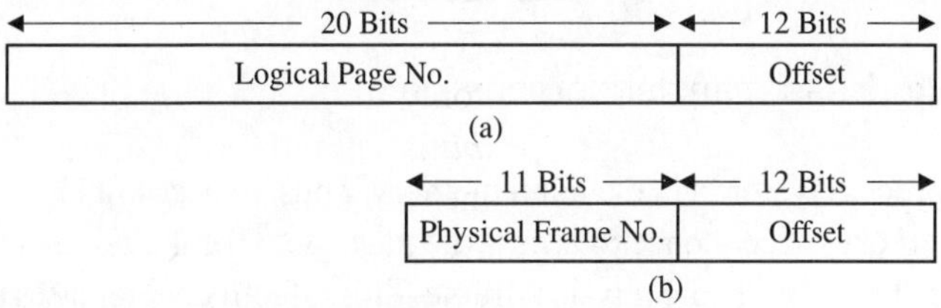

FIGURE 12.9 (a) Logical address (b) Physical address

main memory, it leads to a *page fault*. A page fault is served by the disk scheduler to read the corresponding page into main memory. Till the time the page gets loaded in the main memory, the process remains blocked. The execution of the process resumes only after that. The analogy between Cache memory and virtual memory has been enumerated in Table 12.2.

The address generated by the processor is of size determined by the processor designer. For example, the processor may generate a 32-bit address, capable of accessing 2^{32} bytes. The address generated by the processor is called the *logical address*. On the other hand, the address applied to the physical memory is called the *physical address*. The physical address consists of the frame number and the offset. The disk space being large, it may be considered infinite for the process. However, the physical memory (main memory) is limited in size. For example, the main memory may be 8MB with a 23-bit address. Assuming the page size to be 4KB, the configuration of logical and physical addresses is as noted in Figure 12.9.

12.6.1 Address Translation

The logical to physical address translation is performed by means of a structure, called *page table*. The page table for a process has one entry for each page of the process. The entries for a page are as follows.

- Valid bit: It is set to 1 if the corresponding page is available in the physical memory.
- Frame number: The frame number of the physical memory at which the logical page has been loaded.

Given a logical address, the logical page number is used to index into the page table. The valid bit is checked. If the bit is '0', the page is not available in the physical memory, and hence leads to a page fault. On the other hand, if the valid bit is '1', the frame number is retrieved from the page table. The offset of the address is appended to the frame number to generate the physical address to be applied to the physical memory. The scheme has been shown in Figure 12.10.

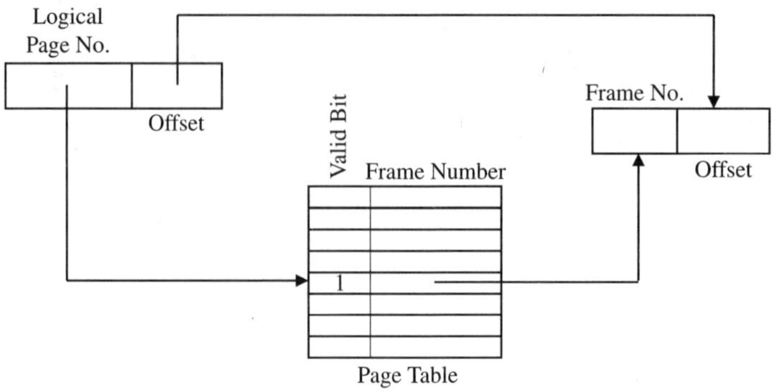

FIGURE 12.10 Address translation scheme

The page table for a process is stored in the main memory. Thus, to access a logical address, the processor accesses the physical memory twice – once to access the page table for the process to get the frame number and then to access the desired offset in the physical frame. This effectively increases the memory access time significantly. A mechanism called *Translation Lookaside Buffer* (*TLB*) is often used to reduce this access time. The TLB is a Cache containing few entries of the page table. For a logical address, the page number part is searched parallel in the TLB. Search is also initiated in the page table. Following the principle of locality, there is a good chance that frequently referred page entries will be found in the TLB. Once a page is located in the TLB, its frame number is retrieved almost immediately. This effectively reduces the memory access time, as in Cache memory hits. The overall strategy has been explained in Figure 12.11.

It may be noted that in the case of virtual memory, the problems related to page fault are similar to the Cache miss. Similar replacement algorithms may be followed to select the victim page to be over-written. However, as the restoration of page from the secondary storage is a time-consuming process, in general, the operating system puts the process into a blocked state till the page is brought into the main

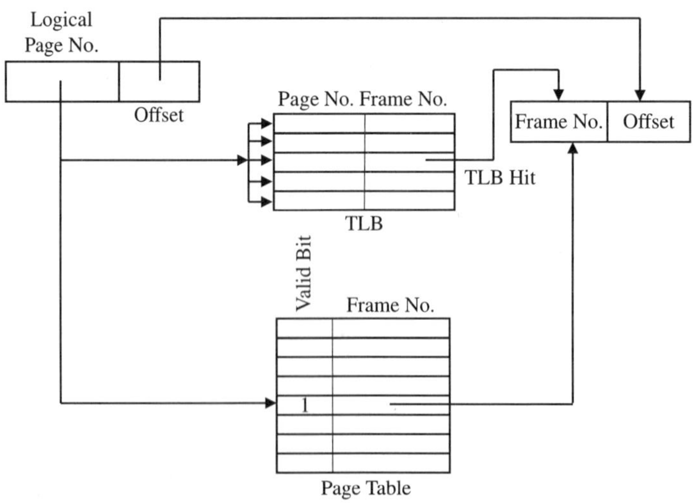

FIGURE 12.11 Virtual memory with TLB

memory. The disk scheduler is informed to carry out the I/O operation and bring the referred page into the main memory. Many of the advanced processors support the virtual memory concept in their hardware directly by incorporating the *Memory Management Unit* (*MMU*) within them. In the following, as we look into the processor architectures, the features will be clarified further.

12.7 The 80286 Microprocessor

LO 6

The 80286 is a microprocessor evolved from 8086 and its successor, 80186. This is the first microprocessor with memory management and protection abilities. It has a 24-bit address bus with the ability to address up to 16 MB of physical memory and 1 GB of virtual memory. Each process can be designed to utilize up to 1 GB of virtual memory. The versions of 80286 can operate at 8/10/12.5 MHz clock frequencies. The 808286 is upward compatible with 8086 in terms of instruction set. The 80286 instruction set includes all instructions for 8086 and 80186, with some more additional instructions. Table 12.3 presents a high-level comparison between 8086, 80186 and 80286.

The 80286 has two modes of operation – *real address mode* and *virtual address mode*. The real address mode operation is just a faster version (upto 6 times) of 8086 with 1MB physical memory, just like 8086. In this mode, 80286 is *object code compatible* with 8086 (that is, code generated for 8086 works directly on 80286, without reassembly). In the protected virtual address mode, the 80286 supports up to 16 MB physical memory and 1 GB virtual memory. In this mode, 80286 is source code compatible with the 8086 (that is, 8086 program needs to be reassembled, targeting 80286).

12.7.1 Internal Architecture

Figure 12.12 shows the internal architecture of the 80286 microprocessor. It can be divided into four functional parts – *Address Unit*, *Bus Unit*, *Instruction Unit* and *Execution Unit*. The units have been discussed briefly in the following.

1. **Address Unit:** This unit is responsible for calculating the physical addresses of the instruction and data to be accessed by the CPU. The address is handed over to the bus unit for onward transmission.
2. **Bus Unit:** This unit performs all read/write operations with the memory and the I/O devices. The physical address is transmitted over a 24-bit bus A_0-A_{23}. The prefetcher module is responsible for the prefetching of instructions. Fetched instructions are kept in a 6-byte queue.
3. **Instruction Unit:** The unit receives instructions from the prefetch queue. The instruction decoder decodes the instructions. Up to 3 decoded prefetched instructions are put into the decoded instruction queue.

TABLE 12.3 Comparison between 8086, 80186 and 80286

Microprocessor	Data Bus Width	Address Bus Width	Memory Size
8086	16 bit	20 bit	1 MB
80186	16 bit	20 bit	1 MB
80286	16 bit	24 bit	16 MB

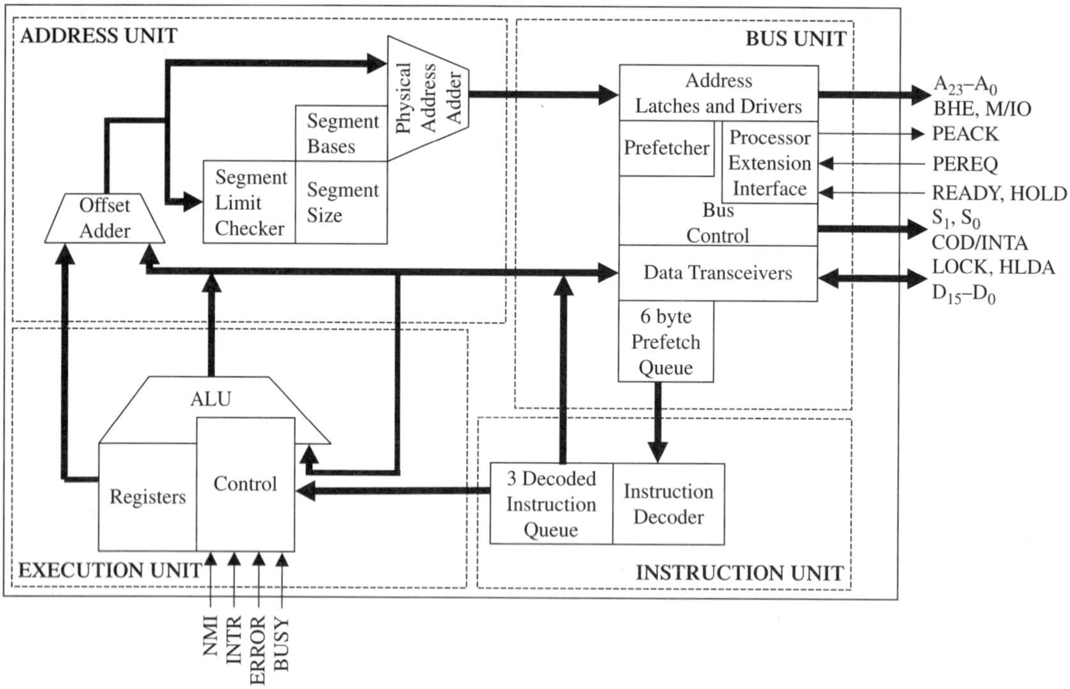

FIGURE 12.12 Internal architecture of 80286

4. **Execution Unit:** The decoded instructions are executed sequentially by the unit. An ALU has been included in this unit for computation. The unit contains a number of registers, as noted next. The registers are mostly similar to the 8086.
   - Eight 16-bit general-purpose registers
   - Four 16-bit segment registers
   - Status and control registers
   - Instruction pointer
   - Stack pointer

Real Address Mode vs. Virtual Address Mode: In the real address mode, the 80286 acts as a fast 8086 only. It remains object code compatible with the 8086. The physical memory accessed is only 1 MB, using the address lines A_0-A_{19}. The BHE signal is used for memory access. The 20-bit physical address is formed in the same way as in 8086, combining the content of a segment register and offset. Due to the presence of extra pipelining stages and improvements at the circuit level, the 80286 works much faster than the 8086, though working in the same fashion. On reset, the 80286 enters into the real address mode only. In this mode, it initializes the IP and other registers and prepares for entering into the protected virtual address mode.

In the protected virtual mode, the 80286 can address 1 GB of virtual memory per task. The virtual memory is mapped onto 16MB of physical memory. The parts of the program code and data are brought into the physical memory, whenever referred to. The processor needs a reset to return to the real address mode.

12.8 The 80386 Microprocessor

LO 6

The 80386 is a high-performance 32-bit microprocessor that was introduced to handle CAE/CAD workstations, high-resolution graphics, office and factory automation, and so on. It provides many facilities to the system designers while maintaining the object code compatibility with its predecessors 8086 and 80286. The important features of 80386 are as follows.

1. It has a 32-bit ALU that can handle 32-bit data at a time.
2. The address bus is 32-bit wide, supporting 4GB of physical memory.
3. Supports 64TB virtual memory with on-chip address translation via dedicated hardware. Translation Lookaside Buffer (TLB) is available to store recently used 32 sets of entries of the page directory.
4. Pipelined architecture, supporting simultaneous instruction fetching, decoding, executing and memory management. Up to 16 bytes of instruction code can be prefetched.
5. Allows the user to switch between different operating systems.
6. Can operate in Real, Protected and Virtual modes.
7. Dedicated hardware for important operations. Multiply/divide unit can perform 32-bit multiplications in 9-41 clocks depending upon the number of significant digits. The unsigned 32-bit division could be done in 38 clocks while signed division in 43 clocks. The barrel shifter can shift 1-64 bits in a single clock.

Internally, the 80386 consists of six functional units – Bus Interface Unit, Code Pre-fetch Unit, Instruction Decoder Unit, Execution Unit, Segmentation Unit and Paging Unit. The interconnection between the units at the block level has been shown in Figure 12.13.

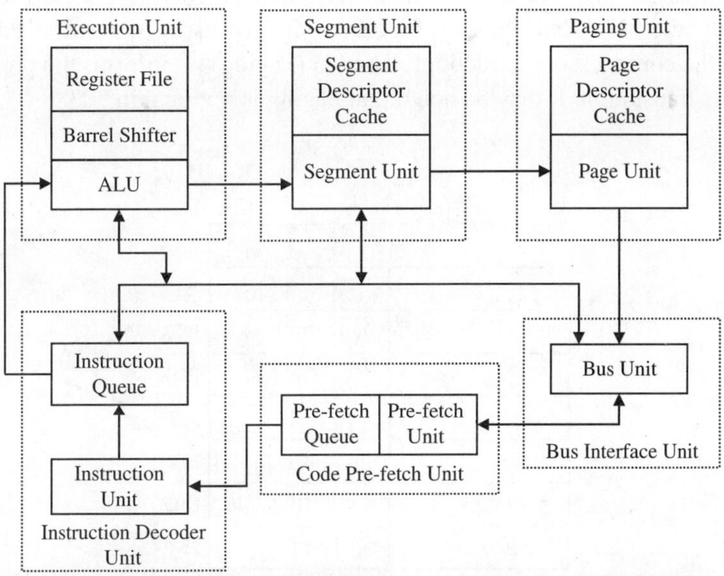

FIGURE 12.13 Functional units in 80386

The Bus Interface Unit is for interfacing the 80386 with the memory and I/O devices. Based on the transfer request, the Code Prefetch Unit prefetches instructions and stores them in a 16-byte instruction queue for decoding. The decoding takes place via the Instruction Decoder Unit. Decoded instructions are stored in the Instruction Queue for execution. The Execution Unit executes the instruction with the help of the ALU and the Barrel Shifter. The Segmentation Unit converts the logical address to a linear address, while the Paging Unit optionally converts the linear address further into a physical address.

12.8.1 Registers

The 80386 has 8 32-bit general purpose registers, as shown in Figure 12.14. The registers can be used as 32-bit or 16-bit ones. Four of the registers can also be used as 8-bit versions. Registers are named in line with those in 8086 and 80286.

The 80386 has six segment registers (compared to 4 for 8086 and 80286). Each of them is of size 16 bits in size. The registers are named as CS (Code), DS (Data), ES (Extra), SS (Stack), FS and GS. The new segment registers FS and GS are available in 80386 and above microprocessors to allow two additional memory segments access by the programs. The IP register also has a 32-bit version. While operating in the real mode, the IP register is 16-bit wide. However, for a protected mode of operation, the register is of 32 bits and is named as EIP. The 16-bit FLAGS register of 8086 has been extended to a 32-bit EFLAGS register.

There are four control registers (CR0-CR3) in 80386, as detailed next.

- **CR0:** The register contains control flags affecting the operation of the whole system, rather than an individual task. The important bits in the register are as follows. The PE (Protection Enable) bit, if set, causes the processor to start executing in protected mode. Resetting the bit returns execution to real mode. Setting the MP (Math Present) bit causes the processor to execute WAIT instructions to coordinate with the coprocessor instructions. The PG (Paging) bit informs the processor to use page tables for address translation. It may be noted that paging is optional in 80386.

31	15	7	0	
	AH	AX	AL	EAX
	BH	BX	BL	EBX
	CH	CX	CL	ECX
	DH	DX	DL	EDX
		SI		ESI
		DI		EDI
		BP		EBP
		SP		ESP

FIGURE 12.14 General registers in 80386

- **CR1:** The register is reserved by the designer.
- **CR2:** This is used to handle page faults. The linear address triggering the fault gets copied into CR2.
- **CR3:** The register is useful in a virtual memory environment. It contains the physical address of the first-page directory entry.

To aid in the debugging process, the 80386 contains six debug registers, out of which four are used for setting breakpoints, one for breakpoint control options, and one for the breakpoint status. While executing instructions, the 80386 continually compares the linear address with the contents of four breakpoint registers. If any breakpoint is reached, an interrupt is generated to initiate debugging. Apart from that, two test registers (TR6 and TR7) are used for page caching.

12.8.2 Memory Management

The 80386 has a dedicated memory management unit in it that performs address translation in hardware. The logical address generated by the 80386 CPU is 48-bit wide and consists of two parts – a 16-bit selector and a 32-bit offset, as shown in Figure 12.15. The logical-to-physical address translation takes place in two steps:

- *Segment translation*: This step transforms the logical address into a linear address (shown in Figure 12.15).
- *Page translation*: It transforms a linear address into the corresponding physical address. This stage is optional and depends upon the system software designers. If the page translation is disabled, the processor outputs the linear address to the address pins directly.

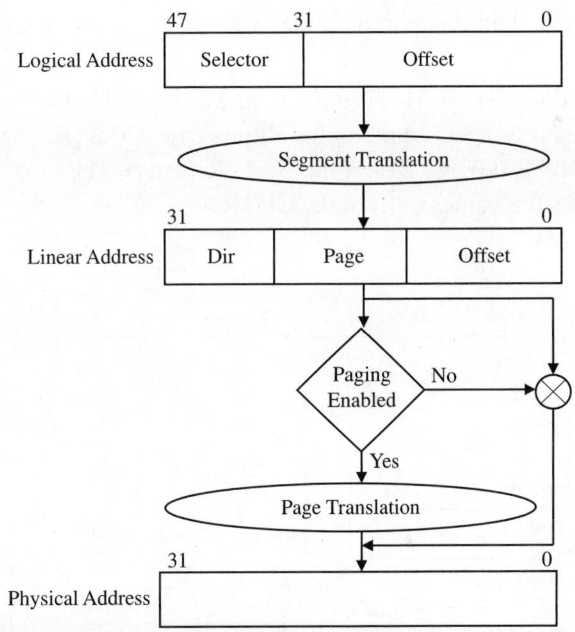

FIGURE 12.15 Logical to physical address translation

12.8.2.1 Segment Translation

The segment translation process converts a logical address into a corresponding linear address. Figure 12.16 presents the segment translation mechanism. For this, the processor uses the following data structures – Descriptors, Descriptor tables, Selectors and Segment registers. The data structures have been enumerated in the following.

1. **Descriptors:** A descriptor contains the data needed for address translation. Descriptors are created by the system software modules and not by the application programmer. The fields of a general segment descriptor has been shown in Figure 12.17 and discussed in the following.
   - **BASE:** It defines the location of the segment in the 4 GB linear space. The base parts distributed over the descriptor are concatenated into a 32-bit value.
   - **LIMIT:** The 20-bit limit field defines the size of the segment. The interpretation of the field is dependent on the value of the granularity bit G. If the granularity bit is clear, the limit is taken to be in bytes. For G = 1, the unit of limit is taken as 4 KB.
   - **TYPE:** Identifies the type of the descriptor. Details of this can be found in the 80386 manual.
   - **DPL:** Defines the *Descriptor Privilege Level*, used by the memory protection mechanisms.
   - **Segment Present (P):** If the bit is 0, the descriptor is invalid, and cannot be used for address translation. The bit is cleared for the following two cases – either the linear space spanned by the segment is not mapped by the paging mechanism, or the segment is not present in the memory.
   - **Accessed (A):** The bit is set when the segment is accessed.
2. **Descriptor Tables:** The segment descriptors are stored in the descriptor tables. A descriptor table is an array of descriptors, containing up to 8192 descriptors. There are two kinds of descriptor tables, as noted next.
   - *Global Descriptor Table* (*GDT*): This is one for the system and holds descriptors for the operating systems, code and data snippets common for all processes. The GDT is located by the processor using the register GDTR (Global Descriptor Table Register). The instructions LGDT and SGDT are used to load and store values to/from the GDTR.

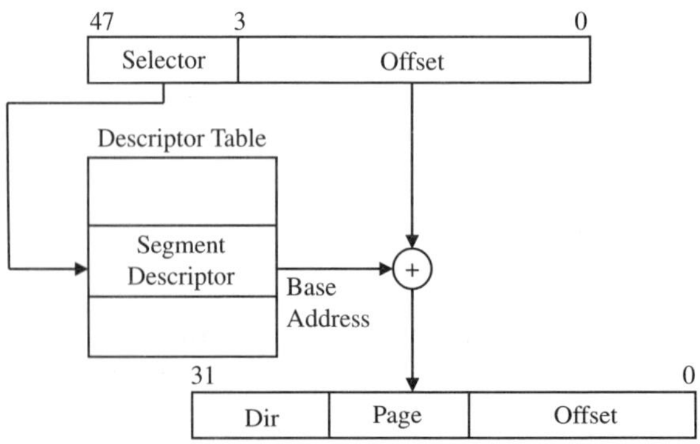

FIGURE 12.16 Segment translation

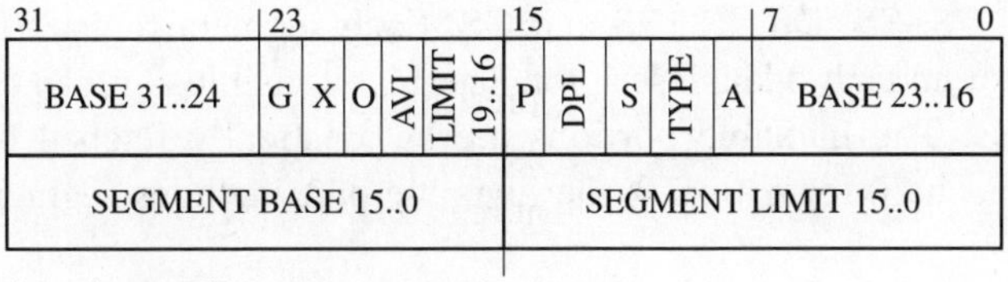

A: Accessed
AVL: Available for use by system programmers
DPL: Descriptor privilege level
G: Granularity
P: Segment present
S: 1 for application code/data segments, 0 for system segments

FIGURE 12.17 Segment descriptor structure

- *Local Descriptor Table* (*LDT*): There can be multiple LDTs, one for each process, holding the process specific information. The LDT is located by the register LDTR (Local Descriptor Table Register). The instructions LLDT and SLDT are used to load and store values to/from the LDTR.

3. **Selectors:** The selector part of the logical address specifies a descriptor table (GDT or LDT) and provides an index to access a particular descriptor within the table. The format of a selector has been shown in Figure 12.18. It has three fields in it – *index*, *TI* and *RPL*. The index field selects one of the 8192 descriptors in the table. The index value is multiplied by 8 (length of a descriptor) and added to GDTR/LDTR register content to access the appropriate segment descriptor. The TI (Table Indicator) field marks the table type – 0 for GDT and 1 for LDT. The RPL (Requested Privilege Level) field is used in conjunction with the protection mechanism.
4. **Segment Registers:** In order to avoid the overhead of consulting the descriptor table every time the 80386 accesses memory, the 80386 stores information from the descriptors into the segment registers. The structure of segment registers has been shown in Figure 12.19. There are six

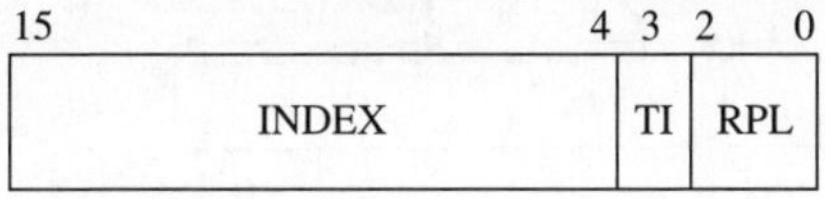

FIGURE 12.18 Selector format

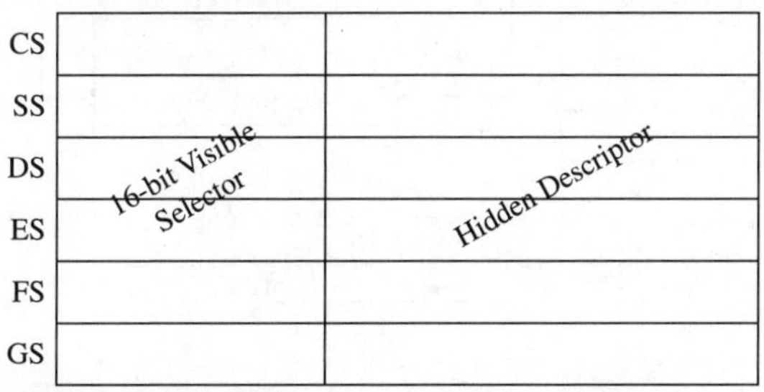

FIGURE 12.19 Segment registers

segment registers – CS, SS, DS, ES, FS and GS. Each segment register has two parts in it – a 16-bit visible selector and a hidden descriptor. The 16-bit visible part can be loaded using normal program instructions. The invisible portions are automatically fetched by the processor from the descriptor table. The information in the invisible part includes segment base address, limit, type, etc.

12.8.2.2 Page Translation

This is the second and optional stage of address translation from linear address to physical address. The stage is in effect only if the PG bit of the CR0 register is set, which is typically done by the operating system during software initialization. The page size in 80386 is 4 KB. As noted in Figure 12.20, the linear address consists of three parts – 10-bit DIR, 10-bit PAGE and 12-bit OFFSET. The figure also shows the linear-to-physical address translation procedure. The DIR field indexes into a page directory. The base of the page directory is available in the register CR3. The DIR entry identifies a page table. The PAGE entry indexes into the page table determined by the page directory. Finally, OFFSET is used to address a byte within the page determined by the page table.

The structure of a page table entry has been shown in Figure 12.21. It consists of the following components.

- **Page Frame Address:** It spans over bits 12 to 31 of the entry. Thus, it gives the 20-bit physical starting address of a page frame. As the pages are of 4KB in size, the lower order 12 bits of the address are always 0.
- **Present bit P:** This bit indicates whether the page is present in memory or not. If P = 0, the page table entry is invalid. An attempt to access this entry causes a page-exception signal to be generated. In response to that, the page-not-present exception handler can bring the required page to physical memory.

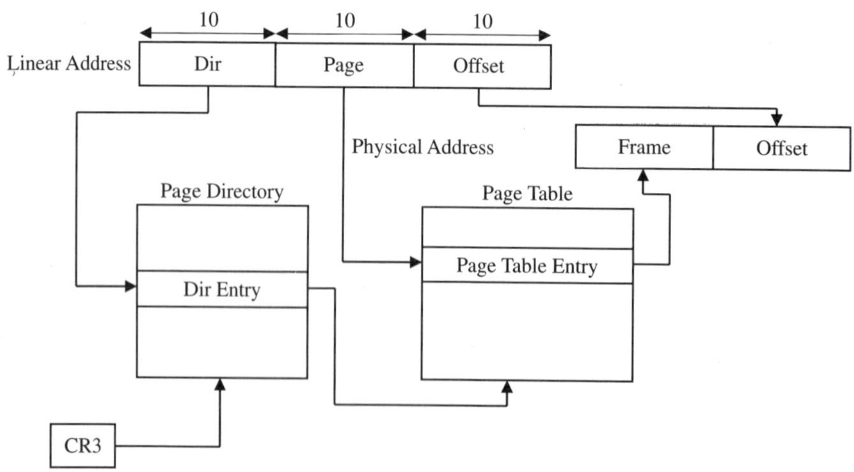

FIGURE 12.20 Linear-to-physical address translation

31 … 12	11 … 0							
Page Frame Address	AVAIL	0 0	D	A	0 0	U/S	R/W	P

P: Present
R/W: Read/Write
U/S: User/Supervisor
D: Dirty
AVAIL: For system programmer use
0: Reserved

FIGURE 12.21 Page table entry format

- **Access bit A:** This bit is set if the page is accessed by the processor, after the page has been loaded. The bit is useful to implement different page replacement strategies.
- **Dirty bit D:** The bit is set if any byte in the page has been written. This bit is also useful to implement page replacement policies.
- **R/W and U/S bits:** The read/write (R/W) and User/Supervisor (U/S) bits are used for memory protection.

As far as the modes of operation are concerned, similar to the 80286, 80386 also supports the real and virtual modes. Further, the 80386 has a protected virtual address mode (PVAM). In this mode, the address is 32-bit wide, and paging is enabled. The segment registers contain an index into a table of segment descriptors. Read/write permissions in the descriptor ensure memory protection. The system in PVAM contains one code segment descriptor and a data segment descriptor with each segment of length 4 GB. PVAM supports memory management, virtual memory, multitasking, protection, debugging, segmentation and paging.

12.9 The 80486 Microprocessor Enhancements

LO 6

The 80486 (also called *i486*) is a higher performance follow-up of 80386. Introduced in the year 1989, the 80486 is binary compatible with 8086, 80286 and 80386. It has a typical frequency of operation of 50 MHz. The significant improvements in 80486 over its predecessors are as follows.

- The 80486 contains 8 KB on-chip level-1 (L1) SRAM Cache. It may be noted that 80386 does not have any on-chip Cache.
- Enhanced external bus for Cache coherency and burst mode fill up of Cache line.
- Improved tightly-coupled pipelined architecture makes it feasible to execute simple instructions in one clock cycle, while the 80386 needed two clock cycles.
- The integrated floating-point unit allows faster algorithms for floating-point computation. In 80386, the coprocessor 80387 needs to be connected.
- The memory management unit has enhanced performance.
- A few instructions have been added.

12.10 The Pentium Microprocessor

LO 6

The Pentium is one of the powerful microprocessors belonging to the X86 series of Intel. It is a 32-bit, superscalar (explained later) microprocessor introduced in the year 1993. It is an advancement over the 80386 and 80486 processors with modifications in Cache structure, data bus width and faster numeric coprocessor. It also provides a dual integer processor for handling two instructions simultaneously in each cycle. The main features of the Pentium processor are as follows.

- 32-bit address bus offering 4GB physical address space.
- 64-bit data bus. The data bus serves the Cache, not registers. Internal and external data buses are connected via the Caches.
- 8KB instruction Cache, 8KB data Cache.
- Dynamic branch prediction.
- Internal parity checking.
- Superscalar architecture – more than one instruction gets executed in one clock cycle using different execution units at the same time.

12.10.1 Pentium Architecture

The internal architecture of the Pentium processor has been shown in Figure 12.22. It consists of units – Bus Unit, Paging Unit, Control ROM, Prefetch Buffer, Execution Unit with two pipelines (U and V), Code and Data Caches, Instruction Decoder, Branch Target Buffer, Dual Processing Logic and Advanced Programmable Interrupt Controller. Description of the important units follows.

Code and Data Caches: The Pentium contains two separate 8KB Cache for code and data, respectively. The Cache line is 32-bit, with each Cache having its TLB. The presence of separate Cachcs ensures that no conflict occurs in simultaneous instruction and data fetches. The Caches are 2-way set associative. Thus, each Cache has 128 sets, each set has two lines.

Branch Target Buffer: It is used to store the target address and branch operation statistics. This helps the Pentium to use the most likely target address for instruction fetch. This reduces the requirement of *pipeline flush* (discarding prefetched instructions) in case of conditional branch execution.

Control Unit: This unit controls the operations of five-stage integer pipelines U and V, as well as the eight-stage floating-point unit. Two integer operations can be performed by the Pentium, giving it a superscalar architecture.

Integer Pipelines U and V: Pentium has the capability to have two instructions issued in parallel – it is known as *pairing*. Though both U and V pipelines are five-stage ones, they are not equally powerful. While the U pipeline can handle the full instruction set of the Pentium, the V pipeline has limited capacity. A set of *pairing rules* determine the instruction that could be sent to the V pipeline. Initially, an instruction is loaded into the U pipeline. The next instruction could be loaded into the V pipeline if supported by the pairing rules. Otherwise, the next instruction must wait for the availability of the U pipeline. The instruction decoding and pairing decisions are made in the hardware.

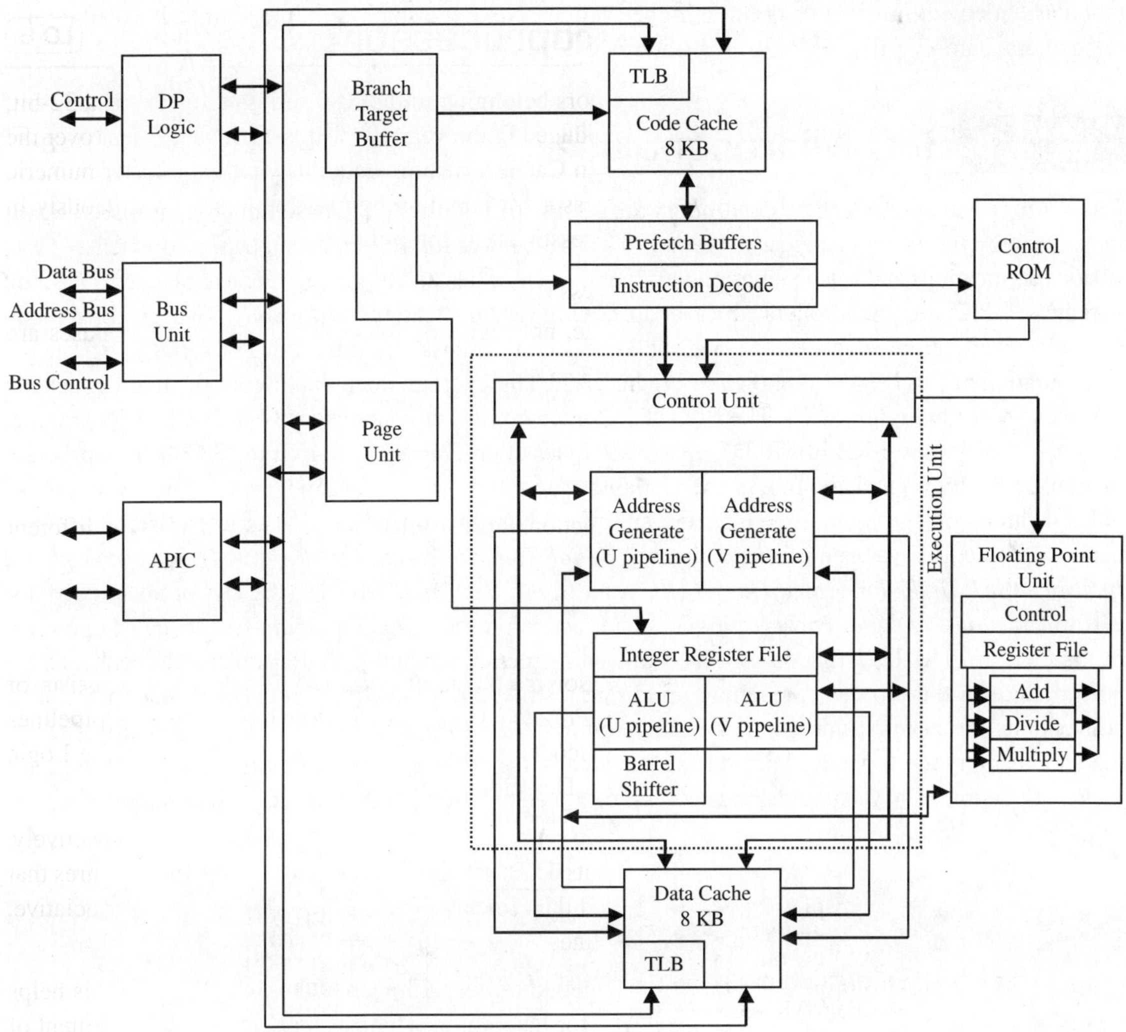

FIGURE 12.22 Pentium architecture

Floating Point Unit: Compared to the 80486 that integrates the floating point coprocessor on-chip, the floating point processor of the Pentium is a pipelined one with eight stages.

APIC: This is the *advanced programmable interrupt controller* that manages the interrupts, similar to the chip 8259A.

It may be noted that Intel has introduced many other advanced processors in the Pentium series. These include Pentium Pro, Pentium II, Pentium III and Pentium 4 processors. However, the Pentium series of processors have been replaced with Core i3, i5, i7 and i9 processors in recent years. While Pentium may be good for low-performance applications (such as, office works, web browsing, etc.) with a lower budget, the i3, i5, i7 and i9 are good for more computationally intensive applications (such as, video editing) with a definite higher budget. Next, we shall look into some advanced micro-controllers, namely PIC, MSP430, AVR and ARM.

12.11 PIC Microcontroller

LO 7

PIC – Programmable Interface Controllers are a class of popular RISC based microcontrollers. It was introduced by the *General Instruments* in 1993. These microcontrollers have been used extensively in various applications, such as in smartphones, audio accessories and advanced medical devices. Different variants of PIC microcontrollers are available with varying features and capabilities. Among them, PIC18F constitutes an advanced one. We shall restrict our discussions to the PIC18F series only. The pin configuration of PIC18F452 is shown in Figure 12.23.The internal block diagram of the microcontroller has been shown in Figure 12.24. The PIC18F452 microcontroller contains an 8-bit CPU, 2 MB program memory, 256 bytes to 1 KB EEPROM, up to 3968 bytes of on-chip SRAM, 4KB to 128 KB flash program memory. Four high-precision timers are available. The timers can aid in input capture, output compare, pulse width modulation, real-time interrupt and watchdog timer operations. A number of serial communication protocols have been included, namely, *Serial Communication Interface* (*SCI*), *Serial Peripheral Interface* (*SPI*), *Inter-Integrated Circuit* (I^2C) and *Controller Area Network* (*CAN*). For analog signals, a 10-bit *Analog-to-Digital converter* (*ADC*) interface is present. The maximum frequency of operation of the CPU is 40 MHz. The processor has built-in instruction pipelining that can be used to enhance the performance of the processor, reaching up to one instruction per cycle. As a major deviation from the conventional processors, this microcontroller follows *Harvard architecture* in which the program and the data memory are separate. The address and data buses of the two memories are separate. The policy makes it possible to have simultaneous access to data and instruction during program execution.

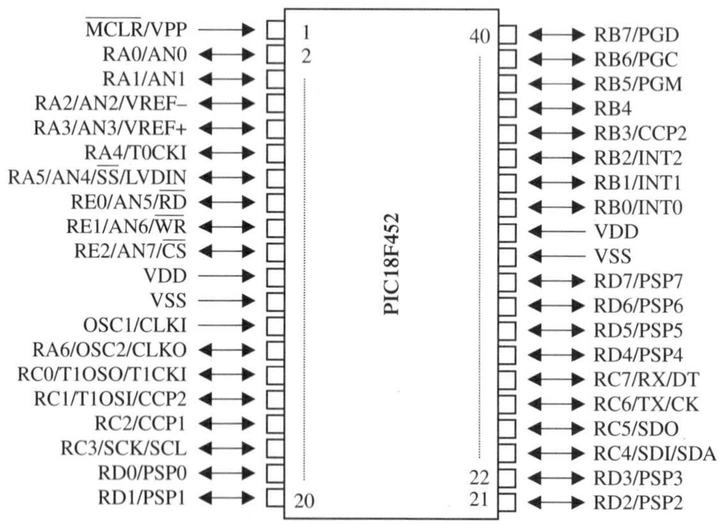

FIGURE 12.23 Pin configuration of PIC18F452

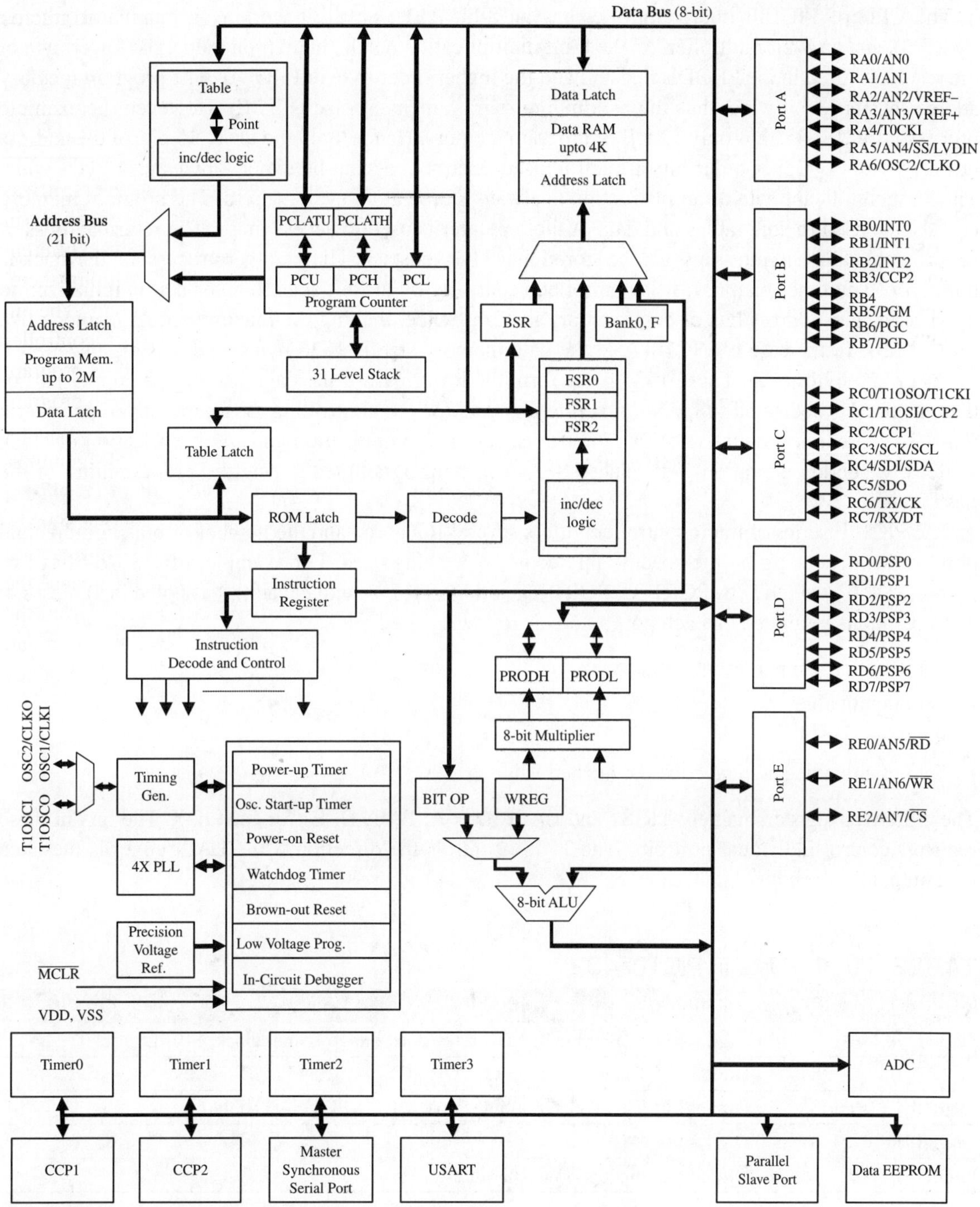

FIGURE 12.24 Block diagram of PIC18F452

The CPU of PIC18F microcontroller has an 8-bit ALU, an 8-bit working accumulator register (WREG), and an 8-bit multiplier. As the 8-bit multiplication results in a 16-bit value, the lower byte of the result is stored in the PRODL register, and the higher order byte in PRODH. The program memory address bus is 21-bit wide, thus the maximum program memory size is 2 MB. However, the example processor PIC18F452 has only 32 KB of program memory. Hence, only the lower 15 bits of the address bus are used – 6 higher order bits are left unused. Instructions can be either 2 bytes or 4 bytes wide. The least significant byte of an instruction is always stored at an even address. The program memory can also be used to hold tables and data. A table pointer is used to access those. The return address of subprogram calls and interrupts can be stored in a 31-level stack. Unlike other processors, the stack is not part of any other memory; it is controlled separately by a 5-bit stack pointer that is initialized to 0, on reset. The address bus of data memory is 12-bit wide, making the maximum data memory size to be 4 MB. In the case of PIC18F452, the data memory size is 1536 bytes, and it is organized as 6 banks of 256 bytes each. These 1536 bytes form the lower part of the data memory, the upper part of data memory is dedicated to *special function registers* (*SFR*s) controlling the operations of peripheral units (such as, timers/counters, ADC, interrupts, serial communication and so on), selection of clock sources, watchdog operations, etc. These SFRs can be programmed during the programming of the flash memory.

The PIC18F series of microcontrollers offers several I/O ports, and the number of pins in individual ports varies depending on the specific microcontroller being used. For example, PIC18F452 has five ports—PORT_A, PORT_B, PORT_C, PORT_D, and PORT_E—with capacities as noted in Table 12.4. The operations permitted on each port bit are as follows:

- Set direction – input or output
- Set output value
- Read input value
- Set output value and read back the latched value

There are two registers namely TRIS*x* and LAT*x* (x = A, B, C, D, E) for each port. The bits in these registers control individual port bits. The TRIS*x* controls the direction while LAT*x* controls the latch data output for each bit.

TABLE 12.4 Ports in PIC18F452

Name	Bits	Function
PORT_A	7	Digital I/O, analog input, ADC reference voltage, Timer clock, SPI slave select
PORT_B	8	Digital I/O, External interrupts, CCP, PGM, PGC and PGD interface
PORT_C	8	Digital I/O, Timer, External interrupts, CCP, SPI, USART interface
PORT_D	8	Digital I/O, PSP
PORT_E	3	Digital I/O, analog input, PSP interface

12.11.1 Timers

There are four timers in PIC18F452, named Timer0, Timer1, Timer2 and Timer3. A brief overview of the timers has been presented next.

Timer0: This timer can be configured as an 8-bit timer or a 16-bit timer. An 8-bit programmable prescaler is also supported. The clock source to the timer can be internal or external. With an external clock source, the module can also act as a counter. On overflow, an interrupt can be generated. The operation of the timer is controlled via the special function register T0CON.

Timer1: This is a 16-bit timer that can act as both timer and counter. The module is controlled by T1CON. Compared to Timer0, the prescalar value can be 1, 2, 4 or 8 only.

Timer2: This is an 8-bit timer. It has a *Period* register. Whenever the timer value becomes equal to the *Period* register value, an interrupt is generated. The timer also has programmable prescalar and postscalar.

Timer3: This timer is similar to Timer1 in operation.

12.11.2 Capture/Compare/PWM Modules (CCP)

There are two CCP modules present in PIC18F452. The modules work in association with timers 1, 2 and 3 to provide the capture, compare and pulse width modulation (PWM) operations. There are two registers – CCPR*x*L and CCPR*x*H (x = 1, 2) in each CCP module, explained later. The registers CCP1CON and CCP2CON control the CCP1 and CCP2 modules, respectively. The operations that can be performed by the individual CCP modules are as follows.

Capture: In *capture mode*, a CCP module works like a stopwatch. On occurrence of an event, the time instant corresponding to the occurrence is recorded, the clock continues to run (unlike stopwatch that stops on occurrence of the event). Depending upon the control register setting, the 16-bit time value of Timer1 or Timer3 gets captured into CCPR*x*H:CCPR*x*L.

Compare: The 16-bit content of CCPR*x*H:CCPR*x*L is compared continually with the Timer1 or Timer3 value. A match in the time value gets reflected in the status of some hardware pin. For example, for the CCP1 module, on the occurrence of a match, the RC2/CCP1 pin gets affected. The change in the pin is also programmable – it can be programmed to become high, low, toggle, no change, or to generate an interrupt. For the CCP2 module, the RB3/CCP2 pin gets affected.

PWM: The CCP modules can be utilized to produce a pulse width modulated output at CCP1 or CCP2 pin. The operation is controlled by Timer2. The period and duty cycle of the output are programmable. For this, CCPR*x*L and CCP*x*CON registers are to be loaded accordingly.

12.11.3 Analog-to-Digital Converter (ADC) Module

A 10-bit ADC with multiple analog channels is available in the PIC18F family of microcontrollers. Specifically, in PIC18F452, there are eight channels AN0-AN7. The channels share the pins of ports PORTA and PORTE. The operation of the ADC is controlled by the SFRs, ADCON0 and ADCON1.

The resulting digital values are put into the registers ADRESH (higher order 2-bits) and ADRESL (remaining 8-bits). An interrupt can also be generated at the completion of the conversion process. While connecting an analog source to any of AN0 to AN7 pins, the source resistance is recommended to be limited to 2.5 kΩ.

12.11.4 Interrupts

There are two categories of interrupt sources in the PIC18F452 – core interrupt sources and peripheral interrupt sources. The following are the core interrupt sources.

- External interrupt pins – INT0, INT1 and INT2.
- Changes in the states of any of the PORTB pins (also includes INT lines).
- Timer 0 overflow.

On the other hand, the following interrupt sources are considered as peripheral interrupts.

- Parallel slave port read/write
- ADC conversion completion
- USART receipt/transmit
- Synchronous serial port
- CCP1 and CCP2
- Timer1, Timer2 and Timer3 overflows
- EEPROM/Flash write
- Low-voltage detection
- Bus collision

The task of defining priorities of interrupts is left to the programmer. Accordingly, the programmer can group the interrupts into two groups – *high priority* and *low priority*. The processor does not distinguish between the interrupts belonging to the same priority group. All high-priority interrupts vector to address 00008H, whereas all low-priority interrupts jump to the address 00018H. A number of special function registers control the operation, enabling/disabling of interrupts.

12.11.5 Addressing Modes

Following are the addressing modes supported by PIC18F. Based on RISC principle, only four addressing modes are available to specify the operand.

- *Register Direct*: An 8-bit value is used to directly identify the register. For example, the instruction "movwf 0x20, A", copies the WREG register to the data register identified by 0x20.
- *Immediate*: The operand value is specified in the instruction itself. For example, the instruction "addlw 0x10" adds the value 0x10 to the WREG register.
- *Inherent*: Here, the operand is implied in the instruction. For example, in the instruction "andlw 0x3C", the value 3CH is ANDed with the WREG, the operand WREG being inherent here.
- *Indirect*: In this case, a special function register acts as a pointer to the actual data register. For example, in the instruction "movwf INDF0", INDF0 gives the indirect operand.

12.12 MSP430

LO 7

MSP430 is an ultra-low-power, 16-bit, mixed-signal microcontroller designed on RISC principle. Introduced in the year 1992 by *Texas Instruments*, the microcontroller is targeted toward low-power, low-cost embedded applications. It has six different low-power modes that can disable clocks and the CPU. Wakeup time is less than 1μS. However, the memory capacity is limited to 512 KB flash and 66 KB RAM. External memory buses are absent, limiting the total memory capacity. It is difficult to accommodate applications requiring large buffers and data tables. The maximum CPU speed is 25 MHz. MSP430 has six generations in it, differing over the capacity, size of memory and on-chip ADC variations. The generic structure of the microcontroller at the block diagram level is shown in Figure 12.25.

12.12.1 Peripherals

The on-chip peripherals available in MSP430 can be grouped into (i) General Purpose I/O Ports and (ii) Integrated Peripherals.

1. *General Purpose I/O Ports*: There are 11, 8-bit I/O ports. The port bits have functions multiplexed with other peripheral devices. Thus, when these devices are not used, the corresponding pins can be used for general-purpose I/O. MSP430 family defines the 11 ports as P0-P10. However, no chip implements more than 10 of them. Each port is controlled by a set of programmable registers corresponding to data input, data output, port direction (input or output), port bit selection, etc.
2. *Integrated Peripherals*: Both analog and digital peripherals are available as integrated components in the MSP430 series of microcontrollers. In the following, we shall look into such peripherals, starting with the analog ones followed by others. The important analog peripherals are ADC, Analog Pool, Comparator, DAC and Op-Amps.
   - *Analog-to-Digital Converter (ADC)*: Two types of ADCs are available in the MSP430 family. These are 10-bit and 12-bit successive approximation converters. A 16-bit sigma-delta converter is also available. A 16-word conversion-and-control buffer makes it possible to convert and store up to 16 analog samples without CPU intervention, minimizing the power consumption.

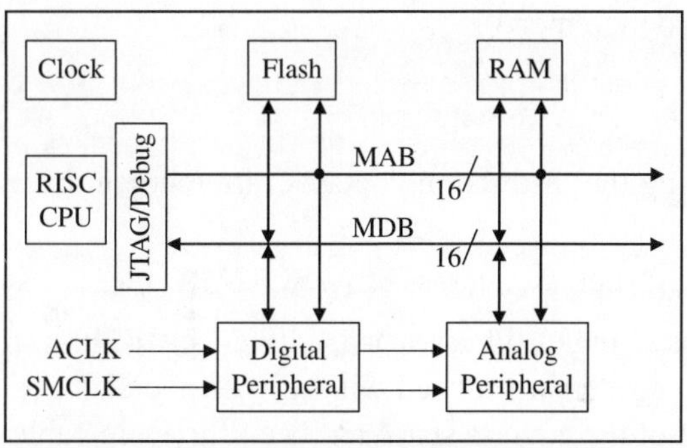

FIGURE 12.25 Generic MSP430 block diagram

- *Analog Pool* (*A-Pool*): This is a module that can be configured to perform various analog functions – ADC, DAC, Comparator, Temperature Sensor or Supply Voltage Supervisor (SVS). With this module, the user program gets the flexibility to program a series of analog functions with a single setup.
- *Comparator_A+*: An analog comparator comparing voltages on two input terminals V_+ and $V_?$. Output is high if $V_+ > V_?$ and low if $V_+ < V_?$. It acts like a one-bit ADC with a precision slope.
- *Digital-to-Analog Converter*: It is a 12-bit DAC that can be configured in 8-bit or 12-bit mode.
- *Op-Amps*: Operational amplifiers with single-supply, low-current operation having rail-to-rail output. Configuration is software selectable to be unity-gain mode, comparator mode, inverting and non-inverting programmable gain amplifiers, differential and instrumentation amplifiers, etc.
- *Timers*: There are two independent 8-bit timers that can be cascaded into a 16-bit timer, as well. Two 32-bit Real-Time Clock (RTC) have been included with one of them having battery-backup facility. There are three more 16-bit, high-precision timers. A Watchdog timer is available to detect software problems. If a selected time interval, estimated for a software task expires, a system reset can be generated with the help of the Watchdog timer.
- *Brownout Reset* (*BOR*): It detects low supply voltage and resets the device by triggering a power-on reset.
- *Communication*: UART, SPI, I^2C and USB are supported for communication.

12.12.2 CPU Architecture

MSP430 follows a *von Neumann* architecture with a single address space, and byte-addressed memory. The processor contains 16, 16-bit registers R0-R15. The register R0 is the program counter, R1 stack pointer, R2 status register and R3 is a constant generator. Registers R4 to R15 are available for general use.

There are 27 instructions – each having a byte and a word version. The following are the addressing modes for the operands in the instructions.

- Register
- Indirect Register
- Indirect Register with Auto increment
- Indexed
- Immediate
- Symbolic – Word following the instruction contains difference in bytes to the destination address, relative to PC.
- Absolute – Fixed addresses, such as, address of ADC, UART, etc.

As noted earlier, MSP430 has a single address space, without any paging structure. The specialty of this architecture is that even the registers associated with the peripherals have been mapped onto memory locations. The general layout of the address space has been shown in Table 12.5.

TABLE 12.5 Memory map of MSP430

Address Range	Usage
0000H-0007H	Processor Special Function Registers (Interrupt Control Registers)
0008H-00FFH	For 8-bit peripherals. Must be accessed as 8-bit data
0100H-01FFH	For 16-bit peripherals. Must be accessed as 16-bit data
0200H-09FFH	Up to 2048 bytes of RAM
0C00H-0FFFH	1024 bytes of bootstrap loader ROM (Flash)
1000H-10FFH	256 bytes of data (Flash ROM)
1800H-19FFH	512 bytes of data FRAM
1100H-38FFH	Extended RAM
1100H-FFFFH	Up to 60 KB of program ROM. Last 16 or 32 bytes are interrupt vectors

12.13 Advanced Virtual RISC (AVR) Microcontroller

LO 7

The AVR (*Advanced Virtual RISC*) microcontrollers are a series of processors introduced by Atmel Corporation in the year 1996. The first one in the series is AT90S8515. There are three varieties of AVR microcontrollers – *TinyAVR*, *MegaAVR* and *XmegaAVR*. Out of these three, MegaAVR is the most popular one. Table 12.6 presents a summary of the three variants.

AVR is an 8-bit RISC based microcontroller. Most of its instructions are executed in one cycle. It is about four times faster than PIC along with different power-saving modes. The key features of the ATMega32 AVR are noted next.

- 2 KB Static RAM
- 32 KB Flash
- Programmable USART
- SPI interface
- Four PWM channels
- Programmable watchdog
- 32, 8-bit general purpose registers
- 1 KB EEPROM
- 10-bit ADC with 8 analog channels
- One 16-bit, two 8-bit timer/counter
- 32, programmable I/O limes
- On-chip boot program

TABLE 12.6 Comparison of AVR series microcontrollers

Series	Pins	Flash Size	Remarks
TinyAVR	6-32	0.5-8 kB	Small
MegaAVR	28-100	4-256 kB	More peripherals
XmegaAVR	44-100	16-384 kB	Direct memory access, Event system

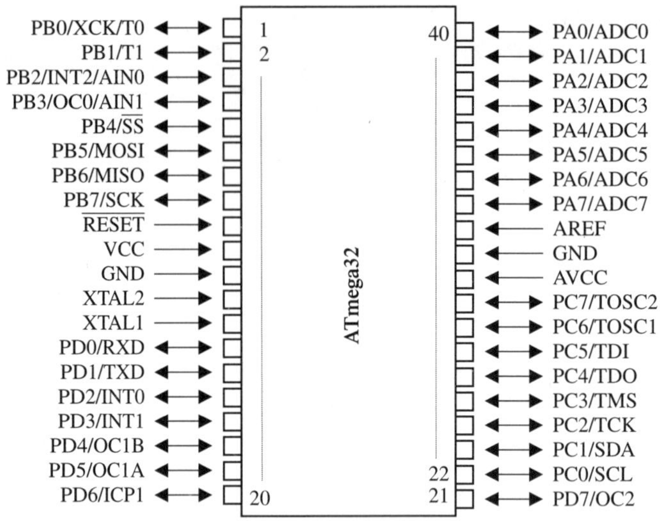

FIGURE 12.26 Pin configuration of ATmega32

A pin diagram of ATMega32 has been shown in Figure 12.26. The chip has 40 pins. Out of that, two pins (Pin 10, 11) are for power supply and ground. Pins 12 and 13 are for oscillator. Pin 9 is the reset pin. For the operation of the internal ADC, three pins are dedicated for power supply and reference voltage. For digital I/O operations, 32 I/O pins are arranged into 4 groups of 8-bit ports. Apart from being used as digital I/O bits, Port A pins can also be used as eight analog input channels. Out of the three timers present, Timer 0 and Timer 2 are of 8-bit, while Timer 1 is of 16-bit. For communication, the standards SPI, I^{2}C and USART are supported. A block diagram of ATmega32 has been shown in Figure 12.27.

There is an on-chip analog comparator available with ATmega32. Based on the results of comparison, an interrupt can be generated. To support external interrupts, three pins have been dedicated. Sensing of the interrupts are programmable. ATmega32 can operate at a wide range of frequencies, spanning over 1 MHz to 16 MHz. Most of the instructions take one clock cycle for execution, multiplication takes two clock cycles. *JTAG Boundary Scan* facility is available for on-chip debugging.

The processor programming model of ATmega32 contains 32, 8-bit general purpose registers, named R0-R31. Three 16-bit registers, named X, Y and Z are also available. These 16-bit registers overlap with the general purpose registers R26 to R31. The overlapping facilitates using the three registers either as address pointers, or to contain data values wider than 8 bits. For program memory access, the Z register can be used. Table 12.7 lists all the registers available in the ATmega32.

The registers are also mapped to some of the SRAM locations. The 32 general purpose registers correspond to the SRAM locations 00H to 1FH. Apart from that, the I/O ports span over the locations from 20H to 5FH while the user data can be kept from 60 H to 85 H.

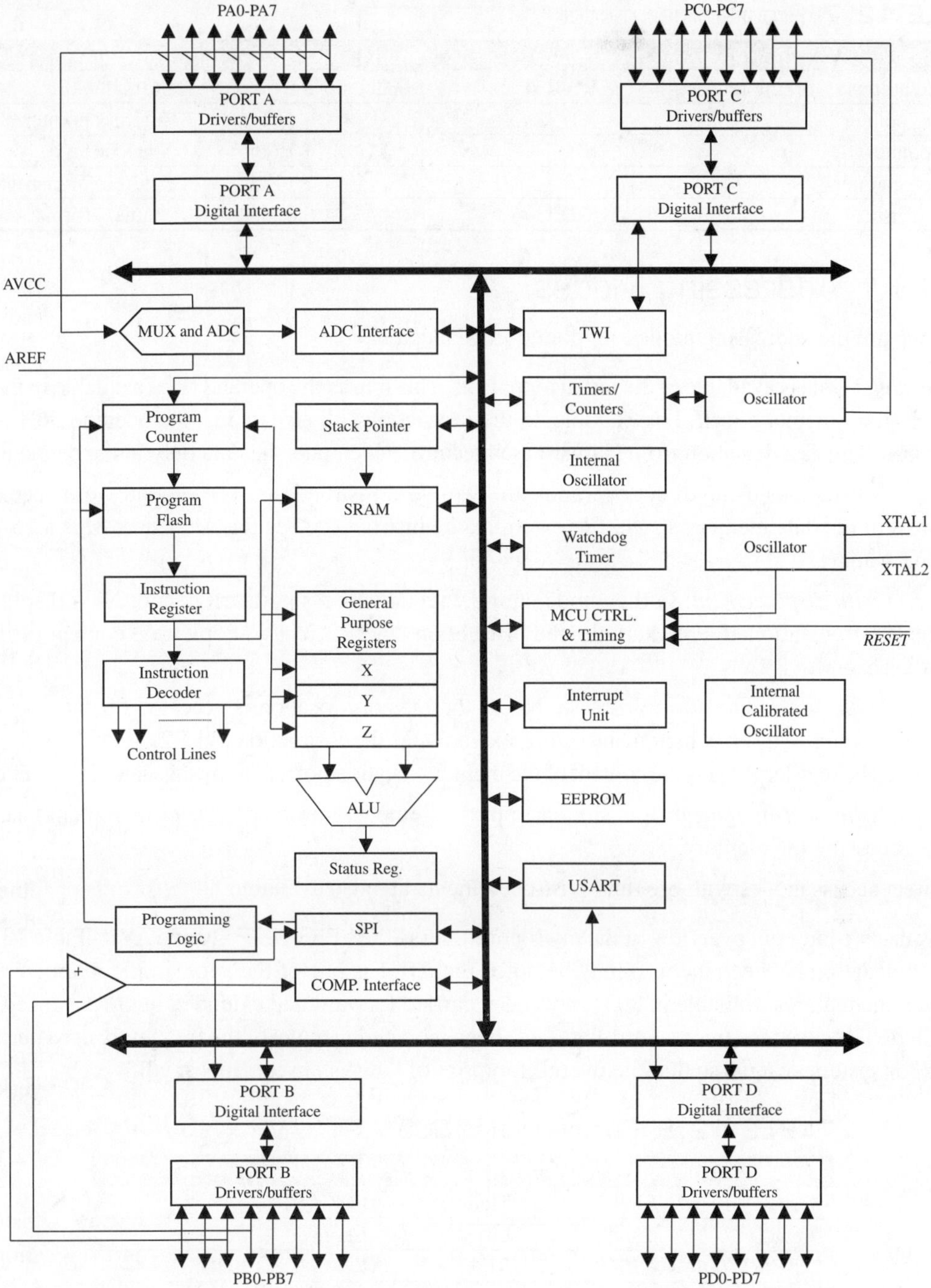

FIGURE 12.27 Block diagram of ATmega32

TABLE 12.7 Programmer model

Category	Name	Number	Size	Function
General purpose working register	R0-R31	32	8 bit	Store data
Address pointer	X,Y, Z	3	16 bit	Address pointer
Stack pointer	SP	1	16 bit	Stack access
Program counter	PC	1	14 bit	Address of instruction
Status register	SREG	1	8 bit	Status information

12.13.1 Addressing Modes

Following are the addressing modes supported in ATmega32.

- *Immediate*: Data is available in the instruction itself. This makes the operand to be available in the Flash, the program memory itself. For example, in the instruction "ldi r16, 0x56", the operand 56H is put as part of it. This is a deviation from Harvard architecture, which puts the data only in the data memory.
- *Direct*: Instructions using direct operands are two-word instructions. The second word specifies the address of the data memory space. For example, the instruction "lds R1, *M*" where *M* is a 16-bit data memory address.
- *Indirect with displacement*: In this mode, along with an address pointer (X, Y or Z), a displacement value is also mentioned. For example, the instruction "ldd R2, Y + *y*" copies the content of memory location Y + *y* to R2.
- *Indirect with pre-increment*: In this case, before the memory location is accessed for the operand, the address pointer register is incremented. For example, in the instruction "ld R2, +X", X is first incremented. The register R2 gets the content of the memory location pointed to by this new X register content.
- *Indirect with post-increment*: It is similar to pre-increment. However, register increment takes place after accessing the memory.
- Indirect access modes with pre- and post-decrements are also available.

Now that we have got overview of the microcontrollers 8051, PIC, MSP430 and AVR, Table 12.8 presents a comparison between them. It may be noted that 8051 is one of the very simple microcontrollers. However, the features available in 8051 have been carried forward and extended in the advanced microcontrollers. The advanced microcontrollers have been included in many of the frequently used peripherals needed for system designs so that the overall footprint of the system remains small.

TABLE 12.8 Comparison of 8051, PIC, AVR

Feature	8051	PIC	MSP430	AVR
Speed	Slow	Moderate	Fast	Fast
Memory	Small	Large	Large	Large
CISC/RISC	CISC	RISC	RISC	RISC
ADC	No	Yes	Yes	Yes
Timers	Yes	Yes	Yes	Yes
PWM	No	Yes	No	Yes

12.14 ARM Microcontroller

LO 8

One of the most advanced microcontrollers, ARM is a 32-bit RISC processor developed by the *ARM Corporation*. It was previously known as the *Acron RISC Machine*. Unlike other processor chips, ARM Corporation does not produce any chips by themselves. Rather, the CPU design is licensed to various companies for the production of systems built around ARM architecture. Some such companies are NXP, STMicroelectronics, Texas Instruments, to name a few. The licensee uses the ARM architecture to develop their own processors that are compliant with the ARM instruction set. Some notable features of the ARM architecture are as follows.

- The architecture is very simple requiring less number of transistors in the chip. For an entire system built on a chip (called, *System-on-Chip*, *SoC*), more silicon die area is kept available for realizing memory and other peripherals, such as, ADC, Timers/Counters, SPI, I2C, USART, and so on.
- The instruction set architecture is very carefully designed and supports pipelined design, optimizing the power consumption. Due to its low power consumption, ARM has become almost the default choice for applications with severe power constraints.
- Two different instruction sets are supported. The default one, called ARM, consists of 32-bit instructions. Another 16-bit instruction set, called THUMB, has been designed to result in better code density and even further power reduction.
- Due to architectural novelties, the performance is often much higher than that of other contemporary advanced processors, even those with CISC architecture.
- The architecture is highly modular. The designer can choose the modules to be included in the processor implementation. Only the integer pipeline, responsible for integer arithmetic, is mandatory. Depending upon the requirement of the target application, the designer may include other modules (for example, real arithmetic).
- Full debugging of the system is possible through the built-in JTAG debug port and on-chip embedded in-circuit emulator.

Looking back in history, ARM1 was introduced in 1985 as a 26-bit RISC processor. In 1992, the 32-bit RISC version ARM7 came into existence. Thereafter, several processor architectures were introduced over the years, namely ARM8, ARM9TDMI, ARM10, ARM11, Cortex, etc. In the following section, ARM7 is discussed in more detail to provide a basic understanding of the ARM architecture.

12.14.1 The ARM7 Architecture

A block diagram level architecture of ARM7 CPU has been presented in Figure 12.28. Both data bus and address bus are of 32-bit width. It may be noted that the direction of the ALE signal is reversed, compared to other processors like 8085.This has been done as unlike 8085/8051, the address and data buses are totally separate; there is no necessity for demultiplexing. Instead, the ALE signal has been used for the external memory/device to inform the CPU to keep the content valid on the address bus.

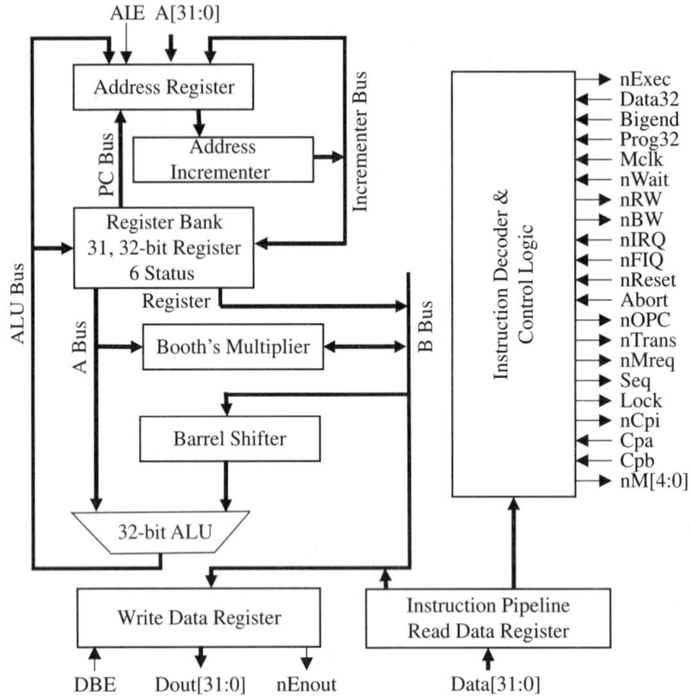

FIGURE 12.28 Block diagram of ARM7

This helps in interfacing devices that need the address to be valid for more than one clock cycle. A large number of registers are present in the CPU, following the RISC philosophy. It has 31, 32-bit registers accessible by the users, spanning over operating modes. Apart from that, there are six status registers used in different modes of operation (explained later). An address incrementer module increments the address after every access to the memory. An efficient 32-bit ALU has been included to make the arithmetic operations fast. Moreover, the ALU has been relieved of complex operations like multiplication and logical shifting. For multiplication, a dedicated multiplier module implementing the Booth's multiplication algorithm has been incorporated. A barrel shifter module performs rotate and shift operations. One of the operands to the ALU comes through the A bus directly from the registers. However, the operand coming on the B bus passes through the barrel shifter module first. This architectural feature makes it possible to shift/rotate the second operand before carrying out the operation. The execution of instructions is carried out through a three-stage pipeline, similar to the classical *fetch-decode-execute* pipeline. The uniformity in the size and execution time for the instructions helps in the smooth realization of the pipeline. It may be noted that in the ARM7 pipeline, the execute stage is heavily loaded as it has to first read the operands, then perform ALU operation and finally write back the result. The later versions of ARM processors contain 5, 6 and 8 stage pipeline structures with more balancing between the pipelined stages.

12.14.2 Instruction Set Architecture

The ARM7 is a typical RISC architecture. However, it includes several enhancements that are more like CISC features to enhance performance further. The RISC features of ARM7 are as follows.

- There are 16 general-purpose registers accessible by the user, in any mode of operation.
- ARM follows a *load/store* architecture model. Instructions performing data operations operate only on registers. A memory operand cannot directly take part in arithmetic and logical operations. It must first be loaded into a CPU register.
- Only simple addressing modes are used, simplifying the instruction decoding stage.
- All ARM instructions are 32-bit wide, and most of them have a regular three-operand encoding. This uniformity enhances the balance between the pipeline stages.

The CISC features adopted in ARM are as follows.

- Each instruction has the capacity to control the ALU and the shifter. For example, a simple ADD instruction can add two operands. However, with the help of the shifter, the instruction can also get one of its operands shifted or rotated by a number of bits.
- Two complex addressing modes, autoincrement and autodecrement, have been included.
- For block data movements, multibyte load/store instructions have been introduced. Multiple bytes of data can be transferred between memory and registers. Execution time for the instruction depends upon the number of bytes moved. This is a violation of equal execution time requirement of RISC instructions.
- There are four condition code bits attached to each instruction. The instruction gets executed only if the specified condition is satisfied (elaborated later).
- Instructions involving the ALU operations may or may not affect the status flags. A pure ADD instruction does not affect the status flags. If the user wants the status flags to get affected, ADDS instruction has to be used.

12.14.3 Registers and Mode of Operation

There are six modes in which the ARM processor may operate, namely, *User mode, Fast Interrupt Processing (FIQ) mode, Normal Interrupt Processing (IRQ) mode, Supervisor (SVC) mode, Undefined (UNDEF) mode* and *Abort mode*. In user mode, normal user programs are executed. While executing in user mode, a program can make a system call to reach the supervisor mode. The ARM processor supports two types of interrupts – fast and normal. In response to the fast interrupt, the processor switches to the FIQ mode. The normal interrupts are processed in IRQ mode. If the code part of a received instruction is not understandable by the processor, the undefined mode is entered. Abort mode corresponds to memory fault.

In ARM, each mode has got its associated set of registers. The user mode has 16 general-purpose registers R0-R15. Among them, R15 is the program counter. Unlike other processors, the R15 register can also be utilized as a general purpose register. ARM does not have any defined default stack. If required, the user can implement a stack as part of the program. Conventionally, the register R13 is used as a stack pointer. Due to the absence of a stack, subprogram branching becomes an issue. For subprogram call and return, the register R14 is used as the *link register*. To perform a procedure call, the current program counter value (that is, the return address) is automatically copied into the R14 register. To return from the procedure, the user may simply copy R14 into R15. Each operating mode has got its own copy of the R13 and R14 registers. Each mode has a status register of its own, as well. All modes, except the user mode, have a register to save the content of the status register of its previous mode. As a special case, the FIQ mode has an additional set of registers for R7-R12 as well. The availability of these extra registers helps in writing the service routines using these extra registers only. This eliminates the necessity of saving and restoring registers while writing an interrupt service routine, saving considerable context-switching time.

12.14.4 ARM Instructions

There are two instruction sets supported by ARM processors.

1. ARM – standard 32-bit instruction set
2. THUMB – 16-bit compressed instruction set

The processor executes the ARM instructions directly, while the THUMB instructions are first converted into their corresponding ARM versions and then executed by the processor. The THUMB instructions are stored in a compressed form in the memory, which first is decompressed and converted to equivalent ARM instructions for execution by the processor. As the THUMB instructions are only 16-bit wide, two such instructions occupy one memory word. The resulting code density is better than many of the CISC processors. However, the decompression introduces additional delay, and hence the execution speed suffers. Table 12.9 presents a summary of the ARM instruction categories.

ARM instructions can be executed conditionally. For conditional execution, a condition code needs to be prefixed with the instruction. The instruction gets executed provided the status flags satisfy the specified condition. The instruction "EQADD R2, R3, R4" has the effect of R2 getting R3 + R4, provided the equality condition is met. In case the condition is not satisfied, even if the instruction reaches the CPU and proceeds to the execution phase, it is treated in a manner similar to "no operation". Incorporation of this feature has the potential to eliminate small branchings in the program. The resulting code becomes more of a straight line one, augmenting the pipeline performance. For illustration, consider the following high-level code fragment.

```
if R1 = R2 then
   R3 = R4 + R5
else
   R3 = R4 – R5
```

We show two tentative assembly codes for this high-level code fragment. The left one has been created without any conditional instruction feature. It has created some additional branch instructions. The code on the right side has used conditional instructions. It has resulted in a straight-line code. For the code on the left, assuming that the equality condition is satisfied, after executing "JE L1", the processor has to wait for the instruction at L1 to get loaded into the CPU. The normal prefetched instruction "SUB

TABLE 12.9 ARM instruction set summary

Category	Description	Example	Remarks

Data Processing	32-bit Addition, Subtraction, Multiplication	ADD R1, R2, R3; R1 = R2 + R3
		ADD R1, R2, R3, LSL #2; R1 = R2 + R3*4
		ADDS R1, R2, R3; R1 = R2 + R3 with status flags affected
			Registers must be used for 1st operand and destination
			Immediate value or register as 2nd operand, may be shifted or rotated
			Permitted immediate values are limited

Byte Data Transfer	Transfer data between memory and registers	LDR R0, [R8]; R0 = Mem[R8]
		LDR R0, [R1,-R2]; R0 = Mem[R1 – R2]
		LDR R0, [R1,#4]; R0 = Mem[R1 + 4]
		LDR R0, [R1,#4]!; R0 = Mem[R1 + 4], R1 = R1 + 4
			Data transfer can be 1, 2 or 4 bytes
			Base plus offset addressing can be used
			Auto-increment mode supported

Block Data Transfer	Multi-byte data transfer between memory and registers	LDMIA R12!, {R0-R11}; Load 48 bytes of data from memory buffer pointed to by R12 to registers R0-R11 STMIA R13!, {R0-R11}; Store 48 bytes of data from registers R0-R11 to memory buffer pointed to by R13	Data transfer between to/from memory and 1-16 registers Lowest memory address correspond to lowest register number Used for stack or bulk data movement
Multiplication	Integer multiplication with 32/64 bit results	MUL R0, R1, R2; R0 = R1 * R2 MULA R0, R1, R2, R3; R0 = R1 * R2 + R3	Separate source and destination registers R15 cannot be used 1 clock pulse for each pair of bits in source, plus 1 clock to start *Early terminates* if source is 0 or all 1
Software Interrupt	Interrupts the CPU via program instruction	SWI #*n*, *n* is a 24-bit number	Processor jumps to SWI interrupt vector Value of *n* is ignored A total of 2^{24} calls possible
Branching	Branch to a different location	B <*target*>; Branch to *target* BL <*target*>; Branch to *target* with PC+4 saved in R14	Relative branching Jump range ±32 MB BL for subroutine For Subroutine return, copy R14 to R15

R3, R4, R5" has to be discarded in that case. This creates an instance of the problem of *pipeline stall*. As it can be noted, the pipeline stall does not occur with the right side code.

```
     CMP R1, R2              CMP R1, R2
     JE L1                   EQADD R3, R4, R5
     SUB R3, R4, R5          NESUB R3, R4, R5
     JMP L2
L1:  ADD R3, R4, R5
L2:
```

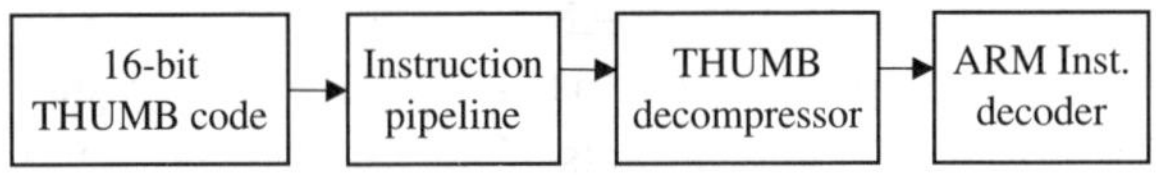

FIGURE 12.29 THUMB to ARM conversion

To implement conditional branching, conditional execution of branch instructions can be utilized, in a similar fashion.

The THUMB instruction set contains 16-bit instructions that are stored in the memory in a compressed form. The instructions belonging to the THUMB instruction set are executed unconditionally, except the branch instructions. However, with the THUMB instruction set, the user has unlimited access only to the subset of registers R0-R7 and R13-R15. The number of software interrupts is restricted to 256. PUSH and POP instructions are included in the THUMB instruction set. On reset, the processor enters into ARM instruction set mode. The user program has to execute a "BX <*target*>" or "BLX <*target*>" instruction to switch between the ARM and the THUMB instruction sets. However, the THUMB instructions are not directly executed by the ARM processor. Once they reach the instruction pipeline, the THUMB instruction goes through a decompressor and gets converted into its equivalent ARM instruction. It goes into execution thereafter. The situation is shown in Figure 12.29. Programs written in THUMB require less memory space due to compression. Power consumption is also up to 30% less in the THUMB code. However, the ARM code is about 40% faster than THUMB. Keeping these observations in view, while designing a typical system, ARM instruction set should be used for on-chip memory containing small speed-critical routines, while THUMB should be used with off-chip memory for non-critical routines.

12.15 LPC214x – An ARM7 Implementation (LO 8)

Several implementations of microcontrollers are available that use ARM7 as the processor. In the following, a typical such implementation, LPC214x has been discussed. These microcontrollers are very popular due to their tiny size and low power consumption, particularly for devices that require miniaturization. There are five different microcontrollers in the series, namely LPC2141/42/44/46/48. Table 12.10 presents a comparison between them.

TABLE 12.10 Comparison across LPC214x devices

Device	SRAM (kB)	Flash (kB)	No. of ADC Channels	No. of DAC Channels
LPC2141	8	32	6	-
LPC2142	16	64	6	1
LPC2144	16	128	14	1
LPC2146	40	256	14	1
LPC2148	40	512	14	1

It comes in a quad-flat package. The size of SRAM varies between 8 to 40 KB, along with 32-512 KB Flash. For interfacing, USB 2.0 is supported. One to two on-chip ADCs with 10-bit resolution and a total of 6-14 analog channels are available. Some variants have 10-bit DAC also as on-chip peripherals. There are two 32-bit timers/counters having four capture and four compare channels. A 6-channel pulse width modulation module and a watchdog timer are included. Interfaces like SPI, SSP, I²C and USART can be utilized for communication. A real-time clock module with an independent power supply has been included in it. Up to 21 external interrupt pins are available for general-purpose digital I/O. The block diagram of the LPC214x microcontroller is shown in Figure 12.30.

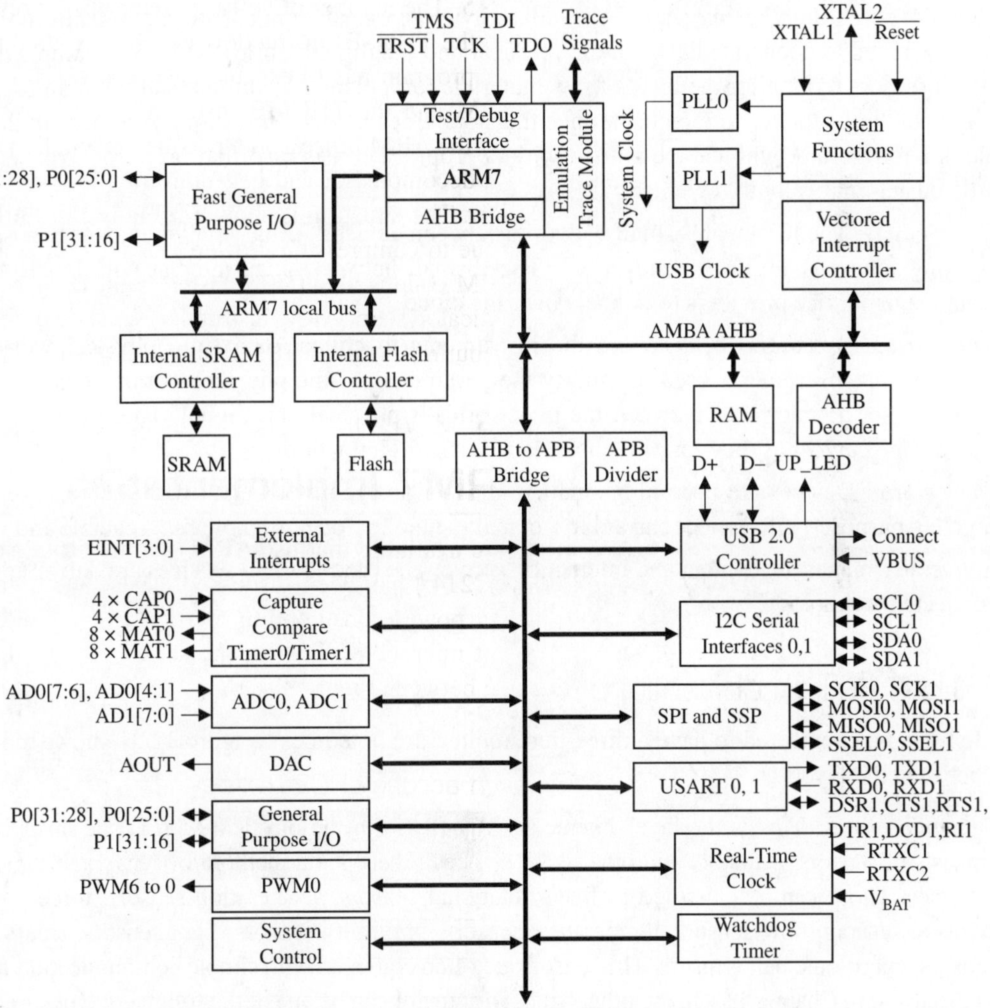

FIGURE 12.30 Block diagram of LPC214x

12.16 Internet of Things (IoT) LO 9

Internet of Things (*IoT*) is the interaction between the physical world and the digital world. With the advancements in the electronic design and the availability of devices for sensing physical world parameters, signal transmission, processing and actuation of control commands, the internet has become the part-and-parcel of today's world. The IoT paradigm attempts to connect almost all devices and appliances to the internet, so that complex tasks with a high degree of intelligence can be carried out in a collaborative manner. It has flourished immensely due to the developments in software, falling hardware prices and the social acceptance of the digital wave. The key features of IoT are as follows.

- *Use of Artificial Intelligence (AI)*: The availability of a large amount of information aids in the use of *AI* algorithms to cooperate between the physical and the digital world. A typical example can be to attach sensors to monitor the level of raw materials for a plant and automatically put an order to the corresponding supplier to replenish, once the stock goes below a threshold. Another similar example is a refrigerator with cabinets fitted with appropriate sensors, so that preferred grocers may be ordered automatically to replenish.
- *Connectivity*: The IoT creates small networks between devices. Thus, smaller and cheaper-scale networking solutions are in demand in IoT. These small networks may interact with larger networks from major service providers to connect over the cloud.
- *Sensors and Actuators*: *Sensors* are the instruments to collect data from physical world, while *Actuators* are instruments used to modify the parameters of the physical world. Thus, sensors and actuators act as interfaces between the two worlds – physical and digital. These devices are also called *edge devices* as they form the boundary of the digital world.
- *Device Size*: Devices are becoming smaller, cheaper and more powerful with the passage of time. Smaller, purpose-built devices can aid IoT in making the IoT operation precise, scalable and versatile.
- *Active*: IoT makes the content and interaction active. The products and services are delivered through connected technology.

12.16.1 IoT Architecture

The IoT can be considered to have a three-tier architecture noted next. Figure 12.31 shows the interaction between the layers.

1. *Perception layer*: This is the layer interacting directly with the physical world. The layer primarily consists of sensors, actuators and other edge devices. There are a plethora of sensors that may act as interface. Those can be classified as linear/rotational sensors, acceleration sensors, force torque and pressure sensors, flow sensors, temperature sensors, proximity sensors, light sensors, smart-material sensors, micro- and nanosensors. The sensors are often equipped with simple communication facilities, as discussed in Chapter 11. On the other hand, an actuator can be one of the following types – electrical, electromechanical, electromagnetic, hydraulic, pneumatic, smart materials, micro- and nano-actuators.

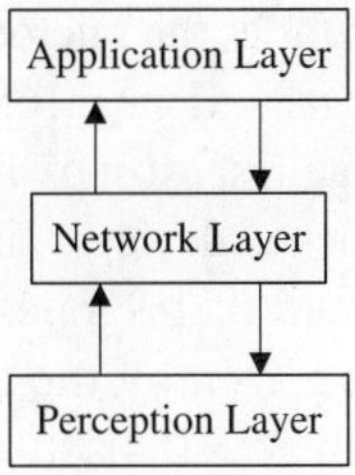

FIGURE 12.31 Three-tier IoT architecture

Most of the components of IoT system at this level are aptly called *Devices* or *Things*. The newer IoT devices come ready with basic communication facilities (such as low-power Bluetooth, Ethernet, Zigbee, Wi-Fi, etc.). However, many of the *pre-IoT* devices use traditional analog or serial communications. To interface these devices, generally a microcontroller or single-board computer is used. The devices are attached to the processor as peripherals. The processors, responsible for collecting information, must provide connectivity with a master gateway.

2. *Network layer*: This layer consists of routers and gateways. The layer is responsible for the discovery, connection and translation of devices over a network. It coordinates the operation of the perception layer with the application layer. The IoT gateways typically work from the close proximity of edge sensors and other devices. A multitude of gateways may be used to serve high-volume edge devices. The IoT gateway performs data normalization, connection and data transfer between the devices at the perception layer and the cloud, forming the application layer. Additionally, the gateways may support some computing and peripheral functionality, such as telemetry, protocol translation, pre-processing and filtering of massive raw sensor data sets, etc. Data encryption is often provided to prevent security attacks. Certain gateways also support Real-Time Operating System (RTOS).
3. *Application layer*: This layer is responsible for data processing and storage. It provides specialized services to the users. Typically implemented over the cloud, the layer communicates with gateways through wired or cellular internet.

12.16.2 IoT Platforms

In order to help the IoT designers build autonomous interactive objects, it is essential to have some hardware platforms. Examples of such platforms are *Arduino* and *Raspberry Pi*. In the following, we shall be looking into the Arduino platform.

Arduino is an open-source hardware platform containing a programmable circuit board that can be integrated with other digital/analog boards and devices. It was introduced in the year 2005 primarily to facilitate non-engineers access a low-cost, simple tool for hardware projects. The boards use a variety of microcontrollers and are equipped with digital/analog I/O pins, and SCI, including USB (on some models). The microcontroller can be programmed in C/C++ programming languages (*Embedded C*) using a standard API called *Arduino Programming Language*. Arduino provides an *Integrated Development Environment* (*IDE*) that runs on many of the standard operating systems. As far as the microcontroller is concerned, most Arduino boards contain an 8-bit AVR microcontroller – ATmega8, ATmega168, ATmega328 or ATmega2560 with varying amounts of flash memory, I/O pins and other features. Most of the boards include a 5V *linear regulator* (capable of generating a regulated 5V supply for attached devices) and a 16MHz oscillator.

Though many different Arduino boards and their clones are available, one of the highly versatile boards is the *Arduino UNO* that can be used even by the novice designers. It uses an *ATmega328P* microcontroller (operating at 16MHz) with 32KB flash and 2KB SRAM. Another microcontroller, ATmega16U2, present in the board, incorporates a USB-to-serial converter acting as a bridge between the computer and Arduino UNO. Data received from a computer over USB is converted into a serial data format understandable by the ATmega328P. The major components on the board, apart from these two microcontrollers, are as follows.

- *Digital Pins*: There are 14 digital input/output pins, labelled 0 to 13. Out of them, Pin 0 acts as a *data receive* (*RX*) line and Pin 1 acts as a *data transmit* (*TX*) line for serial data transmission and may be connected to the *TX* and *RX* pins of a *UART*, respectively. Pin 13 is connected to a tiny built-in LED on the board that may be used to test simple programs. Pins 3, 5, 6, 9, 10 and 11 support *pulse width modulation* (*PWM*). Pins 2 and 3 can also be configured to trigger as an interrupt on a low value, rising/falling edge or a change in value. Pins 10, 11, 12 and 13 can also be used for SPI communication as follows.

    Pin 10 – Slave Select (SS)

    Pin 11 – Master Out Slave In (MOSI)

    Pin 12 – Master In Slave Out (MISO)

    Pin 13 – Serial Clock (SCK)

- *Analog Pins*: There are six analog input pins A0 to A5 with voltage range 0-5V. The pins can monitor analog voltage levels in 1024 steps – 10-bit resolution.
- *Power Pins*: There are several power pins available on the Arduino UNO board.
    - *Vin*: Used to power the board with an external power source of 7-12V DC.
    - *GND*: Three ground pins can be used to ground the board and its connected devices.
    - *5V*: It provides a regulated 5V DC output voltage that can be used to power other devices requiring 5V DC.
    - *3.3V*: It provides regulated 3.3V DC output voltage.
- *Reset Pin*: Used to reset the board by setting it to low.
- *ICSP Header*: *In-Circuit Serial Programming Header* can be used to program the microcontroller externally.
- *USB Connection*: USB port to establish a communication bridge between the board and a computer. It can also act as a power source.
- *Power Jack*: It is an alternative power input source.
- *I2C Pins*: Clock and Data pins are included in I^2C communication.
- *AREF Pin*: Analog Reference pin provides a reference voltage from an external power supply.

Arduino UNO boards have been used extensively in developing IoT systems that can be found in the literature. For small IoT systems, they work as the ideal one. Raspberry Pi is another series of small single-board computers. They have an ARM-compatible CPU. Processor speed ranges from 700 MHz to 2.4 GHz, on-board memory 256 MB to 6GB, along with USB-attached SSD cards. Some models of Raspberry Pi have built-in Wi-Fi, Ethernet and Bluetooth connectivity. Many peripherals are also supported.

SUMMARY

This chapter has discussed about some of the important features of advanced microprocessors and micro-controllers. All these architectures could be classified as CISC or RISC. A clear discussion on the features of RISC and CISC processor architectures has been presented, which also distinguishes between the two. The memory hierarchy of these processors contains a Cache at its top level. The Cache memory organization techniques have been discussed for direct mapping-, set associative- and fully associative Cache memories. The locality of reference helps in the fruitful utilization of Cache. This has been followed by the concepts of virtual memory. The address translation mechanism has been detailed. Advanced microprocessors, like 80286, 80386, 80486 and Pentium have been enumerated. Among the advanced microcontrollers, a number of examples have been taken up. The first one is PIC. In this category, the advanced series PIC18F has been discussed. Other advanced microcontrollers include MSP430 and AVR. Among the recent microcontrollers targeting low-power applications, ARM has become the choice by default. Again, ARM has a number of processor architectures. The most advanced is the Cortex series. However, for a better understanding, ARM7 architecture has been developed. Several companies have implemented the ARM7 differently. The one, LPC214*x* by the NXP is a very popular processor. More detailed discussions on each of these processors and series are beyond the scope of the book. The readers are requested to look into the literature for the same. The microcontrollers are being used extensively in designing systems based on the IoT philosophy. The IoT features, architectures and platforms have been enumerated in the chapter.

EXERCISES

Multiple Choice Questions

1. CISC processors attempt to minimize
   (a) Instructions per program
   (b) Cycles per instruction
   (c) Both A and B
   (d) Neither A nor B
2. RISC processors attempt to minimize
   (a) Instructions per program
   (b) Cycles per instruction
   (c) Both A and B
   (d) Neither A nor B
3. Which one is not a CISC feature
   (a) Large number of registers
   (b) Large number of instructions
   (c) More addressing modes
   (d) None of the above
4. Which one is not a RISC feature
   (a) Large number of registers
   (b) Hardwired controller
   (c) More addressing modes
   (d) None of the above

5. In a load-store architecture, memory is accessed by
   (a) Load
   (b) Store
   (c) Both Load and Store
   (d) None of the above
6. Which architecture has more dependency on software for better performance
   (a) RISC
   (b) CISC
   (c) Both RISC and CISC
   (d) None of the above
7. Size of the address bus in bits for the PIC18F series is
   (a) 16 (b) 21
   (c) 24 (d) None of the above
8. PIC18F can work up to crystal frequency of
   (a) 10 MHz (b) 20 MHz
   (c) 30 MHz (d) None of the above
9. The bit width of port A of the PIC18F processor is
   (a) 7 (b) 8
   (c) 15 (d) 16
10. The number of timers in PIC18F is
    (a) 2 (b) 3
    (c) 3 (d) 4
11. The mode in which a timer works as a stopwatch is
    (a) Capture (b) Compare
    (c) PWM (d) None of the above
12. The high-priority interrupt vector in PIC18F is
    (a) 00008 H (b) 00010 H
    (c) 00016 H (d) 00018 H
13. The width of instructions in the THUMB instruction set, in number of bits, is
    (a) 16 (b) 32
    (c) 48 (d) None of the above
14. ALE signal in ARM processor is
    (a) To the processor
    (b) From the processor
    (c) Not present
    (d) Bidirectional
15. The register that acts as a program counter in ARM is
    (a) R13 (b) R14
    (c) R15 (d) None of the above
16. The register useful for making subprogram calls in ARM is
    (a) R13 (b) R14
    (c) R15 (d) None of the above
17. Conditional execution of instructions in ARM helps in improving
    (a) Pipeline performance
    (b) Stack overflows
    (c) Loop implementation
    (d) None of the above
18. In LPC2148, the register responsible for setting the direction of I/O ports is
    (a) IO*x*DEP
    (b) IO*x*PIN
    (c) IO*x*SET
    (d) None of the above
19. ARM instruction to change the instruction set is
    (a) BX (b) XCH
    (c) THUMB (d) None of the above
20. The Cache closest to the CPU is
    (a) L1 (b) L2
    (c) L3 (d) None of the above
21. If the Cache hit ratio is 85%, the Cache access time is 0.01μS, and the main memory access time is 0.2μS, the effective memory access time will be (in μS)
    (a) 0.02 (b) 0.04
    (c) 0.015 (d) None of the above
22. Direct mapping Cache organizations map a memory block to number of Cache lines
    (a) 1 (b) 2
    (c) All lines (d) None of the above
23. In virtual memory organization, processes and main memory are divided respectively into
    (a) Page and Frame
    (b) Frame and Page
    (c) Both Page
    (d) Both Frame

24. The size of the address bus (in bits) of 80286 is
(a) 16 (b) 20
(c) 24 (d) 32

25. The number of segment registers in 80386 is
(a) 4 (b) 5
(c) 6 (d) 8

26. With G=1, the unit of limit for segment size in 80386 is
(a) 1 byte (b) 1 KB
(c) 2 KB (d) 4 KB

27. In 80386 memory management, the optional phase is
(a) Segment translation
(b) Page translation
(c) Both Segment and Page Translation
(d) None of the above

28. Which integer pipeline in Pentium is more powerful?
(a) U
(b) V
(c) U and V equally powerful
(d) X

29. Pairing rules determine the instructions that could be sent to the pipeline
(a) U (b) V
(c) Both U and V (d) None of the above

Answers of Multiple Choice Questions

1. a	2. b	3. a	4. c	5. c	6. a	7. b	8. d	9. a	10. d
11. a	12. a	13. a	14. a	15. c	16. b	17. a	18. d	19. a	20. a
21. b	22. a	23. a	24. c	25. c	26. d	27. b	28. a	29. b	

Long Answer Type Questions

1. What are the features of a CISC processor?
2. List the features of a RISC processor.
3. What role does a compiler play in the choice of RISC vs. CISC processors?
4. How do overlapping register windows help in subprogram calls?
5. Distinguish between RISC and CISC processors.
6. Elaborate the Cache memory organization policies.
7. Explain the address translation process for a virtual memory organization.
8. Enumerate the internal architecture of the 80286 microprocessor.
9. State the memory management policy of the 80386 microprocessor.
10. What are the I/O ports of the PIC18F series of processors?
11. How are the *Capture* and *Compare* operations carried out by the timer modules?
12. List the non-RISC features of ARM processors.
13. Discuss the architecture of the PIC18F452 microcontroller.
14. Compare the ARM7 architecture and with the 8051.
15. Discuss the internal structure of the LPC2148 microcontroller.
16. Describe the features of IoT architecture.
17. Enumerate the modules present in Arduino UNO.

Index

E

F

G

H

I

U

V

W

X

Z